An Introduction to

Soil Mechanics
and
Foundations

THIRD EDITION

C. R. SCOTT
B.A., M.I.C.E., M.I.Struct.E.

Senior Lecturer in Civil Engineering, The City University, London

APPLIED SCIENCE PUBLISHERS LTD
LONDON

APPLIED SCIENCE PUBLISHERS LTD
RIPPLE ROAD, BARKING, ESSEX, ENGLAND

First edition 1969

Second edition 1974

Reprinted 1979

Third edition 1980

British Library Cataloguing in Publication Data

Scott, Charles Robin
    An introduction to soil mechanics and
    foundations.—3rd ed.
    1. Foundations—Design and construction
    2. Soil mechanics
    I. Title
    624′.1513   TA775

    ISBN 0-85334-873-1

WITH 16 TABLES AND 205 ILLUSTRATIONS

© APPLIED SCIENCE PUBLISHERS LTD, 1980

Printed in Great Britain by Galliard (Printers) Ltd, Great Yarmouth

Preface

This book is mainly intended to meet the needs of undergraduate students of Civil Engineering. A certain amount of factual information, in the form of design charts and tables, has been included. I hope, therefore, that the book will prove to be of use to the students when their courses are over, and will help to bridge the awkward gap between theory and practice.

In preparing the first edition of this book, I had two principal aims: firstly to provide the student with a description of soil behaviour—and of the effects of the clay minerals and the soil water on such behaviour—which was rather more detailed than is usual in an elementary text, and secondly to encourage him to look critically at the traditional methods of analysis and design. The latter point is important, since all such methods require certain simplifying assumptions without which no solution is generally possible. Serious errors in design are seldom the result of failure to understand the methods as such. They more usually arise from a failure to study and understand the geology of the site, or from attempts to apply analytical methods to problems for which the implicit assumptions make them unsuitable. In the design of foundations and earth structures, more than in most branches of engineering, the engineer must be continually exercising his judgement in making decisions. The analytical methods cannot relieve him of this responsibility, but, properly used, they should ensure that his judgement is based on sound knowledge and not on blind intuition.

In this third edition, a few minor alterations have been made to take account of recent research. The main changes, however, concern the core of the book, consisting of Chapters 5 to 8. These have been largely rewritten to place the study of soil strength and consolidation more firmly in the context of plasticity theory. Substantial changes have also been made in Chapters 9, 10 and 11 to allow a fuller discussion of the methods of analysis of lateral earth pressure, bearing capacity of foundations and the stability of slopes.

I am grateful to the following for permission to reproduce diagrams and other material from their publications: the

Councils of the Institutions of Civil, Structural and Water Engineers, the Director of Road Research, the Clarendon Press, Butterworths and Co., McGraw-Hill Book Co., John Wiley and Sons, Addison-Wesley, Prof. A. Casagrande, and the Controller of H.M. Stationery Office. Extracts from British Standards and Codes of Practice are reproduced by permission of the British Standards Institution, 2 Park Street, London W1, from whom copies of the complete Standards and Codes of Practice may be obtained.

Finally, I should like to thank my colleagues in The City University for their help and encouragement, and in particular Mr R. H. Foster, Mr J. D. Coleman, Mr G. A. Watt and Mr J. S. Evans for their generous assistance in the revision of the text and their many helpful suggestions.

Contents

List of symbols

The symbols used in the text conform, as far as possible, with the recommendations of BS 1991: Part 4: 1961. Other symbols have been chosen to conform generally with established usage. The reader should notice that, in conformity with BS 1991, the bulk density (γ) has, throughout this book, been defined as the weight density or unit weight (that is, the gravitational force per unit volume, measured here in kN/m^3). The significance of the more important symbols is given below.

A air void ratio: area: pore pressure coefficient.

A_f pore pressure coefficient A at failure.

a area.

B breadth: pore pressure coefficient.

b breadth.

C_c compression index: coefficient of curvature.

C_s swelling index: compressibility of the soil skeleton.

C_u uniformity coefficient.

C_v compressibility of the pore fluid.

c apparent cohesion (in terms of total stress).

c' apparent cohesion (in terms of effective stress).

c'_e Hvorslev's effective cohesion.

c'_r residual value of apparent cohesion (in terms of effective stress).

c_u apparent cohesion under undrained conditions (in terms of total stress).

c_v coefficient of consolidation.

c_s coefficient of swelling.

c_w adhesion between soil and a retaining wall.

D depth: diameter.

D_f depth factor.

D_{10} effective particle size.

d depth: diameter: length of drainage path.

E Young's modulus.

e void ratio.

e_c critical void ratio.

F factor of safety.

F_c factor of safety with respect to cohesion.

F_ϕ factor of safety with respect to friction.

f plastic potential function.

G_s specific gravity of soil particles.

H height.

H_c critical height of an unsupported vertical bank.

h height: total hydraulic head.

I_σ influence coefficient for stress below a foundation.

I_ρ influence coefficient for settlement of a foundation.

i hydraulic gradient.

i_c critical hydraulic gradient for unstable upward flow.

K coefficient of lateral earth pressure.

K_a Rankine's coefficient of active earth pressure.

$K_{ac}, K_{aq}, K_{a\gamma}$ general coefficients of active earth pressure.

K_p Rankine's coefficient of passive earth pressure.

$K_{pc}, K_{pq}, K_{p\gamma}$ general coefficients of passive earth pressure.

K_0 coefficient of earth pressure at rest.

k permeability.

L length.

LI   liquidity index.

LL   liquid limit.

l length.

m_v coefficient of volume change.

N Taylor's stability number for earth slopes.

N_c, N_q, N_γ bearing capacity factors.

n porosity.

P force.

P_a active thrust of earth on a retaining wall.

P_p passive resistance of earth to a movement of a retaining wall.

PI   plasticity index.

PL   plastic limit.

p pressure: mean normal stress $(=\sigma_1 + \sigma_2 + \sigma_3)$.

p' mean effective stress $(=\sigma'_1 + \sigma'_2 + \sigma'_3)$.

p_a active lateral earth pressure.

p'_c preconsolidation pressure (spherical consolidation).

p'_e equivalent consolidation pressure (spherical consolidation).

pF soil suction.

p'_i initial value of p'.

p_p passive lateral earth pressure.

q rate of flow: deviator stress $\left(= \dfrac{3}{\sqrt{2}} \tau_{oct} \right)$.

q_a allowable bearing pressure of a foundation.

q_f gross ultimate bearing capacity of a foundation.

q_n net foundation pressure.

q_0 surcharge load on the ground surface.

RD   relative density.

R_p overconsolidation ratio (spherical consolidation).

R_σ overconsolidation ratio (one-dimensional consolidation).

r radius.

r_u	pore pressure ratio.
S_r	degree of saturation.
SL	shrinkage limit.
T	torque: time factor: surface tension.
T_v	time factor (one-dimensional consolidation).
t	time.
U_v	degree of consolidation (one-dimensional).
$\bar{U}$	mean degree of consolidation.
u	pore pressure.
u_a	pore air pressure.
u_w	pore water pressure.
V	volume.
V_s	volume of solids.
V_v	volume of voids.
v	velocity: specific volume ($= 1 + e$).
W	weight.
w	water content.
w_{sat}	saturated water content.
X	body force in the (negative) x direction.
Z	body force in the (negative) z direction.

β	slope angle: inclination of x axis to horizontal: dilatancy angle.
Γ	specific volume at the critical state for $p' = 1\cdot0$.
δ	angle of wall friction.
γ	bulk density (weight density): shear strain.
γ'	submerged density (weight density).
$\dot{\gamma}$	shear strain rate (with respect to time).
γ^p	plastic shear strain.
γ_d	dry density (weight density).
γ_f	density of fluid.
γ_p	density of soil particles.
γ_{sat}	saturated density (weight density).
γ_w	density of water (weight density).
ε	normal strain.
ε^p	plastic component of normal strain.
$\dot{\varepsilon}$	normal strain rate (with respect to time).
η	dynamic viscosity: efficiency of a pile group.
κ	slope of an overconsolidation line on the $v : \log_e p'$ plane.
λ	slope of the normal (spherical) consolidation line on the $v : \log_e p'$ plane.
M	slope of the critical state line in a plane of constant specific volume.
M_0	slope of the state boundary surface in a plane of constant specific volume.
μ	coefficient of friction.

N specific volume for normal (spherical) consolidation at $p' = 1{\cdot}0$.

v Poisson's ratio.

ρ surface settlement.

σ total normal stress.

σ' effective normal stress.

$\sigma_1, \sigma_2, \sigma_3$ principal normal stress components.

σ_a, σ_r axial and radial stress components in a triaxial compression test.

σ'_c preconsolidation pressure.

σ'_e equivalent consolidation pressure $\left.\right\}$(one-dimensional).

σ'_i initial consolidation pressure.

σ_n component of stress normal to a surface of sliding.

σ_{oct} octahedral normal stress.

σ_v vertical component of normal stress.

τ shear stress components.

τ_f shear strength.

τ_{max} maximum shear strength.

τ_{oct} octahedral shear stress.

Φ potential function

ϕ angle of shearing resistance (in terms of total stress).

ϕ' angle of shearing resistance (in terms of effective stress).

ϕ'_{cv} angle of shearing resistance at constant volume (in terms of effective stress).

ϕ'_e Hvorslev's effective angle of internal friction.

ϕ_u angle of shearing resistance for undrained conditions (in terms of total stress).

ϕ'_r residual angle of shearing resistance (in terms of effective stress).

ϕ_μ angle of friction at an intergranular contact.

χ coefficient defining effective stress in partially saturated soils.

Ψ stream function.

CHAPTER 1

The analysis and classification of soils

SOILS AND SOIL FORMATION

1.1 Engineering soils: Most of the Earth's land surface, and parts of the bed of the sea, are covered with a layer of granular sediments, mainly derived from the breakdown and decomposition of rocks. Where such sediments either remain entirely uncemented, or are so lightly cemented as not to change their essentially particulate nature, or where cemented materials have been broken up by excavation, they are described as *soils*. It is this particulate or granular nature which distinguishes soils, in the general engineering sense, from rocks.

1.2 Soil formation: The processes of soil formation are complex, but they need only be considered here in so far as they directly affect the engineering properties of the resulting material. Most soils have been formed by the disintegration of rock as a result of weathering processes which may be classified as mechanical or chemical. *Mechanical weathering* is the fragmentation of the parent rock by physical forces, such as those resulting from temperature stresses or from the formation of ice. Temperature stresses, caused by cooling of the rock mass or as a result of daily temperature changes near the surface, lead to cracking. If water percolates into these cracks and subsequently freezes, the resulting expansion opens the cracks further, until eventually pieces are broken from the rock mass. By the same processes, these pieces may then be broken down into smaller and smaller fragments. In dry climates, the impact of sand grains carried by high winds may also cause rapid erosion of rock surfaces.

Mechanical weathering leaves the crystal structure of the material unchanged and clearly identifiable with that of the parent rock. The products of *chemical weathering*, on the other hand, are the result of attack on the rock minerals by water or oxygen or by alkaline or acid materials dissolved in the soil water. Carbon dioxide from the air and organic matter in the top soil are common sources of such dissolved acids.

The clay minerals are a group of complex crystalline materials consisting mainly of aluminium silicates, but containing other materials as well. They are mainly the products of chemical weathering, although there is evidence

1

that they are sometimes formed by hydrothermal action (that is, by the combined action of sub-surface heat and water). Their crystal structures are mainly determined by the climatic conditions under which they have been formed, and do not necessarily reflect the crystal structure of the parent material.

Mechanical and chemical weathering processes take place simultaneously, so that most natural soils contain both primary (rock) minerals and clay minerals, although the proportion of the latter may be small. The weathered material may be carried away into streams and rivers, whence it is deposited in lakes, in the sea, or along the course of the river itself. The size of particle that can be carried by moving water depends critically on the speed of the flow, so that particles of different sizes may be deposited in different places as the speed of the river changes. This results in some sorting of the particles, so that those found in any one deposit may be predominantly of one size. During transport in this way, there is considerable abrasion of the particles, which are broken down into smaller sizes, and often acquire a smoother surface and more rounded shape.

Glaciers and moving ice sheets may also transport soil particles, but the effect is rather different from that in rivers. On the one hand, very large boulders may be plucked from the rock surface and carried great distances: at the other extreme, stones picked up in the bottom layer of the moving ice will, by abrading the rock over which they pass, produce a rock flour of very fine particles. Soil carried by the moving ice is deposited as the ice melts, forming *moraines*. Where these deposits remain as laid down, they contain materials of a great range of sizes, intimately mixed together. On the other hand, the soil may be carried away by the melt water and so pass into the river system.

Leaching, by the action of water passing through the ground, may—particularly under tropical conditions—produce extensive beds of weathered material without subsequent erosion and transportation. Such material is called a *residual soil*. Near the surface, the native rock material may be almost entirely altered in mineral form, but at greater depths the alteration is less and less complete.

SOIL CONSTITUENTS AND THEIR PROPERTIES

1.3   Soil constituents: Any sample of soil will be found to contain some or all of the following:

      (a)   Solid phase:
            (i) Primary rock minerals

        (ii) Clay minerals
       (iii) Intergranular cement
      (iv) Organic matter
  (b) Liquid phase:
        (i) Water
       (ii) Dissolved salts
  (c) Gaseous phase:
        (i) Air (and sometimes other gases)
       (ii) Water vapour.

All these are constituent parts of the soil, and all will, to a greater or lesser extent, affect the engineering properties.

1.4 Primary rock minerals: These are pieces broken from the parent rock. They are generally relatively large, being seldom less than 0·002 mm in diameter, although some soils of glacial origin contain very small particles of rock flour. The particles are generally rounded or angular in shape.

Where particles of this type form the principal part of the soil minerals (as in gravels and sands), the engineering properties depend mainly on the grading (that is, the variation in particle size) and on the closeness of packing. These two factors are to some extent interdependent, for, if the particles are all of about the same size, it is impossible to pack them closely. If, however, there is a good gradation of size from the smallest to the largest, the smaller particles can be made to pack the spaces between the larger. A densely packed material of this type has a low compressibility and a high resistance to shearing, on account of the interlocking of the particles. The shape and texture of the particles also have some effect on the properties, but the mineral composition is irrelevant.

1.5 Clay minerals: These are mainly the products of chemical weathering. The particles are very small, their main dimension being seldom more than 0·002 mm, and frequently very much less. They are commonly in the form of flat plates (although needles, tubes or rods may occur), and in some cases are only a few molecules thick. They therefore have a

TABLE 1.1 *Typical values of the specific surface of soil particles.*

Clay minerals		m^2/g
	Kaolinite	5 to 30
	Illite	50 to 100
	Montmorillonite	200 to 600
Clean sand		2×10^{-4}

very large surface area. The surface area is most conveniently expressed by the *specific surface*—that is, the surface area of a unit mass of the material. Table 1.1 shows typical values of the specific surface for a number of clay minerals.

1.6   *Intergranular cement:*   In some soils, a considerable quantity of cementing material (such as calcite, iron oxide or silica) is deposited on the surfaces of the soil particles. This material may originate in dissolved salts, introduced from elsewhere by the soil water, or may be the residual product of the disintegration of the soil minerals by leaching. In either case, the mineral forms a cement between the particles, increasing the shear strength and reducing the compressibility of the soil.

1.7   *Organic matter:*   Organic matter in the soil is derived from plant or animal remains. It is generally concentrated in the top 0·3 to 0·5 m of the soil, but leaching may carry it down much further in permeable soils, while peat deposits may occur at considerable depths, where the normal processes of decomposition are arrested due to lack of air.

Fresh organic matter in the soil is liable, in the presence of air, to attack by bacteria. The end product of this attack is a group of very complex organic compounds collectively known as *humus*. All these organic materials have properties which are very undesirable in engineering structures. These properties are summarised below.

(a)   Organic material will absorb large quantities of water (up to five times its own weight). Increases in pressure applied to the material cause large volume changes, as the water is expelled. Thus, a bed of peat 3 m thick might settle 0·5 m under quite a modest increase in load. There would also be considerable swelling if the load were removed. Lowering the ground water level by drainage may also cause a reduction in the volume of the soil and a general settlement of the land.

(b)   The material has a very low shear strength and will adversely affect the strength of any soil of which it forms a considerable part.

(c)   Humus has a very large base exchange capacity. (The significance of this is explained in Chapter 2.)

(d)   The presence of organic matter inhibits the setting of cement. Highly organic soils cannot be stabilised with cement.

The quantities of organic matter which can be tolerated depend on its nature and on the purpose for which the soil is required. Humus derived in very acid conditions usually has the most serious effects in inhibiting the setting of cement. In general, less than $\frac{1}{2}\%$ of organic matter is unlikely to affect the setting of cement: 2% to 3% could seriously alter the strength and compressibility of the soil.

1.8   Water: Change in the water content of the soil is the greatest single cause of variation in the engineering properties. Shear strength, compressibility and permeability are all, directly or indirectly, related to the water content.

In conjunction with the electro-chemical forces at the crystal boundaries, the pore water plays a large part in determining the special properties of the clay minerals. The reasons for, and effects of, these special properties are considered in detail in the next chapter.

It must be remembered that the water is as much part of the soil as are the solid particles. Although incapable of carrying shear stresses, the water can support a normal pressure, and this pressure is often a significant part of the stress carried by the soil. An increase in the total stress applied to the soil results generally (unless the soil can drain freely) in changes both in the contact pressure between the soil particles and in the pressure in the pore fluid. Both effects will have to be considered when discussing the stresses in the soil.

In partially saturated soils, there is surface tension at the air/water interfaces within the pores. The pressures in the pore air and in the pore water are not generally the same.

1.9   Dissolved salts: Wherever water passes through the soil, it can transport salts in solution. From an engineering standpoint, the most important of these are the sulphates, because of their destructive effect on concrete. Calcium sulphate occurs fairly generally in British clays, but is only moderately soluble. Sodium and magnesium sulphates are less common, but are more dangerous because of their greater solubility.

The tolerable concentration of the sulphate ion in the soil water depends on the nature of the concrete structures with which it will come in contact. Detailed recommendations are given in reference [1.1].

1.10   Air: Not all soils are fully saturated (that is, the voids between the soil particles are not entirely filled with water). Even heavy clay soils may contain as much as 1% or 2% of air voids. Where the proportion of air in the soil is very small (less than about 5% of the voids), it is mainly in the form of very small bubbles, held in position under high pressure by surface tension. These bubbles are not easily expelled or compressed, and have relatively little effect on the compressibility of the soil. A rather larger proportion of air (up to 15%) results in the formation of larger pockets at lower pressures, and these have a significant effect on the volume changes and pore pressure changes which result from changes in externally applied loads.

Where the voids contain a higher proportion of air, this is mostly continuous throughout the soil mass, and may be fairly easily expelled. This expulsion may take place as a result of compaction of the soil (that is, by the forcible rearrangement of the soil particles into a smaller volume). This can lead to large settlements of the ground surface. Alternatively, the air may be expelled by an influx of water into the voids, causing a reduction in shear strength, particularly in the case of clay soils.

1.11   *Water vapour:*   In partially saturated soils, the relative humidity of the air in the pore spaces is high. The vapour pressure may vary from place to place, because of differences in temperature or for other reasons. If the degree of saturation is low, so that the air spaces are mainly continuous, there may be considerable migration of water in the form of water vapour.

INDEX TESTS

1.12   *Particle size analysis and the Atterberg limit tests:*   In view of the great diversity of soils and soil materials, engineers have long recognised a need to characterise soils simply, so that their engineering properties may be quickly assessed. Tests developed for this purpose are called 'Index Tests', and are as follows:

(a)   *Particle size analysis* to determine the grading. This allows an estimate to be made of the engineering properties of coarse-grained soils.

(b)   *The Atterberg limit tests.* These are a series of empirical tests, from the results of which it is possible to estimate the engineering properties of fine-grained soils. The tests are described in Section 1.16 below.

1.13 *Particle classification by size:* Soil particles may be classified on the basis of their size. The system adopted in the Code of Practice for Site Investigations, CP 2001:1957 [1.2] is shown in Table 1.2.

TABLE 1.2 *Particle size classification.*

Description		Nominal diameter (mm)
Cobbles		more than 60·0
Gravel	Coarse	60·0 to 20·0
	Medium	20·0 to 6·0
	Fine	6·0 to 2·0
Sand	Coarse	2·0 to 0·6
	Medium	0·6 to 0·2
	Fine	0·2 to 0·06
Silt	Coarse	0·06 to 0·02
	Medium	0·02 to 0·006
	Fine	0·006 to 0·002
Clay		less than 0·002

[After 1.2]

Glossop and Skempton [1.3] have shown that the limits of the different groups in this system correspond to real differences in the physical properties of the soils.

It is unfortunate that the terms clay, silt, sand and gravel are used here to define the particle sizes of soil fractions. Elsewhere they are used to define types of complete soil. Although a gravel soil will consist mainly of gravel sized particles, it will generally contain a proportion of material smaller than 2 mm. A clay soil may contain a great deal of material (perhaps more than half) which is larger than 0·002 mm, and which is not formed of clay minerals.

1.14 *Particle size analysis* [1.4]. The coarse particles may be separated by sieving. After treatment to remove organic matter and intergranular cement, the gravel and sand sized particles are separated by passing the soil over a series of sieves of decreasing mesh size. The weight of soil retained on each sieve is recorded.

Material passing the 75 μm sieve is too fine for further division by sieving, and is separated by a process of sedimentation. The fine soil, passing the 75 μm sieve, is first treated with a deflocculating agent to disperse the particles, so that they will settle individually in the suspension. A dilute suspension of the treated soil is then shaken up to disperse the particles evenly, and is allowed to stand.

Stokes's law states that the velocity of a spherical particle sinking in a still fluid is

$$v = \tfrac{2}{9}r^2\left(\frac{\gamma_p - \gamma_f}{\eta}\right)$$

where r is the radius of the particle

γ_p, γ_f are the unit weights of the particle and the fluid respectively

η is the viscosity of the fluid.

Then

$$D = \left(\frac{18h\eta}{(\gamma_p - \gamma_f)t}\right)^{\frac{1}{2}}$$

where D is the particle diameter

h is the distance fallen in time t.

If the suspension were to contain spherical particles, all of the same diameter D, all the particles would sink at the same rate. If the particles were originally distributed uniformly, and were allowed to settle, the particle which started from the surface would be at a depth h after time t. Above this level, the suspension would be clear, but below this point, since all particles sink at the same rate, the concentration of particles would be uniform and unchanged from the beginning of the test.

If a uniformly distributed suspension is prepared, containing particles of various sizes, and if a sample is taken at a depth h below the surface after the suspension has settled for time t, the sample will contain no particles larger than D. All particles smaller than D will be present in the sample in the same proportions as at the beginning of the test. The effect is therefore the same as if the sample had been divided on a sieve of mesh size D.

The concentration of particles remaining in the suspension at any level and at any time may be determined as follows:

(a) *By the pipette method.* Samples are drawn off (usually about 10 ml at a depth of 100 mm) at specified times. The samples are dried, and the weight of the solid residue is recorded.

(b) *By the hydrometer method.* The specific gravity of the suspension is measured at specified times, using a hydrometer. The effective depth of the measurement (h) is assumed to be at the centre of volume of the hydrometer.

Stokes's law assumes that

(a) the particles are spherical
(b) the flow around the particles is laminar, and
(c) the particles are much larger than molecular size.

The assumptions (a) and (c) are not really valid in the case of soil particles, and considerable errors may result, particularly in the case of fine-grained soils. Departure from spherical shape, and molecular influence, both cause particles to settle more slowly than a sphere free from molecular influence. The particle size calculated from Stokes's expression really represents an 'equivalent spherical diameter' and the results are often expressed in this form. Further errors may arise because the dispersion of the particles may not always be complete, and the sedimentation process is therefore of rather uncertain accuracy.

However, in estimating the engineering properties of fine-grained soils the grading is usually of much less significance than the chemical activity determined by the mineral type. It is usually sufficient if we can determine the approximate proportions of fine, medium and coarse silt, and of clay. The methods described above have been found to be sufficiently accurate for this purpose.

1.15 *Particle size distribution curves:* The results of the particle size analysis are plotted on a curve of the type shown in Fig. 1.5. The particle size is shown on the horizontal axis, on a logarithmic scale. The vertical scale shows the proportion (%) of the whole sample which is smaller than any given size. From the shape of the grading curve, it is possible to classify coarse-grained soils, and to make an estimate of their engineering properties.

1.16 *The Atterberg limits* [1.4]: Particle size analysis discloses very little about the engineering properties of very fine-grained soils in which the clay minerals predominate. A better indication of their properties is obtained by measuring the water contents at which certain changes in the physical behaviour can be observed.

The *shrinkage limit* (SL) is the water content below which no further shrinkage takes place as the soil is dried. If the water content is above the shrinkage limit, drying causes a loss of water without a corresponding increase in the air content of the voids, and the volume decreases. Further drying, at a water content below the shrinkage limit, causes no appreciable reduction in volume, the lost water being replaced by air drawn into the voids.

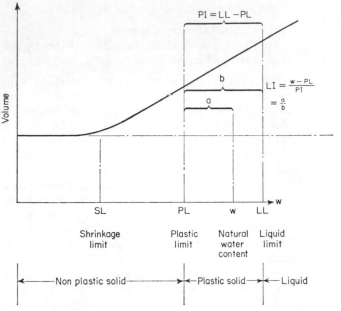

Fig. 1.1   The Atterberg limits.

The *liquid limit* (LL) is the minimum water content at which the soil will flow under a specified small disturbing force. The disturbing force is defined by the conditions of the test. The soil sample is placed in the cup of the standard apparatus (Fig. 1.2), and a groove is cut with a standard tool. The cup is lifted and dropped 10 mm on to a block of rubber of specified hardness. The liquid limit is defined as the water content such that 25 blows will just close the groove for a length of 13 mm.

The *plastic limit* (PL) is the minimum water content at which the soil can be deformed plastically. It is defined as the minimum water content at which the soil can be rolled into a thread 3 mm thick.

The *plasticity index* (PI) is the range of water content over which the soil is in the plastic condition.

$$PI = LL - PL$$

The *liquidity index* (LI) expresses the natural water content in terms of the liquid and plastic limits.

$$LI = \frac{w - PL}{LL - PL} = \frac{w - PL}{PI}$$

where w is the natural water content. The liquidity index varies from zero for soils at the plastic limit to 1·0 for soils

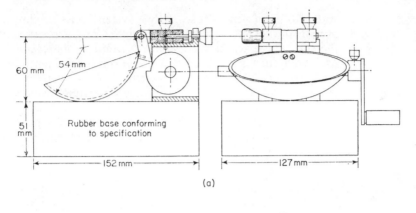

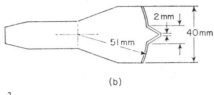

(b)

After BS 1377 [1·4]

Fig. 1.2 (a) Liquid limit apparatus. (b) Standard grooving tool.

at the liquid limit. Exceptionally, certain 'quick' clays have a liquidity index greater than 1·0. These clays are discussed in Section 2.26 below.

1.17 Activity: Even in clay soils, the proportion of clay minerals may be no more than half. The Atterberg limits measure the behaviour of the whole soil, but are mainly determined by the clay minerals present. The results of the tests, therefore, depend on the proportion of clay mineral in the soil, as well as on its nature. It is sometimes necessary to measure the behaviour of the clay mineral fraction only. The *activity* of the soil is therefore defined [Skempton and Bishop, 1.8] as

$$\frac{\text{PI}}{\% \text{ by weight of clay sized particles } (<0.002 \text{ mm})}$$

SOIL CLASSIFICATION SYSTEMS

1.18 The purpose of classification systems: The aim of any classification system is to provide a set of common definitions which will permit useful comparisons to be made between different soils. To be of general use, such a system must be simple, precise and directly related to the engineering properties of the soil. There are, unfortunately, a number of systems in use. Many of these are only of value for special purposes, while others were originally derived from agricultural practice, and are not really suitable for engineering use.

1.19   Analysis of grading curves: The important features of a particle size grading curve may be expressed in terms of the *effective*

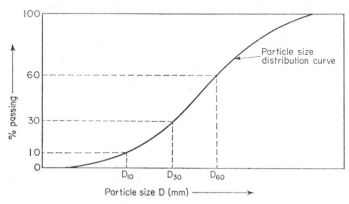

Fig. 1.3   Analysis of grading curves.

size, the *uniformity coefficient* (C_u) and the *coefficient of curvature* (C_c). Three sizes are determined from the grading curve (Fig. 1.3):

(a) D_{10} = the size such that 10% (by weight) of the sample consists of particles having a smaller nominal diameter

(b) D_{30}, D_{60} = the equivalent sizes for 30% and 60% of the sample respectively.

Then:

(a)   The effective size is defined as D_{10}
(b)   The uniformity coefficient (C_u) is D_{60}/D_{10}
(c)   The coefficient of curvature (C_c) is

$$\frac{(D_{30})^2}{D_{60} \times D_{10}}$$

Low values of the uniformity coefficient imply a uniform close grading. Ideally, the coefficient of curvature should be about 2.

1.20   Casagrande's extended classification system (CP 2001 [1.2]): In this system, each soil is allotted two letters; a prefix depending

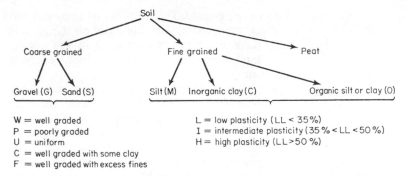

W = well graded
P = poorly graded
U = uniform
C = well graded with some clay
F = well graded with excess fines

L = low plasticity (LL < 35%)
I = intermediate plasticity (35% < LL < 50%)
H = high plasticity (LL > 50%)

Fig. 1.4   Casagrande's extended classification system.

on the predominant particle size, and a suffix related to the engineering properties (Fig. 1.4).

Coarse-grained soils are classed as gravels (G) or sands (S), depending on whether the predominant size is greater or less than 2 mm. The suffix is allotted according to the shape of the grading curve (Fig. 1.5).

(a) W = *well graded* material. The smaller particles will pack the spaces between the larger, giving a dense mass of interlocking particles, with high shear strength and low compressibility.

(b) U = *uniform* material. As the grains are almost all of one size, this material cannot be tightly compacted. The shear strength is generally low.

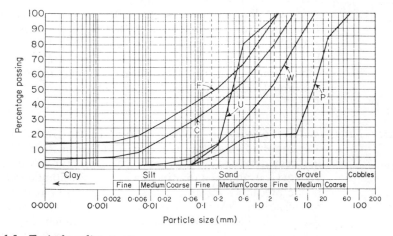

Fig. 1.5   Typical grading curves.

(c)   P = *poorly graded*. This implies that there is a gap in the grading. In the example (Fig. 1.5) there is almost no material in the coarse sand range. Such a soil will not generally compact well.

(d)   C = *well-graded material with some clay*. This is an ideal fill material. It can be closely compacted, and the small proportion of clay acts as a binder, giving a tough dense material with a high shear strength and low compressibility. The permeability is also low, as the clay fraction blocks the pores between the coarse grains.

(e)   F = *well-graded material with an excess of fines*. Where the fine material is more than sufficient to fill the spaces between the larger particles, the latter are no longer in contact with each other, but are embedded in a matrix of fine particles. The fine material begins to control the properties of the whole soil.

Fine-grained soils are classified from the results of liquid and plastic limit tests, plotted on Casagrande's plasticity chart (Fig. 1.6). Soils with plasticity indices above the 'A' line are usually inorganic clays (C). Values below this line indicate silts (M) or organic soils (O).

The suffix is allotted on the basis of the liquid limit.

H = high plasticity (LL > 50%)

I = intermediate plasticity (35% < LL < 50%)

L = low plasticity (LL < 35%).

Soils with liquid limits less than 20% are generally sands. Comparing soils with equal liquid limits, high values of the plasticity index indicate toughness and dry strength, but low permeability and consequently a low rate of volume change under load. Both shear strength and compressibility are directly related to the plasticity index.

1.21   *The Unified Soil Classification System* [1.6]:   In the United States, a system known as the 'Unified Soil Classification System' has been developed. This is based on Casagrande's original system [1.5], and is similar in many respects to that described in the preceding section. It has been adopted—with trivial modifications—in several other countries, and it is likely that a similar classification will be included in future versions of CP 2001 [1.2]. A comparison of the systems is

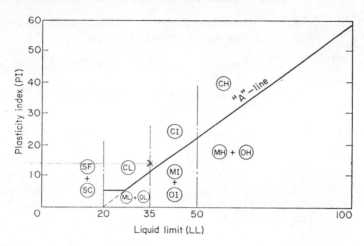

Fig. 1.6   Casagrande's plasticity chart.

contained in reference [1.7]. In this system, prefix letters are allotted as follows:

(a)   Coarse-grained soils (more than half retained on the 75 μm sieve):

G = Gravel (more than half the coarse fraction in the gravel range)
S = Sand (less than half the coarse fraction in the gravel range).

(b)   Fine-grained soils (less than half retained on the 75 μm sieve):

C = Inorganic clay (PI above the 'A' line and PI > 7)
M = Silt   ⎫ (PI below the 'A' line or
O = Organic clay ⎭   PI < 4)

(c)   Highly organic soils:

Pt = Peat.

Suffix letters are allotted as follows:

(a)   Coarse-grained soils having less than 5% passing the 75 μm sieve:

W = Well graded (C_u > 4 (for gravel) or > 6 (for sand); 1 < C_c < 3).
P = Poorly graded (not meeting all the grading requirements for suffix W).

    (b) Coarse-grained soils having more than 12% passing the 75 μm sieve:

        C = with clay (PI above the 'A' line and PI > 7)

        M = with silt (PI below the 'A' line or PI < 4).

    (c) Fine-grained soils:

        H = high plasticity (LL > 50%)

        L = low plasticity (LL < 50%).

Soils having 5% to 12% passing the 75 μm sieve are given a dual classification (*e.g.* GW–GM). Soils having PI above the 'A' line and 4 < PI < 7 are given a dual classification (*e.g.* GC–GM, CL–ML).

WATER CONTENT AND DENSITY

1.22   Effect of changes in density and water content: Shear strength, permeability and compressibility are all greatly affected by changes in the water content and in the closeness of packing of the particles.

1.23   Water content: This is defined as

$$\frac{\text{weight of water in the soil}}{\text{weight of the solid matter}}$$

and is usually expressed as a percentage.

       by
      weight

Air		
Water	}	w
Solid	}	1

The standard method for determining the water content is to dry the soil to constant weight in an oven maintained at 105°C to 110°C. A change in the oven temperature makes a considerable difference to the amount of adsorbed water driven off, particularly in the case of fine-grained soils.

Wherever possible, therefore, the standard temperatures should be used, so that the results may be directly comparable.

Some clay minerals (*e.g.* halloysite) lose crystal water at temperatures well below 110°C. Such materials should not be heated but should be dried at normal laboratory temperature in a desiccator. Organic soils may lose considerable weight due to oxidation at temperatures approaching 100°C, and lower oven temperatures (50°C to 60°C) must generally be used for these soils.

Example 1.1   Water content by oven drying.
Given that
the mass of the wet soil sample and container
$$= 0.317 \text{ kg}$$
the mass of the dry soil sample and container
$$= 0.276 \text{ kg}$$
the mass of the container $= 0.090 \text{ kg}$
Then

$$w = \frac{(0.317 - 0.276) \times 9.81}{(0.276 - 0.090) \times 9.81}$$

$$= \frac{0.040}{0.182}$$

$$= 22.0\%$$

Rapid determinations of water content may sometimes be made in the field by cruder methods—such as heating the soil over a sand bath, or soaking the sample in alcohol and igniting it. These methods are quick, but they involve high temperatures: they are not suitable for soils containing appreciable amounts of clay minerals, since most of these lose considerable quantities of crystal water at temperatures well below 500°C.

1.24   *Bulk density* (γ):   This is defined as the *total weight of soil per unit volume*. The volume of the sample may be determined

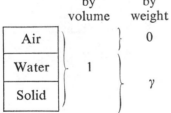

by volume   by weight

(a)   by direct measurement, for regularly shaped undisturbed samples, or

(b)   by measuring the volume of the hole from which the sample has been taken, where undisturbed samples cannot be obtained.

1.25   *Bulk density by direct measurement:*   This is best seen in an example.

Example 1.2
A soil sample, 0.1 m in diameter and 0.15 m long, has a mass of 2.35 kg
Then

$$\gamma = \frac{2.35 \times 10^{-3} \times 9.81}{(\pi/4) \times 0.1^2 \times 0.15}$$

$$= 19.6 \text{ kN/m}^3$$

1.26 Bulk density by the sand replacement method: Where an undisturbed sample cannot be obtained (as in the case of coarse-grained fill materials), the bulk density must be measured *in situ*. A small hole is made in the ground, and the soil taken from it is weighed. The hole is then filled with dry sand, deposited from a standard container (Fig. 1.7) operated in a specified manner. The weight of sand needed to fill the cone

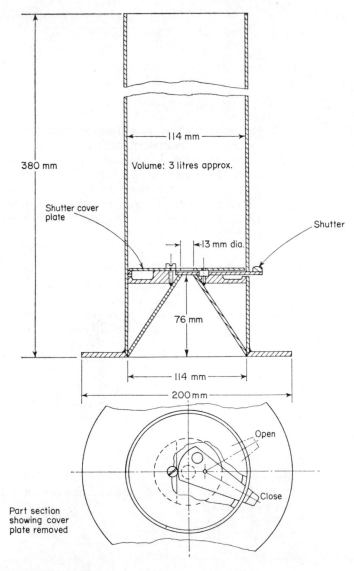

Fig. 1.7 Sand pouring cylinder.

at the base of the container is determined by operating the container on a flat surface. The bulk density of the sand, when deposited from the standard container, is determined by filling a canister of known volume.

Example 1.3
    Given that
        the initial mass of the container and sand
$$= 3\cdot114 \text{ kg}$$
        the final mass of the container and sand
$$= 1\cdot390 \text{ kg}$$
        the mass of soil taken from the hole $= 1\cdot850 \text{ kg}$
        the mass of sand required to fill the
          cone of the container $= 0\cdot321 \text{ kg}$
        the bulk density of the dry sand $= 14\cdot9 \text{ kN/m}^3$

Then the mass of sand required to fill the hole

$$= 3\cdot114 - 1\cdot390 - 0\cdot321$$
$$= 1\cdot403 \text{ kg}$$

and the bulk density

$$\gamma = \frac{1\cdot850}{1\cdot403} \times 14\cdot9$$

$$= 19\cdot6 \text{ kN/m}^3$$

1.27 Dry density (γ_d): This is defined as the weight of solid material per unit volume of soil. (This does not imply that the soil is, or ever has been, dry. Because the soil may shrink during drying, γ_d is generally not equal to the bulk density of the dried soil.)

From the diagram on the left, it may be seen that,

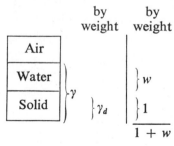

$$\frac{\gamma}{\gamma_d} = \frac{1+w}{1}$$

or

$$\gamma_d = \frac{1}{1+w} \cdot \gamma$$

Example 1.4
    Given that

$$\gamma = 19\cdot6 \text{ kN/m}^3$$
$$w = 22\%$$

Then

$$\gamma_d = \frac{19 \cdot 6}{1 \cdot 22}$$

$$= 16 \cdot 1 \ \text{kN/m}^3$$

1.28   Specific gravity of the soil particles (G_s):   For fine-grained soils, this may be determined using a density bottle of the normal type. For larger grained soils, a *pycnometer* is used. This is simply a glass screw-top jar (Fig. 1.8) with a modified cone-shaped top, designed to prevent air bubbles being trapped under the lid while ensuring that the volume is exactly maintained. Removing the air from the sample is the most difficult part of the procedure, particularly from samples of fine-grained soils.

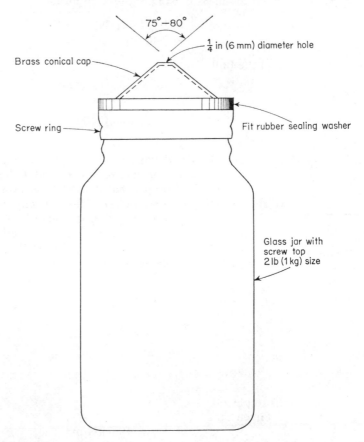

Fig. 1.8   A pycnometer.

The density bottle or pycnometer is filled with water and is weighed (W_1). A sample of dry soil of known mass (W_s) is placed in the bottle, which is weighed again (W_2). Then the mass of water displaced by the soil is

$$(W_1 + W_s) - W_2$$

and

$$G_s = \frac{W_s}{(W_1 + W_s) - W_2}$$

Example 1.5

Given that

the mass of the pycnometer full of water (W_1)
$$= 1{\cdot}734 \text{ kg}$$
the mass of the dried soil sample $(W_s) \quad = 0{\cdot}391 \text{ kg}$
the mass of the pycnometer containing the soil sample
and filled with water $(W_2) \qquad = 1{\cdot}980 \text{ kg}$

Then,

$$G_s = \frac{0{\cdot}391}{(1{\cdot}734 + 0{\cdot}391) - 1{\cdot}980}$$

$$= 2{\cdot}70$$

Most of the commoner primary minerals have specific gravities in the range 2·55 to 2·75 with a mean value of 2·65. The clay minerals generally have slightly higher values of G_s, ranging from about 2·70 to 2·85 with a mean value of 2·75. A few rock minerals have high specific gravities in the range 4·0 to 5·0, but such materials are not common and are seldom found in sediments. A high value of G_s often denotes iron-containing minerals in the soil.

1.29 Void ratio (e): This is defined as $\dfrac{\text{volume of voids}}{\text{volume of solids}}$

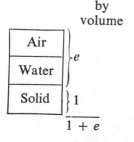

by volume

Example 1.6

Given that

$$\gamma = 19{\cdot}6 \text{ kN/m}^3$$
$$w = 22\%$$
$$G_s = 2{\cdot}70$$

Then

$$\gamma_d = \frac{19 \cdot 6}{1 \cdot 22}$$

$$= 16 \cdot 1 \ \text{kN/m}^3$$

Then the volume of solids in each m³ of soil

$$= \frac{16 \cdot 1}{2 \cdot 70 \times 9 \cdot 81} \ \text{m}^3$$

$$= 0 \cdot 608 \ \text{m}^3$$

and

$$e = \frac{1 - 0 \cdot 608}{0 \cdot 608}$$

$$= 0 \cdot 645$$

1.30 Porosity (*n*): This is sometimes a more convenient expression than the void ratio. It is defined as

$$\frac{\text{volume of voids}}{\text{total volume}}$$

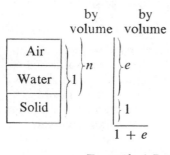

by volume   by volume

From the diagram on the left, it may be seen that

$$n = \frac{e}{1 + e}$$

or

$$e = \frac{n}{1 - n}$$

Example 1.7

If

$$e = 0 \cdot 645$$

$$n = \frac{0 \cdot 645}{1 \cdot 645}$$

$$= 0 \cdot 392$$

1.31 Degree of saturation (S_r): This is defined as

$$\frac{\text{volume of water}}{\text{volume of voids}}$$

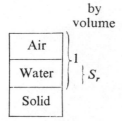

by volume

Alternatively, it may be thought of as

$$\frac{\text{water content of the soil}}{\text{water content of the soil if saturated at the same void ratio}}$$

1.32   Relation between S_r, e, w, and G_s: Consider a sample containing unit volume of solids.

		Then the weight of solids	$= G_s\gamma_w$
by	by	and the weight of water	$= S_r e\gamma_w$
volume	weight		

Therefore

$$w = \frac{S_r e\gamma_w}{G_s\gamma_w}$$

$$S_r e = wG_s$$

Air
Water
Solid

$\Big\} S_r e$ $\Big\} S_r e\gamma_w$

e

1 $G_s\gamma_w$

Example 1.8
Given that

$$e = 0\cdot645$$
$$w = 22\%$$
$$G_s = 2\cdot70$$

Then

$$S_r = \frac{0\cdot22 \times 2\cdot70}{0\cdot645}$$

$$= 92\%$$

1.33   Saturated water content (w_{sat}) and saturated bulk density (γ_{sat}): The *saturated water content* and the *saturated bulk density* are the water content and bulk density which the soil would have if saturated at the same void ratio.

Example 1.9
Given that

$$\gamma = 19\cdot6 \text{ kN/m}^3$$
$$w = 22\%$$
$$S_r = 92\%$$

Then

$$w_{sat} = \frac{0\cdot22}{0\cdot92}$$

$$= 23\cdot9\%$$

$$\gamma_{sat} = 19\cdot6 \times \frac{(1 + 0\cdot239)}{(1 + 0\cdot22)}$$

$$= 19\cdot9 \text{ kN/m}^3$$

1.34   Air void ratio (A):   This is usually defined as $\dfrac{\text{volume of air}}{\text{total volume}}$

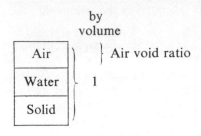

Example 1.10
Given that

$$\gamma = 19\cdot6 \text{ kN/m}^3$$
$$w = 22\%$$
$$G_s = 2\cdot70$$

Then

$$\gamma_d = \frac{19\cdot6}{1\cdot22}$$

$$= 16\cdot1 \text{ kN/m}^3$$

For each m^3 of the soil,

$$\text{volume of solids} = \frac{16\cdot1}{2\cdot70 \times 9\cdot81} = 0\cdot608 \text{ m}^3$$

$$\text{volume of water} = \frac{16\cdot1}{9\cdot81} \times 0\cdot22 = 0\cdot361 \text{ m}^3$$

Then volume of air $= 1 - (0\cdot608 + 0\cdot361) = 0\cdot031 \text{ m}^3$
and the air void ratio $= 3\cdot1\%$
Alternatively, given that

$$n = 0\cdot392$$
$$S_r = 92\%$$

Then the air void ratio $= n(1 - S_r)$
$$= 0\cdot392(1 - 0\cdot92)$$
$$= 3\cdot1\%$$

COMPACTION

1.35   The purpose of compaction: Compaction is the application of
mechanical energy to the soil, to rearrange the particles,
and to reduce the void ratio, usually by driving out air.

When placing any fill material, it is generally desirable to achieve the smallest possible void ratio, for three reasons:

(a) The maximum shear strength occurs approximately at the minimum void ratio

(b) Large air voids may lead to subsequent compaction under working loads, causing settlement of the structure during service

(c) If large air voids are left in the soil, they may subsequently be filled with water, which may reduce the shear strength of the soil. This increase in the water content may also be accompanied by appreciable swelling, and loss of strength in some clays.

Some natural cohesionless soils (particularly some uniform fine sands deposited under water) have a loose structure which is very unstable. These soils must be compacted before loading.

1.36 *Factors affecting compaction:* Compaction is measured in terms of the dry density achieved. This is found to be a function of

(a) the water content

(b) the compactive effort applied to the soil, and

(c) the nature of the soil.

1.37 *The British Standard compaction tests:* In these tests, a sample of soil is compacted, using a standard compactive effort, into a mould whose volume is $9.44 \times 10^{-4}\,\text{m}^3$ ($\frac{1}{30}\,\text{ft}^3$). The soil in the mould is weighed, the water content is measured, and from these measurements the dry density of the soil is calculated.

The test is repeated several times with gradually increasing water contents, until the whole of the relevant range of water content has been covered. The measured values of dry density are plotted against water content, as shown in Fig. 1.9.

In the original form of the test (generally known as the Proctor compaction test), the soil is compacted in three layers, each layer receiving 25 blows from a hammer having a mass of 2.5 kg dropped 0.305 m. As compaction plant increased in weight, a greater compactive effort became necessary for test purposes, and the American Association of State Highway Officials (AASHO) devised a modified test. This is identical with the Proctor test, except that the hammer has a mass of 4.5 kg and is dropped 0.457 m. These two tests have been adopted as the British Standard compaction tests.

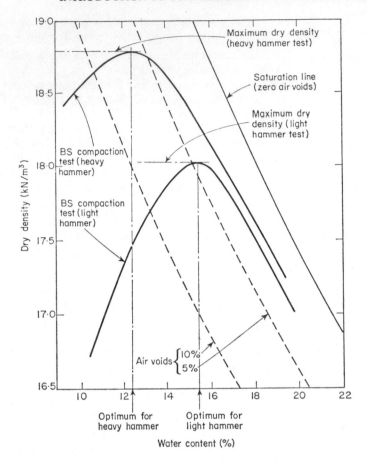

Fig. 1.9   Water content density curves for BS compaction tests.

1.38   The effect of water content on compaction: The shearing resistance to relative movement of the soil particles is large at low water contents. As the water content increases, it becomes progressively easier to disturb the soil structure, and the dry density achieved with a given compactive effort increases. However, if the dry density is plotted against the water content for a given compactive effort, it will be seen that the dry density reaches a peak, after which any further increase in water content results in a smaller dry density.

The reason for this can be readily seen if lines are plotted on the same diagram representing the relation between dry density and water content for complete saturation, and for various air void ratios.

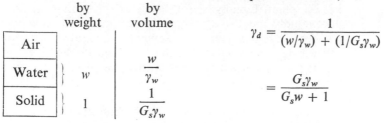

For complete saturation,

$$\gamma_d = \frac{1}{(w/\gamma_w) + (1/G_s\gamma_w)}$$

$$= \frac{G_s\gamma_w}{G_sw + 1}$$

Then for a partially saturated soil, where A is the air void ratio as defined in section 1.34 above, we have

$$\gamma_d = (1 - A) \cdot \frac{G_s\gamma_w}{G_sw + 1}$$

The lines for complete saturation, and for 5% and for 10% air void ratios are shown in Fig. 1.9. It will be seen that the maximum dry density occurs at about 5% air void ratio (or about 85%–90% saturation). At this point much of the air remaining in the soil is in the form of small occluded bubbles, entirely surrounded by water, and held in position by surface tension. As the compaction proceeds, these bubbles become increasingly difficult to displace, and an increasing amount of the compactive effort results only in a momentary increase in pore pressure, so that there is less and less permanent volume change. Complete saturation can seldom be achieved by compaction alone, and, at the higher values of the water content, the curve lies almost parallel to the saturation line.

From the dry density/water content curve, we can determine two quantities;

(a) the *maximum dry density*, and
(b) the *optimum water content* at which this maximum dry density is achieved.

1.39 The effect of variations in the compactive effort: Both the maximum dry density and the optimum water content are found to depend on the compactive effort used. Increasing the compactive effort increases the maximum dry density, but reduces the optimum water content. The air void ratio at the peak density remains very much the same.

It may be seen that, at high water contents, there is little to be gained by increasing the compactive effort beyond a certain point. Since most natural soils in this country have water contents above the optimum for even the lightest plant, and since it is generally impracticable to dry the soil, little improvement in the dry density results from the use of heavier plant. Equally, extra passes of the equipment produce rapidly diminishing returns, once adequate compaction has been achieved. Heavy equipment is generally used, not because it greatly increases the dry density, but for the purely economic reason that it produces a given compactive effort more cheaply.

1.40   The effect of soil type on compaction: The highest dry densities are produced in well-graded coarse-grained soils, with smooth rounded particles. Uniform sands give a much flatter curve, and a lower maximum dry density, as is shown in Fig. 1.10.

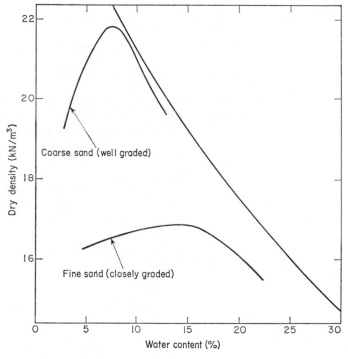

After ref [1·10]

Fig. 1.10   Effect of grading on the compaction of sands.

Clay soils have much higher optimum water contents, and consequently lower maximum dry densities. The effect of increasing the compactive effort is also much greater in the case of clay soils.

Table 1.3 shows typical results of compaction tests for different types of soil.

TABLE 1.3 *Typical values of maximum dry density and optimum water content.* (Based on information in reference [1.10])

(t)	BS Light hammer (Proctor) test		BS Heavy hammer (AASHO) test	
	max. dry density (kN/m³)	optimum water content (%)	max. dry density (kN/m³)	optimum water content (%)
Clay	15·2	28	18·2	18
Silty clay	16·3	21	19·1	12
Sandy clay	18·1	14	20·4	11
Sand	19·0	11	20·6	9
Gravel–sand–clay-mixture	20·0	9	22·0	8

1.41 Relative density (RD): In cohesionless soils of high permeability, such as clean sands and gravels, the compaction is not limited by saturation of the soil, since water can be expelled during the compaction process. It is more useful therefore to compare the actual void ratio of the natural or compacted soil with the void ratio in the densest and the loosest states.

The maximum void ratio (e_{max}) may be determined approximately by pouring the dry soil into a mould, but the value of e_{max} obtained in this way varies somewhat, depending on the rate of pouring and the height from which the soil is dropped. Rapid pouring from a low level produces the loosest state, with the highest void ratio. More consistent results are obtained if the soil is deposited through water, as recommended by Kolbuszewski [1.9].

The minimum void ratio (e_{min}) may be determined by placing the soil in a standard compaction mould under water. The soil should be placed in the mould in three layers, each layer being thoroughly compacted with a vibrating hammer.

The relative density of the soil is defined as

$$RD = \frac{e_{max} - e}{e_{max} - e_{min}}$$

where e is the void ratio of the natural or compacted soil.

REFERENCES

1.1   Building Research Station. 1975. Concrete in sulphate-bearing soils and groundwaters. *Digest* **174**.

1.2   British Standards Institution. Site investigations. CP 2001:1957.

1.3 GLOSSOP, R. and SKEMPTON, A. W. 1945. Particle size in silts and sands. *J. Inst. Civil Eng.* **25.**

1.4 British Standards Institution. Methods of test for soils for civil engineering purposes. BS 1377:1975.

1.5 CASAGRANDE, A. 1947. Classification and identification of soils. *Proc. Amer. Soc. Civil Eng.* **73.**

1.6 US Corps of Engineers. 1960. The unified soil classification system. Waterways Exp. Est., Vicksburg, Miss.

1.7 DUMBLETON, M. J. 1968. The classification of soils for engineering purposes. Road Research Lab. Report LR 182.

1.8 SKEMPTON, A. W. and BISHOP, A. W. 1954. Soils. In *Building materials: their elasticity and inelasticity.* (ed. Reiner, M.) North-Holland Pub. Co.

1.9 KOLBUSZEWSKI, J. J. 1948. An experimental study of the maximum and minimum porosities of sands. Proc. 2nd Int. Conf. on Soil Mech. and Found. Eng. **1.**

1.10 Road Research Laboratory. 1952. *Soil mechanics for road engineers.* HMSO.

CHAPTER 2

The clay minerals

ELECTRO-CHEMICAL FORCES

2.1 Clay mineral crystals and surface forces: The clay minerals are a group of complex alumino-silicates, mainly formed during the chemical weathering of primary minerals. Because of their small size and flat shape, they have very large specific surfaces. There is usually a negative electric charge on the crystal surfaces, and the electro-chemical forces on these surfaces are therefore predominant in determining their engineering properties. In order to understand why these materials behave as they do, it will be necessary to examine their crystal structure in some detail.

2.2 Primary valency bonds: An atom positively charged by the loss of one or more electrons is attracted to atoms negatively charged by the gain of electrons. This force of attraction is responsible for the formation of chemical compounds and for the regular structure of crystalline materials, and is known as an *ionic bond.* Ionic bonds are, compared with some other electro-chemical forces, relatively strong. They decrease in strength inversely as the square of the distance between atoms. Although most of the bonds within clay mineral crystals are of this type, there is evidence that some are covalent, resulting from the sharing of electrons by two or more atoms. This is of limited practical significance, as ionic and covalent bonds are of approximately the same strength.

2.3 Van der Waals forces or molecular bonds: These are forces of attraction caused by the field generated by the spinning electrons surrounding the atomic nucleus. Such forces are considerable at spacings of the same order as the size of the atoms, but they decrease inversely as the 5th or 6th power of the interparticle distance in the case of the very finest clays.

2.4 Hydrogen bonds: Where a hydrogen atom in one molecule or crystal is adjacent to an oxygen atom in another, a particularly strong form of molecular bonding occurs between the two. This attractive force is responsible for some of the weaker bonds between crystal layers (*e.g.* in the mineral kaolinite), and for the adhesion and regular orientation of water molecules at the crystal faces. Hydrogen bonding also affects the structure of liquid water by causing adjacent molecules to associate, and is therefore of great importance in all soil and biological systems. Information about hydrogen bonds is obtained from studies of the infra-red spectra of various materials (including clay minerals).

2.5   *Polar forces:*   In many molecules, the positive and negative charges are not concentric. Where two such molecules approach each other, there is a resultant attractive force, as is shown in Fig. 2.1.

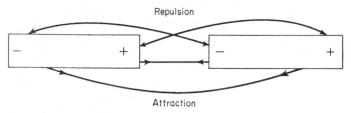

Fig. 2.1   *Forces between polar molecules.*

BASIC STRUCTURE OF THE CLAY MINERALS

2.6   *Tetrahedral and octahedral layers:*   The clay minerals are formed from two basic units, held together by ionic bonds:

(a)   *Tetrahedral units* (Fig. 2.2). The central ion of the tetrahedron is generally, although not invariably, silicon (Si^{4+}) surrounded by ions of oxygen (O^{2-}). The units are formed into layers

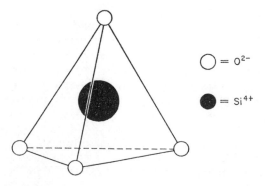

Fig. 2.2   *A tetrahedral unit.*

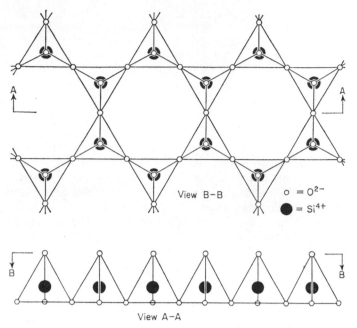

View B–B   o = O^{2-}
                ● = Si^{4+}

View A–A

Fig. 2.3 A tetrahedral layer.

of the form shown in Fig. 2.3 with the general formula $n((Si_2O_5)^{2-})$. The oxygen ion at the apex of each unit carries one unsatisfied valency bond.

(b) *Octahedral units* (Fig. 2.4). These are also formed into layers, with the general formula $n(Al_2(OH)_6)$ or $n(Mg_3(OH)_6)$. Where the central ion is trivalent (*e.g.* aluminium), only two-thirds of the possible central positions must be occupied to balance the positive and negative charges. Such a material is said to be *dioctahedral*. If, on the other hand, the central ion is divalent (*e.g.* magnesium), all positions must be filled for a balanced charge, and the material is said to be *trioctahedral*.

⊘ = $(OH)^-$

● = Al^{3+} or Mg^{2+}

Fig. 2.4 An octahedral unit.

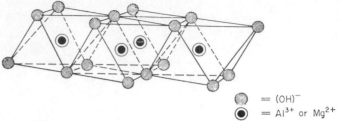

$$\text{\bigcirc} = (OH)^-$$
$$\text{\textcircled{\textbullet}} = Al^{3+} \text{ or } Mg^{2+}$$

Fig. 2.5   An octahedral layer.

If central ions are replaced by others of the same valency, the charge balance is maintained. If, however, ions of different valency enter the lattice, a charge imbalance occurs which can materially increase the surface activity of the clay mineral. The effect of this on the individual minerals is explained later in the chapter.

TWO-LAYER MINERALS—THE KAOLIN GROUP

2.7   *Basic structure:*   The basal spacing in the tetrahedral and octahedral units is almost identical. Hence, a tetrahedral and an octahedral layer will fit together, the unsatisfied bonds in the tetrahedral layer each displacing one hydroxyl group in the octahedral layer. This forms a single sheet of kaolinite with the general formula $n(Al_2Si_2O_5(OH)_4)$.

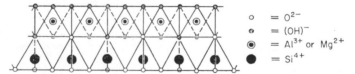

$$\circ = O^{2-}$$
$$\text{\o} = (OH)^-$$
$$\text{\textcircled{\textbullet}} = Al^{3+} \text{ or } Mg^{2+}$$
$$\bullet = Si^{4+}$$

Fig. 2.6   The structure of kaolinite.

2.8   *Stacking of sheets:*   If two such sheets are brought together, hydrogen bonds form between the O^{2-} ions in the tetrahedral layer of one sheet and the $(OH)^-$ ions in the octahedral layer of the next. There is exact conformity between the ions in the two sheets, and they are said to be *regularly stacked*. A particle of kaolinite consists of a large number of such sheets, regularly stacked. Since the hydrogen bonds are weaker than the primary valency bonds, there are distinct planes of weakness between the sheets. However, dislocations (crystal imperfections) are relatively widely spaced, and the particles are therefore fairly large, being commonly up to 1·0 micron in diameter and 0·05 micron thick.

2.9   Other minerals of the kaolin group: Other minerals of this group (*e.g. dickite* and *nacrite*) generally differ only in the stacking arrangements which, for engineering purposes, are irrelevant. The one exception to this is *halloysite*. In this mineral, a layer of water molecules separates each sheet, and there is no regular stacking arrangement. The slight difference in the basal spacing of the two layers causes a curvature of the crystal which can be seen in the tubular form of the particles. From an engineering standpoint, the important feature of halloysite is that the crystals start to lose some of this interstitial water at temperatures below 60°C, forming *metahalloysite*, which has different properties. The process is not reversible. Care is needed when handling samples of halloysite intended for testing.

2.10   Isomorphic substitutions and adsorbed ions: We frequently find that some of the silicon and aluminium ions in the crystal are replaced by other elements, the crystal structure remaining unchanged. Provided that these elements are of the same valency (*e.g.* Fe^{3+} replacing Al^{3+}) this is of little consequence. However, if they are of lower valency (*e.g.* Al^{3+} replacing Si^{4+} in tetrahedral layers, or Mg^{2+} replacing Al^{3+} in a *di*octahedral structure), the crystal is left with a net negative charge. An equivalent number of positively charged ions is attracted to the crystal surface from the soil water. In the kaolins, these isomorphic substitutions are relatively few, and the minerals of this group are therefore nearly inert.

THREE-LAYER MINERALS

2.11   Mica: The prototypes of the three-layer minerals are the primary minerals of the mica group (*e.g.* muscovite and biotite). These are formed of 'sandwiches' of one octahedral layer between two tetrahedral layers, as shown in Fig. 2.7.

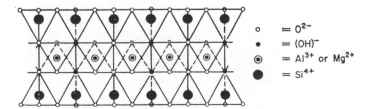

o $= O^{2-}$

• $= (OH)^-$

◉ $= Al^{3+}$ or Mg^{2+}

● $= Si^{4+}$

Fig. 2.7   The structure of mica.

Since the ions in both faces of the sheet are O^{2-}, there are no hydrogen bonds between successive sheets. However, about one quarter of the Si^{4+} ions are replaced by Al^{3+} in the tetrahedral layers, and this leaves about one net negative charge for every four tetrahedral units. This charge deficiency is balanced by the adsorption of potassium ions (K^+) between the crystal sheets. As the potassium ions are of the right size to fit closely into the hexagonal spaces in the tetrahedral layers, they make a good key between successive sheets. This is shown in Fig. 2.8.

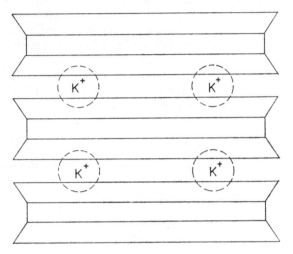

Fig. 2.8   Bonding in mica.

2.12   *Clay micas:*   These minerals are found in almost all temperate clays, and comprise a large part of the clay-sized fraction in most common British soils. They are therefore important, but are much less well understood than the kaolins or the minerals of the montmorillonite group (2.13 below). Starting with the primary mineral muscovite, we can distinguish a continuous series through muscovite–sericite–hydromuscovite–illite–degraded illite, with (in general) a steady decrease in the number of interstitial potassium ions holding the successive layers together. It is possible that the lower (illite) end of the series is continuous with the montmorillonite group of minerals, but this continuity is not fully established, and isomorphic substitutions and different stacking methods for successive layers make identification difficult at this end of the series.

Muscovite is a primary mineral having a low content of combined and adsorbed water, and is nearly inert. The others are secondary (clay) minerals. Towards the lower end of the series, the particles are generally smaller and less well crystallised; the combined, interstitial, and adsorbed water is greater; and soils containing these minerals are progressively poorer as foundation materials.

Existing knowledge of clay minerals in Britain, based on some 3500 X-ray analyses, is summarised by Perrin [2.7]. Few of the minerals seem to match closely the Fithian or Morris illites often used as standards in this work, and much more data are needed [2.8]. Analyses of 19 clay soils from road construction sites are given by Farrar and Coleman [2.9].

2.13   *The montmorillonite group:*   The mineral montmorillonite typically has the formula $n[(Al_{1.67}Mg_{0.33})Si_4O_{10}(OH)_2]$. This means that about one in six of the aluminium ions in the dioctahedral layer are replaced by magnesium. The resulting charge deficiency is made up by the adsorption of cations between the crystal sheets, but these cations are generally not potassium, and are much less effective in keying the sheets together. As a result, water can penetrate between the sheets, causing very large volume changes within the crystal itself. Soils containing much montmorillonite are therefore poor road foundation materials, showing large volume changes between wet and dry seasons. Shallow foundations on such soils may also give trouble.

2.14   *Vermiculite:*   This mineral has a structure similar to that of montmorillonite, except that the interstitial cations are principally magnesium, embedded in a thin layer of regularly arranged water molecules. On heating to about 110°C, about half of this water is driven off, resulting in a reduction in the size of the crystal. This process is rapidly reversed at normal air temperatures, and the material therefore shows expansion properties similar to those of montmorillonite. In the case of vermiculite, however, the extent of the expansion is strictly limited, whereas some montmorillonites may continue to absorb water until the crystal sheets are completely dissociated.

Vermiculite is hardly ever sufficiently abundant in natural soils to affect the soil properties materially, but the absorption properties are used in artificial soils prepared for horticulture.

2.15   *The chlorite group:*   The minerals of this group consist of sheets of trioctahedral mica alternating with single octahedral sheets (brucite). From one quarter to one half of the silicon ions in the tetrahedral layers is replaced by aluminium, leaving a charge deficiency in the mica sheets. This is balanced by an excess charge in the brucite sheets, resulting from the substitution of Al^{3+} for Mg^{2+}.

The minerals of this group differ from each other principally in the stacking arrangements of the successive sheets, and in the location and number of the substituted ions. Well crystallised chlorites have a balanced charge and are nearly inert. Chlorites among the clay minerals may be less well crystallised. They appear to show some random stacking, and probably some hydration of the crystals.

2.16   *Disorder in clay mineral crystals:*   The crystal structures of natural materials always show some deviation from the ideal forms discussed above, the defects being usually more common in the case of montmorillonite and illite and less frequent in the case of the more stable kaolins. As a result of this disorder, considerable diffusion takes place, not merely into the spaces between the sheets, but actually within the crystal lattice itself. This results in a slow process of change, by which the crystal structure may be gradually altered as a result of changes in the surrounding soil water.

Some materials, collectively known as *allophane*, yield no identifiable pattern from X-ray analysis. Such materials do not necessarily have no crystal structure, but they are probably so disordered that no regular arrangement can be discerned. Since most of our information on the crystal structure of the clay minerals is obtained by X-ray diffraction techniques, it is almost impossible to draw any conclusions about the structure of these materials, although electron diffraction studies can sometimes yield information about structures ordered on a very small scale. However, not all amorphous materials are alumino-silicates of the allophane type. Hydrated silica, and oxides of aluminium, iron and manganese, may occur. Other finely divided material may include phosphates, carbonates, sulphates, sulphides, and feldspars (Perrin [2.7], Gillot [2.10]). The poorly crystalline fractions of clay size are attracting increasing attention.

ADSORPTION AND BASE EXCHANGE

2.17   *The adsorption layers:*   Since the ions in the surface layers of the crystals are O^{2-} or $(OH)^-$, water is held against these surfaces by hydrogen bonds and dipolar moments. The exact arrangement of the bound water is disputed, but it is certain that the molecules against the crystal face are tightly bound in a fairly regular pattern. Further away from the surface the bonds become weaker, the arrangement less regular, and the water more fluid and more easily expelled.

An additional factor is the presence of cations, each of which attracts and carries with it its own shell of water molecules. These ions are themselves attracted to the crystal face, partly by the net negative charge on the crystal (resulting from isomorphic substitutions), but also because the surface layers of the crystal are formed of the negative ions O^{2-} or $(OH)^-$. The positively ionised layer has its greatest concentration against the crystal face, and decreases approximately exponentially with the distance from it.

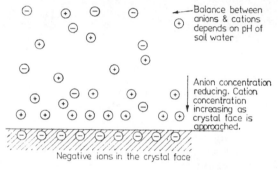

Fig. 2.9 *Adsorption of cations at a crystal face.*

Furthermore, the cations with their water shells will either enhance the degree of order at the particle surface or produce disruption, depending on the perfection of their geometric fit. Ions are constantly diffusing into and out of this layer, and the concentration at any distance from the crystal face is the result of a balance between the number of ions attracted into the layer and the number diffusing out of it.

The thickness of the adsorbed layer depends on two factors:

(a) *The concentration of ions in the soil water.* Increasing the concentration increases the number of cations close to the surface, and therefore reduces the thickness of the layer needed to neutralise the negative ions in the crystal face. Also, where the ion concentration in the free water is higher, the difference between the concentrations at the particle surface and in the free water is less. For a given osmotic gradient, this difference can be accommodated in a thinner adsorbed layer.

(b) *The type of cation in the adsorbed layer*. Monovalent cations (*e.g.* Na^+) naturally lead to thicker layers than divalent ions (*e.g.* Ca^{2+}) since twice as many of the former are required to balance a given charge. Also, the water shell surrounding a sodium cation has a much larger diameter. The rate of diffusion of the ions out of the adsorbed layer depends on their size, so that this will also affect the thickness of the layer. The commonest adsorbed ions, in descending order of layer thickness, are as follows:

$$Na^+$$
$$K^+$$
$$Ca^{2+}$$
$$H^+$$
$$Mg^{2+}$$
$$Fe^{3+}$$

2.18   Base exchange: The type of cation in the adsorbed layers may be altered by altering the type of cation in the soil water. The extent of this alteration (if any) depends on a number of factors, such as the valency and relative concentrations of the cations. The process is called *base exchange*. The base exchange capacity of a mineral is measured in terms of the equivalent weight of the cation available for exchange in unit weight of the mineral. The kaolins, having few isomorphic substitutions and therefore few adsorbed ions, have low exchange capacities in the range 10–40 milliequivalents per 100 g. Montmorillonite and vermiculite have very high exchange capacities (100–150 milliequivalents per 100 g), as the cations adsorbed between the crystal sheets are available for exchange.

STRUCTURE OF CLAY SOILS

2.19   Forces between clay mineral particles: If two particles approach each other in a suspension, the forces acting on them are

    (a)   the Van der Waals forces of attraction, and
    (b)   the repulsion between the two positively ionised adsorbed layers.

At very small separations, the Van der Waals forces are always the larger, and particles which approach sufficiently closely will adhere. However, the Van der Waals forces decrease rapidly with increasing separation. If the adsorbed

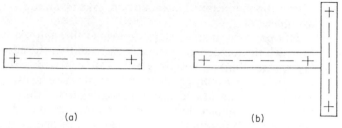

(a)   (b)

Fig. 2.10 (a) Particle with bipolar charge distribution. (b) Edge-to-face arrangement of flocculating particles in a suspension.

layer is thick, the repulsion will be large at distances from the surface at which the Van der Waals forces are small. There will therefore be a large net repulsive force, and particles will tend to settle independently (though very slowly) and will remain dispersed. Contact will only be established if an external force is applied which is large enough to overcome it. On the other hand, if the adsorbed layer is thin, there will be little or no net repulsion at any distance, and random movements of the particles will be enough to bring them into contact. Groups of such particles will form, and will settle together comparatively rapidly, through the suspension. This process is called *flocculation*.

The net forces of repulsion are greatest in the case of particles approaching face to face. As a result, flocculating particles tend to make contact in the edge-to-face arrangement shown in Fig. 2.10. This tendency is accentuated when the particles are flocculating in an acid medium (pH < 7·0). In these conditions the charges on the particles are not evenly distributed: the negative charges are concentrated on the face of the particles, while the positive charges, in the form of hydrogen ions, are concentrated on the broken bonds at the crystal edges. The resulting bipolar charge

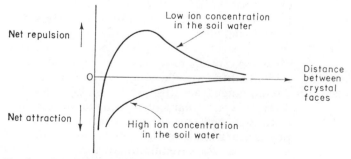

Fig. 2.11 Net force between two particles in a suspension.

distribution increases the tendency to take up an edge-to-face arrangement.

Marine clays (that is, clays deposited through sea water, in which the ion concentration is high, so that the adsorbed layers are thin) generally have a flocculated structure. *Lacustrine clays* (clays deposited in freshwater lakes) generally have a dispersed structure. In this case, few of the particles are in direct contact, most being separated by the adsorbed water layers. A flocculated clay, because of its open structure, generally has a higher liquid limit than one with a dispersed arrangement.

2.20	*Inter-particle bonds:* Bonds may gradually develop at points of contact between particles. These bonds will resist any shearing forces which may be applied subsequently across the points of contact. The nature of these bonds is not quite certain, but they probably have two main causes:

    (a)	*Orientation of the adsorbed water.* Although we talk of particles being in contact, it is probable that the water molecules immediately against the crystal surfaces are never displaced. However, the cations embedded in the adsorbed layers are driven away from the points of contact, and the water molecules are free to develop a stronger and more regular structure than that existing elsewhere in the adsorbed layer.

    (b)	*Cementation.* Cementing material may be deposited at the points of contact, either from solution in the soil water or as the product of chemical weathering.

2.21	*Soil fabric:* The *microstructure* of a clay soil (that is, the geometrical arrangement of the individual particles), and the forces or bonds between those particles, may vary greatly, even between soils with nearly identical mineral compositions. Moreover, superimposed on the microstructure, there is frequently a *macrostructure* consisting of cracks, fissures, partings of different soil material, layering, and other types of discontinuity. This macrostructure, and the microstructure of the soil between the macrostructural discontinuities, are together known as the soil *fabric*. The fabric of a particular soil depends on the conditions during deposition, and on the subsequent history of chemical, physical, and stress changes which have occurred. The fabric is often a very significant feature in determining the behaviour of a particular soil under load.

THE EFFECT ON THE ENGINEERING PROPERTIES

2.22 Permeability: The permeability of a soil is a measure of the resistance to movement of water through the voids. Clearly, this depends to a large extent on the proportion of the voids which is occupied by the adsorbed layers, since the water in these layers is bound in a more or less regular arrangement, and offers a large resistance to displacement. Changing the nature of the adsorbed cations alters the thickness of the adsorbed layer, and therefore alters the permeability of the soil.

2.23 Compressibility: If the normal stress in a soil is increased, and the pore fluid is free to drain from it, the particles are forced together. This causes

    (a)   additional contacts between particles
    (b)   reduction in the thickness of the adsorbed layers, and
    (c)   reorientation and rearrangement of the particles.

If the stress increment is removed, and the pore fluid is free to return to the soil, some expansion takes place as the adsorbed layers return to their original thickness. However, particles which have been forced into contact, or which have been rearranged by rotation, do not return to their original positions. Thus, much of the volume change is irreversible.

The compressibility and swelling properties also depend on

    (a)   the nature of the clay minerals, and
    (b)   the nature and concentration of the cations in the adsorbed layers.

Where clays contain any considerable quantity of montmorillonite, large volume changes accompany changes of stress, as a result of the changes in the thickness of the water layers within the crystals.

The nature and concentration of the ions affects the thickness—and therefore the compressibility and swelling properties—of the adsorbed layers. Soils in which sodium cations predominate in the adsorbed layers are generally more compressible, and swell more, than soils in which the cations are principally calcium. In the case of montmorillonites, the swelling properties of the crystals are also affected by the adsorbed ions. Sodium montmorillonite may adsorb water and swell until the individual sheets in the crystals are almost dissociated.

2.24 Shear strength: Packing the particles more closely increases the interference between the adsorbed layers, and results in more particle contacts. Thus, the shearing resistance of the soil increases with decreasing void ratio.

If the material is physically disturbed, both the structural arrangement and the bonds at the particle contacts are broken down. This results in some loss of shear strength. The *sensitivity* of the soil is defined [Skempton and Northey, 2.1] as the ratio

$$\frac{\text{the shear strength of the undisturbed soil}}{\text{the shear strength of the remoulded soil}}$$

This ratio may be up to about 8 for normally sensitive soils, but may be very much more (perhaps over 100) for very sensitive or 'quick' clays.

In time, diffusion begins to re-establish some of the broken bonds, and there is a gradual regaining of strength. This is known as *thixotropy*. The loss of strength resulting from reorientation of the soil particles is, however, irrecoverable, as is much of that due to broken bonds. Even after a long period, the full strength is not restored.

ORIGIN AND OCCURRENCE

2.25 Clays of normal sensitivity: Clays in marine deposits (*e.g. London clay, Oxford clay*) are predominantly illite, with some additions of other minerals. Thus, the London clay contains about 50% of clay minerals, the remainder being mostly quartz with some feldspar. Of the clay mineral fraction [Grim, 2.2], about 70% is illite, 20% is kaolinite, and 10% is montmorillonite. The montmorillonite results in a fairly considerable shrinkage on drying, although, since the adsorbed ions are principally calcium, the swelling properties are limited.

Kaolins frequently predominate in deposits resulting from the weathering *in situ* of granite. The Cornish *china clay* beds and the *ball clays* of Devon and Dorset have been formed in this way. The ball clays contain an appreciable proportion of organic matter, which considerably increases their plasticity.

In an acid environment and a tropical climate, leaching frequently results in considerable loss of silica, resulting in soils rich in iron and aluminium oxides, and containing halloysite as the principal clay mineral. Table 2.1 shows the constituents of four such clays from Kenya, Java and

Australia. The presence of iron oxide gives many of these soils a characteristic red colour.

In all cases, the iron oxide acts as a cement, forming aggregates of grains. This gives the soil a considerably higher shear strength and lower compressibility than would be expected for a clay: the effect is particularly marked when the halloysite is in the dehydrated form.

TABLE 2.1 *Composition of four tropical red clays*

	Nyeri (Kenya) [2.3]	Sasumua (Kenya) [2.4]	Tjipandoengjang (Java) [2.4]	Silvan dam (Australia) [2.5]
Halloysite	—	60% ± 5%	76%	—
Metahalloysite	50% ± 5%	—	—	80%–90%
Kaolinite	—	4% ± 2%	—	—
Haematite (Fe$_2$O$_3$)	23% (mainly haematite)	—	—	8%–11%
Goethite (hydrated iron oxide)		16% ± 1%	13%	
Non-clay minerals	remainder			

Where weathering has proceeded further, the residual soil may contain a high proportion of the sesqui-oxides of iron and aluminium. These *lateritic soils* occur extensively in Malaysia, India, Africa, and central America. Because of the cementing action of the oxides, they make good road base materials.

Black cotton soils result from the decomposition *in situ* of basalt and similar rocks, under alkaline conditions which are generally the result of poor drainage. They contain a high proportion of montmorillonite, which causes large volume changes on wetting and drying. They therefore present serious problems in the design of foundations. The black colour, at least in some cases, is due to the presence of *nontronite*, a montmorillonite in which the predominant cation is Fe^{3+}.

In many cases, both red clays and black cotton soils have been formed on precisely the same rocks. The red clays have been formed in areas of good drainage on the steeper slopes, while the black cotton soils occur in areas of poor drainage on the hill tops.

Under temperate conditions, soils rich in montmorillonite have generally been formed as a result of the weathering of volcanic ash *in situ*. The best known of these deposits are the *bentonites* of Wyoming. These consist of montmorillonites with only a small admixture of other minerals. The adsorbed ions are principally sodium, and bentonite therefore has a very great swelling capacity. Water can be adsorbed until the individual crystal sheets are completely dissociated, forming a gel with very useful thixotropic properties. The material is used extensively as a drilling mud. *Fuller's earths* which are rich in montmorillonite are found in parts of South Eastern England, although their derivation from volcanic ash is doubtful. The adsorbed cations in this case are mostly calcium, but these can be exchanged for sodium, giving a material with properties similar to bentonite.

Recent very precise analyses (Avery and Bullock, [2.11]) of the clay fractions of 102 British soil profiles have given the following new information:

(a)   Chlorite is commoner in soil clays than earlier work had indicated. Although it is absent from some clays, many contain 10–20% of the mineral, and some contain up to 40%. Although it is not harmful, its presence complicates soil analysis.

(b)   Montmorillonite as such is rare in soils. Most so called montmorillonites in ordinary soils are interstratified mica/montmorillonites of the degraded illite type with low, but not zero potassium content. However, as in the case of true montmorillonites, they have very poor engineering properties.

(c)   The geologically recent rocks and soils on average contain much more of this undesirable mineral than older materials.

2.26   Very sensitive or 'quick' clays: The 'quick' clays, which occur frequently in Scandinavia and northern Canada, were originally deposited in the sea, and consist mainly of illites and chlorites. They have the flocculated structure which is typical of this origin. As a result of the subsequent isostatic uplift of the land, they are now well above sea level and have been leached by fresh water. The reduction in the ion content of the soil water has resulted in a considerable increase in the thickness of the adsorbed layers surrounding the particles. If the soil is disturbed, so that the flocculated structure is broken up, the particles tend to take up a dispersed arrangement. In this condition, the soil flows, since the water content exceeds the liquid limit. Very large slides take place on quite shallow slopes since, once disturbed, the clay will flow out in almost liquid condition [Bjerrum, 2.6].

REFERENCES

2.1 SKEMPTON, A. W. and NORTHEY, R. D. 1952. The sensitivity of clays. *Géotechnique*, **3**.

2.2 GRIM, R. E. 1949. Mineralogical composition in relation to the properties of certain soils. *Géotechnique*, **1**.

2.3 COLEMAN, J. D., FARRAR, D. M. and MARSH, A. D. 1964. The moisture characteristics, composition and structural analysis of a red clay from Nyeri, Kenya. *Géotechnique*, **14**.

2.4 TERZAGHI, K. 1958. Design and performance of the Sasumua dam. *Proc. Inst. Civil Eng.*, **9**.

2.5 LAMBE, T. W. and MARTIN, R. T. 1955. Composition and engineering properties of soil. *Proc. Highways Res. Board*, **34**.

2.6 BJERRUM, L. 1954. Geotechnical properties of Norwegian marine clays. *Géotechnique*, **4**.

2.7 PERRIN, R. M. S. 1971. *The clay mineralogy of British sediments*. The Mineralogical Society, London.

2.8 WEAVER, C. E. and POLLARD, L. D. 1973. The chemistry of clay minerals. *Developments in Sedimentology*, **15**.

2.9 FARRAR, D. M. and COLEMAN, J. D. 1967. The correlation of surface area with other properties of 19 British clay soils. *J. Soil Science*, **18**.

2.10 GILLOT, J. E. 1968. *Clay in engineering geology*. Elsevier, Amsterdam.

2.11 AVERY, B. W. and BULLOCK, P. 1977. *Mineralogy of clayey soils in relation to soil classification*. Soil Survey Technical Monograph No. 10, Rothamsted Experimental Station.

CHAPTER 3

Pore pressure, effective stress and suction

PORE PRESSURE AND EFFECTIVE STRESS

3.1   Pore pressure (u):   The voids between the solid particles of soil are filled with fluid—either water or air or a combination of the two. Except perhaps within the adsorbed water layers, these fluids can offer no resistance to static shear forces. They can, however, support normal pressures, and these normal pressures form part of the total stress in the soil.

3.2   Effective stress (σ'):   Changes in the total stress and in the pore pressures in the soil lead to changes in volume and in shear strength. In considering the stability and compressibility of the soil, the first problem is to determine an *effective stress (σ')* which controls the changes in volume and shear strength. This effective stress will be dependent on the total stress (σ) and on the pore air and pore water pressures (u_a and u_w).

3.3   Terzaghi's expression for effective stress in saturated soils: Terzaghi showed experimentally that, for saturated soils,

$$\sigma' \simeq \sigma - u_w$$

both in respect of compressibility and of shear strength. Subsequent investigation [Skempton, 3.1] has confirmed this to a high degree of accuracy provided that

(a) the area of contact between particles is small, or
(b) the compressibility of the individual particles is small compared with that of the whole soil skeleton.

Even in the case of rock or concrete, where neither of these conditions strictly applies, the expression is generally sufficiently accurate for engineering purposes. In the case of soils, the error is insignificant.

3.4   Pore pressures in partially saturated soils: Where the voids in the soil contain both air and water, surface tension occurs at the air/water interfaces. As a result, the pressures in the water and in the air are not the same.

49

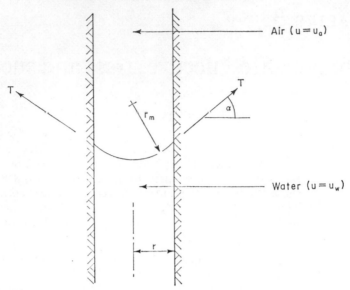

Fig. 3.1   Pressure difference at an air/water interface.

Consider the simple capillary shown in Fig. 3.1. For equilibrium

$$u_a \pi r^2 = u_w \pi r^2 + 2\pi r \sin \alpha T$$

where T is the surface tension (per unit length of air/water interface). Then

$$u_a - u_w = \frac{2T \sin \alpha}{r}$$

$$= \frac{2T}{r_m}$$

where r_m is the meniscus radius.

It may therefore be seen that

(a) u_a is always greater than u_w, and
(b)   the pore pressure difference $(u_a - u_w)$ increases with decreasing radius of the meniscus at the interface.

The meniscus radius depends on the pore size. In sands, therefore, the pressure difference is relatively small, seldom exceeding 5 kN/m². In clay soils, where the mean pore diameter is very much smaller, pressure differences of more than 10 000 kN/m² may occur.

3.5   *Effective stress in partially saturated soils:* Bishop [3.2] has suggested the general expression

$$\sigma' = \sigma - [u_a - \chi(u_a - u_w)]$$

where χ is a parameter dependent on the degree of saturation (S_r). For dry soils, where $S_r = 0$, $\chi = 0$ and $\sigma' = \sigma - u_a$. For fully saturated soils, $S_r = \chi = 1{\cdot}0$ and $\sigma' = \sigma - u_w$. For intermediate values of S_r, the values of χ are to some extent dependent on the proportion of the air which is continuous through the pores, and the proportion which is in the form of occluded bubbles entirely surrounded by water. The occluded air has much less influence on the effective stress than the air in continuous pore spaces.

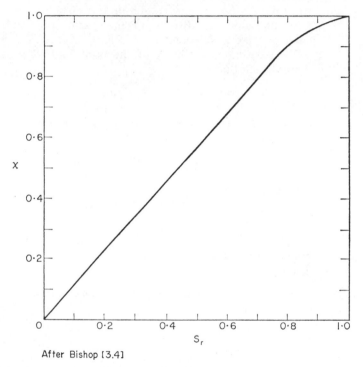

After Bishop [3.4]

Fig. 3.2   Typical relationship between S_r and χ.

Thus, although χ is mainly dependent on the degree of saturation (S_r), it is also influenced by other factors, such as the soil structure and the processes of wetting and drying by which the present degree of saturation has been reached. Fig. 3.2 shows a typical relationship between the degree of saturation and χ.

It follows that the determination of the effective stress in a partially saturated soil is not a simple matter. Moreover, there is evidence that the values of χ (and therefore the values of the effective stresses) are not exactly the same in respect of shear strength and of compressibility. However for soils approaching saturation (that is, where $S_r \geqslant 90\%$) it may be assumed that $\chi = 1\cdot0$ and $\sigma' = \sigma - u_w$ to a sufficient degree of accuracy for engineering purposes.

PORE PRESSURE CHANGES RESULTING FROM TOTAL STRESS CHANGES

3.6   The effect of a uniform increase in total normal stress: Consider an element of soil subjected to a uniform increment of total stress $\Delta\sigma_3(= \Delta\sigma_1 = \Delta\sigma_2)$ under undrained conditions. (The term 'undrained conditions' means that no pore fluid can enter or leave the element.)

Then the compressibility of the pore fluid (C_v) may be defined such that

$$-\frac{\Delta V_v}{V_v} = \Delta u C_v$$

and the compressibility of the skeleton of solid particles (C_s) such that

$$-\frac{\Delta V_s}{V_s} = (\Delta\sigma_3 - \Delta u)C_s$$

where V_v = the volume of the voids
  V_s = the total volume of the element
  $u = u_a - \chi(u_a - u_w)$.

But $V_v = nV_s$ and $\Delta V_v = \Delta V_s$ (if the soil particles are assumed to be incompressible).
Then

$$\Delta u = \frac{1}{1 + n(C_v/C_s)}\,\Delta\sigma_3$$

$$= B\,\Delta\sigma_3$$

where

$$B = \frac{1}{1 + n(C_v/C_s)}$$

In a saturated soil, the pore fluid is almost incompressible, and C_v is small compared with C_s. Then $B \simeq 1\cdot0$. Thus, for a saturated soil, a uniform increase in total stress results in an equal rise in pore pressure and the effective stress remains unchanged.

$$\Delta u = \Delta\sigma_3; \qquad \Delta\sigma'_3(= \Delta\sigma'_1 = \Delta\sigma'_2) = 0$$

In an unsaturated soil, the presence of air in the voids increases the compressibility of the pore fluid.

$$\frac{C_v}{C_s} > 0 \quad \text{and} \quad B < 1\cdot0$$

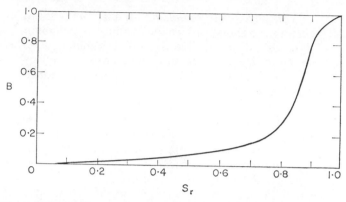

Fig. 3.3   Relationship between the pore pressure coefficient B and the degree of saturation S_r.

Fig. 3.3 shows a typical relationship between B and the degree of saturation (S_r).

3.7   *The effect of a change in the principal stress difference:*   Suppose that the same element is now subjected to a further increment of total stress ($\Delta\sigma_1 - \Delta\sigma_3$) in one direction only. There is a further change in the pore pressure, resulting from the balance of two effects—the effect of the change in mean stress on the one hand, and the effect of shear strains on the other.

(a)   *Increase in mean stress.*   As a result of the increase in mean stress there is an increase in pore pressure, as shown in the preceding section. This increase is equal to the rise in mean stress in the case of saturated soils, but is smaller where air is present in the voids.

(b)   *Shear strains.*   It is a characteristic of particulate materials that shear strains result in changes in volume as the particles are rearranged. After large deformations have taken place, the original soil structure is completely destroyed. The void ratio (e) tends to a critical value (e_c) which depends on the geometry of the soil particles (their shape, grading, etc.), but is unaffected by the original particle arrangement. This critical void ratio is also dependent on the effective normal stress, and any increase in the latter causes a reduction in e_c.

Where particles are tightly packed, with a void ratio $e < e_c$, shear deformation at constant mean effective stress causes an increase in the volume until $e = e_c$. This increase of volume is called *dilatancy*. If the soil is saturated and undrained, almost no volume change is possible, as water is almost incompressible. The effect of dilatancy is therefore to reduce the pore pressure and increase the effective stress until the increase in the latter is sufficient to reduce e_c to the existing void ratio (e).

Dense sands and heavily consolidated clays are generally dilatant, and shear deformations reduce the pore pressure in these soils. Conversely, the void ratio in loose sands and in soft clays is commonly greater than the critical value (e_c), and the effect of shear is to increase the pore pressure.

In unsaturated soils, the effects of shear deformations on the pore pressures are similar, but the actual pore pressure changes are smaller, and some volume change occurs due to the compressibility of the pore fluid.

Thus the change in pore pressure may be expressed as

$$\Delta u = B\{\Delta\sigma_3 + A(\Delta\sigma_1 - \Delta\sigma_3)\}$$

where A and B are Skempton's pore pressure parameters [3.3].

It must be understood that the pore pressure parameter A is not constant. It will be influenced by such factors as the level of normal and shear stresses applied to the soil (including the intermediate principal stress σ_2), and whether the stresses are increasing or decreasing.

Because of this variation, the value of A can only be determined experimentally, but the conditions of the test must be carefully adjusted to the conditions of loading in the ground. The pore pressure parameters will be discussed further in Chapter 6.

SOIL SUCTION

3.8   The water table: In considering the movement of water in the soil, it is convenient to measure all pressures from atmospheric pressure as zero, and to define a *water table*, which is an imaginary surface within the soil joining points at which the pore water pressure is zero (*i.e.* atmospheric). Except where there is rapid downward seepage, the pore water pressure is negative at all points above the water table, and positive at all points below. Most soils are practically saturated below the water table, but above it they may contain considerable quantities of air.

The depth of the water table below the ground surface varies with the soil type, the topography and the climate. There may also be seasonal variations as a result of the changing balance between rainfall and evaporation at different times of the year. Where rainfall exceeds evaporation, the water table is generally close to the surface in soils of low permeability, but may be at depths of 6 m to 10 m in sands. In an arid climate, a high water table is generally only found close to the sea or to some similar source of supply. In the absence of this, the water table may fall to a depth of 30 m or more.

Between the water table and the ground surface, water is held in the soil by *matrix* (or *capillary*) *suction* and by *solute* (or *osmotic*) *suction*.

3.9 *Matrix suction:* The air at the air/water interface may usually be assumed to be at atmospheric pressure. There is, therefore, a negative pore water pressure

$$-u_w = \frac{2T}{r_m}$$

as shown in Section 3.4, above. This negative pressure is the matrix or capillary suction.

3.10 *Solute suction:* If a solution is in equilibrium with a source of pure water through a semi-permeable membrane (that is, a membrane permeable to water but not to the solute), there is a pressure difference across the membrane equal to the pressure exerted on the membrane by the molecules of the solute. This is shown in Fig. 3.4. Thus, when in equilibrium the pressure in the pure water is less than that in the solute by the same amount. Similarly in the pores of a soil, if the ion concentration at some point is less than elsewhere, there is a pressure deficiency at that point. This pressure deficiency is the solute or osmotic suction.

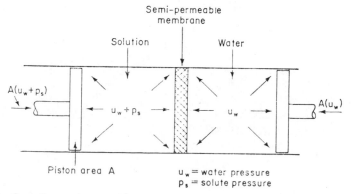

Fig. 3.4 *Osmotic or solute suction.*

3.11 *Total suction:* The total suction is the sum of the matrix and solute suctions. Because of the wide range, it is convenient to measure suction on a logarithmic scale. The *pF* index is therefore defined as

$$\log_{10} \text{ (suction, measured in cm of water)}$$

In clean sands, the matrix suction seldom exceeds 0.5 m of water ($pF = 1.7$) but in oven-dried clays, where the capillaries are very much smaller, suctions up to $pF = 7.0$ are commonly observed.

The solute suction may be as much as one atmosphere ($pF = 3$) between the layers of a montmorillonite crystal, but seldom exceeds one tenth of this value ($pF = 2$) in the free water between the particles. In many cases, therefore, it is of small significance. Many of the methods used to measure soil suctions do not distinguish between the matrix and solute suctions.

3.12 *Desaturation and resaturation:* Consider a soil sample, initially saturated with water at zero pressure, and then subjected to an increasing suction. The maximum matrix suction which can be sustained by the water in any pore is determined by the pore size. Once this suction is exceeded, the pore will drain and air will enter to replace the water.

Since all pores are not of the same size, they do not all drain at the same suction. The water content of the soil is therefore progressively reduced as the suction increases [3.5].

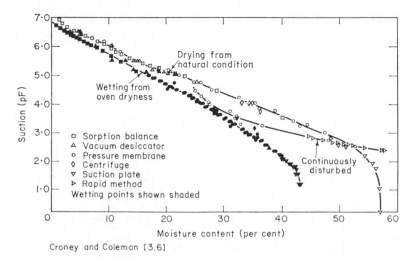

Croney and Coleman [3.6]

Fig. 3.5 *Water content/suction curves for Gault clay.*

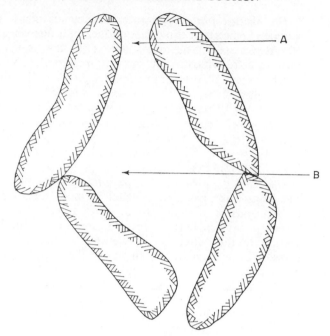

Fig. 3.6   Variation of matrix suction within the pores.

The upper curve in Fig. 3.5 shows the water content/ suction relationship measured in a sample of Gault clay, drained under increasing suction.

Now consider the same sample, drained under a large suction, which is then progressively reduced. As the suction is reduced, water is drawn back into the soil, but the suction at which any particular pore space is resaturated is much less than the suction originally required to drain it. Consider the pore space shown in Fig. 3.6. The suction required to drain the space is governed by the small constriction at *A*, whereas the suction at which water is drawn back to refill the space depends on the much smaller matrix suction at *B*. Thus the water content/suction relationship is not the same for increasing as for decreasing suction. The lower curve in Fig. 3.5 was obtained when the same soil sample was wetted from the oven-dried condition.

3.13   Determination of the relationship between soil suction and water content: Because of the large range of suctions to be measured, no single method is suitable for all cases. A large number of techniques have been developed, of which the suction plate, pressure membrane and vacuum desiccator are the most suitable for routine investigations.

The suction plate apparatus [Croney *et al.*, 3.7]: The sample is placed on the upper surface of a fine porous disc of sintered glass. The under side of the disc is in contact with water to which a known suction is applied. When the water content of the sample has reached equilibrium with the applied suction, the sample is weighed. The applied suction may be varied to give a range of values. However, the suction in the water below the disc is limited to about 1 atmosphere, and the method is restricted to the range $pF = 0$ to $pF = 3$. The suction plate apparatus is now available commercially.

The pressure membrane apparatus [Croney *et al.*, 3.8]: The sample is placed on a cellulose membrane supported on a porous bronze disc which is in contact with water at zero (*i.e.* atmospheric) pressure. Instead of applying a suction to the water, as in the suction plate, a positive pressure is applied to the air above the sample. Once the water content has reached equilibrium, the suction is equal to the applied air pressure. The standard form of the apparatus is suitable for the range of suctions from $pF = 0$ to $pF = 5 \cdot 2$. The pressure membrane apparatus is now available commercially.

The vacuum desiccator: The sample is placed in a desiccator containing a solution of sulphuric acid of known density. The acid solution controls the humidity of the air in the desiccator, the relative humidity depending on the concentration of the solution. When equilibrium has been reached, at a standard temperature of 20°C, the suction in the soil is

$$pF = 6 \cdot 50 + \log_{10}(2 - \log_{10} H)$$

where H is the relative humidity in the desiccator.
The desiccator is evacuated to increase the rate of moisture movement to and from the sample.

The use of strong solutions of sulphuric acid to control the humidity in the desiccator can be very dangerous. The humidity control can be provided just as well, although rather less conveniently, by saturated salt solutions, provided that they are stirred at intervals to maintain their uniformity.

The vacuum desiccator is suitable for large suctions in the range $pF = 5 \cdot 0$ to $pF = 7 \cdot 0$.

3.14 *The effect of normal stress on the soil suction:* In assessing the water content/suction relationship in the ground above the water table, the suction as defined above must be modified to allow for the effect of normal stresses in the soil, since these normal stresses also reduce the water content [Croney, 3.9].

Then we may write

$$\text{soil suction} = -u_w + \alpha \cdot \sigma_m$$

where σ_m is the mean normal stress in the soil, and is equal to

$$\frac{\sigma_1 + \sigma_2 + \sigma_3}{3}$$

In practice, since the horizontal stresses are seldom known with any certainty, σ_m is usually taken to be equal to the vertical component of normal stress σ_z. The coefficient α may be determined from simple tests. Alternatively it may be estimated from the empirical relation

$$\alpha \simeq 0.03 \text{ PI}$$

where PI is the plasticity index.

3.15 *Estimation of equilibrium water content beneath pavements* [Aitchison *et al.*, 3.11]: The construction of an impervious pavement prevents infiltration and evaporation through the surface, and may alter the equilibrium water content of the soil below. Since the shear strength of the soil is dependent on the water content, it is often necessary to estimate the final condition.

Where the water table is close to the surface (that is, at depths less than about 7 m in clays and about 1 m in sands), the equilibrium water content is mainly controlled by the position of the water table, irrespective of the climatic conditions. Since there is no vertical flow, the negative pore pressure $-u_w = \gamma_w \times$ (the height above the water table), and the soil suction can then be computed as described in the last section. From this, and from the water content/suction curves, the final water content profile can be determined.

Where the water table is at a great depth, the equilibrium state is affected by the climate. In arid zones, with a rainfall of less than 0.25 m per year, the effect of the pavement is small, and the ultimate water content beneath the pavement differs little from that of the uncovered soil at the same depth. Where the rainfall is greater, it is usually necessary to use some empirical relation based on the balance between rainfall and evaporation.

3.16   Frost damage in soils: Where both ice and water are present in the soil, the suction is uniquely related to the temperature. By equating the vapour pressure of the ice to that of the supercooled water under negative pressure, it can be shown [Croney, 3.10] that, within certain limits,

$$pF \simeq 4{\cdot}10 + \log_{10} t$$

where t is the depression of the freezing point in °C below zero.

Stated another way, a drop of 1°C in the freezing point is accompanied by a suction of about 12 atmospheres. It follows that very large suctions are possible in freezing soils. If the conditions are right, this suction causes migration of water into the freezing zone, and the formation there of lenses of segregated ice.

This water movement requires

(a)   a high water table, as a source of supply
(b)   a sufficient number of small pores, in which the matrix suction, by depressing the freezing point, will prevent the formation of ice, and
(c)   a sufficiently high permeability.

The initial penetration of frost into the soil is rapid, and the zero isotherm may penetrate 0·1 m below the surface in a few hours. There will generally be no time for significant segregation of ice in the surface layer, but this soil is subjected to cycles of freezing and thawing if the daytime temperature rises above zero. At greater depths, the penetration of the isotherm is much slower. A continuous period of about three days, with subzero temperatures day and night, is required before the zero isotherm penetrates 0·3 m, and about ten days are required for a penetration of 0·4 m. In the British Isles, the maximum penetration is not likely to exceed 0·5 m except in places with exceptional exposure.

Croney and Jacobs [3.12] have examined the frost-susceptibility of a number of soils and road making materials, and their conclusions are summarised below:

(a)   *Cohesive soils* are not seriously frost-susceptible if the plasticity index exceeds 15% for well drained soils and 20% in areas of poor drainage. The liability to frost damage is reduced by better compaction, because the permeability is thereby reduced. Chemical additives may be used to reduce heave in cohesive soils.

(b) *Non-cohesive soils and crushed rocks* (other than limestone) are not frost susceptible if they contain less than 10% of material passing the 75 μm sieve. A small quantity of cement or bitumen additive is sufficient to prevent all heave in non-cohesive soils.

(c) All *crushed chalks* are frost-susceptible. Chemical or cement additives may be used to reduce frost heave, but the quantities required are usually so great that it is cheaper to replace the chalk within 0·5 m of the surface with other non-susceptible materials.

(d) *Oolitic and magnesian limestones* have extensive pores within the particles, and are generally susceptible if the saturated water content of the particles exceeds 3%.

(e) *Burnt colliery shales* and some *pulverised fuel ash* (containing more than 40% of material passing the 75 μm sieve) are frost-susceptible. Tests should be made on representative samples before such materials are used in the top 0·5 m of road construction.

Frost damage takes two forms. Firstly, the expansion of the ice lenses in road bases and subgrades can break up the whole road structure. Secondly, further damage often occurs during a thaw. In Britain, a thaw usually occurs fairly quickly, but (if the permeability is low) some time may elapse before all the water from the segregated ice has drained away, and the equilibrium water content has been re-established. Until this has happened, the strength of the soil may be much reduced, and pavements may be damaged by traffic. In colder climates, where the frost penetration is deeper, the first thaw often affects the surface layer only. Water within this layer cannot drain away, as it is trapped by the still frozen layer below.

REFERENCES

3.1   SKEMPTON, A. W. 1961. Effective stress in soils, concrete and rocks. In *Pore pressure and suction in soils,* Butterworths.

3.2   BISHOP, A. W. 1955. The principle of effective stress. *Teknisk Ukeblad,* **39.**

3.3   SKEMPTON, A. W. 1954. The pore pressure coefficients *A* and *B. Géotechnique,* **4.**

3.4   BISHOP, A. W. 1961. The measurement of pore pressure in the triaxial test. In *Pore pressure and suction in soils.* Butterworths.

3.5   CRONEY, D. and COLEMAN, J. D. 1954. Soil structure in relation to soil suction (pF). *J. of Soil Science,* **5.**

3.6   CRONEY, D. and COLEMAN, J. D. 1961. Pore pressure and suction in soil. In *Pore pressure and suction in soils.* Butterworths.

3.7   CRONEY, D., COLEMAN, J. D. and BRIDGE, P. M. 1952. The suction of moisture held in soil and other porous materials. Road Res. Tech. Paper 24 (HMSO).

3.8   CRONEY, J., COLEMAN, J. D. and BLACK, W. M. P. 1958. The movement and distribution of water in soil in relation to highway design and performance. Highway Res. Board Sp. Rep. No. 40 (Washington).

3.9   CRONEY, D. 1952. The movement and distribution of water in soils. *Géotechnique,* **3.**

3.10   CRONEY, D. 1952. Frost damage to roads in Great Britain. Highway Res. Board Sp. Rep. No. 2 (Washington).

3.11   AITCHISON, G. D., RUSSAM, K. and RICHARDS, B. G. 1965. Engineering concepts of moisture equilibria and moisture changes in soils. In *Moisture equilibria and moisture changes in soils beneath covered areas.* Butterworths.

3.12   CRONEY, D. and JACOBS, J. C. 1967. The frost susceptibility of soils and road materials. Road Research Laboratory.

CHAPTER 4

Permeability and seepage

SOIL PERMEABILITY

4.1 Total head (h) and hydraulic gradient (i): Consider a point B in a mass of saturated uniform porous soil. Let B be at a height z above some arbitrary datum level (A–A in Fig. 4.1), and let the pore pressure at B be u. A column of water in equilibrium with the pore pressure at B would rise to a

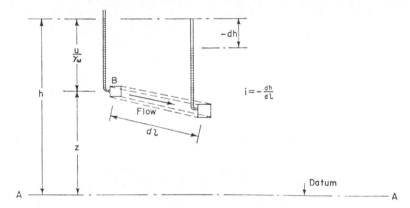

Fig. 4.1   Total head and hydraulic gradient.

height u/γ_w above B. The *total head* (h) at B is defined as the height of this column of water above the datum level. The total head (h) is equal to the sum of the *position head* and the *pressure head:*

$$h = z + u/\gamma_w$$

If the head (h) is everywhere the same, there is no flow of water through the soil. If the head differs in different parts of the soil mass, water flows away from points at which the head is high, and towards points at which the head is lower.

The rate of flow is governed by the *hydraulic gradient*, which is defined as

$$i = -\frac{dh}{dl}$$

where l is the distance measured along the flow path.

63

4.2 Permeability and Darcy's law: Permeability is a measure of the rate at which fluid passes through a porous medium. In the case of water passing through soil, Darcy [4.1] defined a coefficient of permeability (k) such that

$$v = -k\frac{dh}{dl} = ki$$

where v is the *apparent velocity* of flow, and is equal to the average rate of flow of water across unit area in the soil.

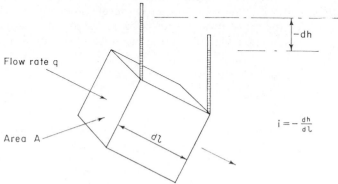

Fig. 4.2   Darcy's law for permeability of soil.

Darcy stated that k is almost independent of the hydraulic gradient (i). The coefficient of permeability has the dimensions of a velocity.

Then, if q is the rate of flow across an area A,

$$v = \frac{q}{A} \quad \text{and} \quad q = Aki$$

4.3 Factors affecting permeability: As defined above, the coefficient of permeability varies with

    (a) the density and viscosity of the soil water
    (b) the turbulence of flow
    (c) the porosity of the soil
    (d) the shape and arrangement of the soil particles
    (e) the degree of saturation, and
    (f) the thickness of the adsorbed layers, in the case of fine-grained soils.

Density and viscosity of the soil water are mainly governed by temperature. Since there is generally little variation in soil temperature, except close to the ground surface, the effect can usually be ignored. Laboratory temperatures, on the other hand, are commonly 5° to 10°C higher than those in the ground, and this may cause a slight increase in the measured permeability, as a result of reduced viscosity.

The permeability is mainly dependent on the shape and size of the voids, and on the length of the flow path. It is therefore greatly influenced by the *porosity* and by the *shape and arrangement of the soil grains*. The *degree of saturation* of the soil also has a considerable effect. For values of S_r greater than about 85% much of the air in the soil is held in the form of small occluded bubbles. Darcy's law is still approximately valid, but the bubbles block some of the pores and reduce the permeability considerably. Within this range, a drop of 1% in the degree of saturation causes a drop of about 2% in the permeability of uniform soils, and about twice this amount in well-graded materials. If the degree of saturation is less than about 85%, much of the air is continuous through the voids, and Darcy's law no longer holds good.

4.4 *Hazen's approximation:* Hazen suggested the following empirical relation for the permeability of filter sands,

$$k = C \times (D_{10})^2 \text{ m/s}$$

where D_{10} = the effective size in mm
C = a coefficient varying between 0·01 and 0·015

The approximation is reasonably good for the type of material for which it was proposed, but the expression makes no allowance for variation in porosity, or in the shape of the particles.

4.5 *The Kozeny-Carman equation* [4.2, 4.3] : It may be shown theoretically that, for laminar flow in saturated soils,

$$k = \frac{1}{k_0 k_T S_s^2} \cdot \frac{n^3}{(1 - n)^2} \cdot \frac{\gamma_w}{\eta}$$

where k_0 is a factor depending on the shape of the pores.
k_T is a factor depending on the 'tortuosity' of the pores (that is, on the mean distance travelled by the water in moving between two points which are unit distance apart in a straight line).
S_s is the surface area of the particles in unit volume of the solid material.

For coarse grained soils, $k_0 k_T$ is about 5·0, so that

$$k \simeq \frac{1}{5·0 S_s^2} \cdot \frac{n^3}{(1 - n)^2} \cdot \frac{\gamma_w}{\eta}$$

4.6 *Fine-grained soils:* Although the Kozeny-Carman equation is reasonably reliable in predicting the permeabilities of coarse-grained soils, it is usually not so useful in the case of silts and clays. There are two reasons for this:

(a) The thickness of the adsorbed layers materially reduces the effective size of the pores in fine-grained soils, and this thickness is influenced by the nature and concentration of the ions in the soil water.

(b) The flat shape of the fine particles makes it impossible to alter the porosity without, at the same time, altering the values of k_0 and k_T.

4.7 *Typical values of the coefficient of permeability* (k): Typical values for natural soils are shown in Table 4.1. Quite small quantities of fine materials tend to block the pores between the coarse particles and greatly reduce the permeability. In any particular case, therefore, the permeability may differ considerably from the typical value quoted. Fissured clays have much higher permeabilities than similar intact materials, because much of the flow takes place through the fissures.

TABLE 4.1 *Typical values of permeability of natural soils.*

	m/s
Gravel	$k > 10^{-2}$
Clean sand	$10^{-2} > k > 10^{-5}$
Silt	$10^{-5} > k > 10^{-8}$
Fissured clay	$10^{-4} > k > 10^{-8}$
Intact clay	$k < 10^{-8}$

PERMEABILITY MEASUREMENTS

4.8 *Laboratory measurements of permeability* [4.4]: Because of the very large range of permeability in soils, no single method is suitable for all cases. The types of equipment most generally used are

(a) the *constant head permeameter*, for permeabilities down to about 10^{-4} m/s, and

(b) the *falling head permeameter*, for values of k between 10^{-4} and 10^{-7} m/s.

Below about 10^{-7} m/s permeability can generally only be measured by indirect means, in a consolidation test (Section 5.11) or in a dissipation test (Section 6.16).

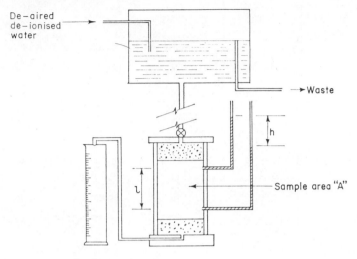

De—aired
de—ionised
water

Waste

h

Sample area "A"

𝓁

Fig. 4.3   A constant head permeameter.

4.9   The constant head permeameter: This apparatus is illustrated diagrammatically in Fig. 4.3. A constant head of water is maintained across the sample during the test. Once equilibrium has been reached, the rate of flow is measured. Filters of coarse material are placed at the top and bottom of the sample to prevent fine material being washed out. Although there is some head loss in these filters, this does not affect the measured value of the permeability, as the hydraulic gradient is measured between tapping points on the side of the sample itself.

Example 4.1
During a constant head permeameter test, a flow of 173 ml was measured in 5 minutes. The sample was 0·1 m in diameter, and the head difference of 0·061 m was measured between tapping points 0·2 m apart.
Then

$$k = \frac{q}{Ai}$$

$$= \frac{173/(5 \times 60 \times 10^6)}{[(0\cdot1^2 \times \pi)/4] \times (0\cdot061/0\cdot2)}$$

$$= 2\cdot4 \times 10^{-4} \text{ m/s}$$

4.10   The falling head permeameter (Fig. 4.4): When the rate of flow through the sample is too small to be accurately measured in the constant head permeameter, the falling head apparatus

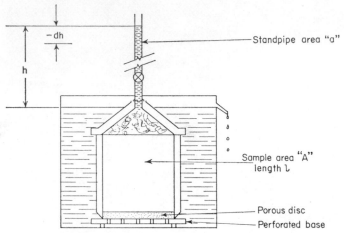

Fig. 4.4   A falling head permeameter.

is used. In this case, the water flows through the sample from a standpipe, and the head loss across the sample at any moment is taken to be the difference between the levels in the standpipe and in the container surrounding the sample. The head loss in the filter above the sample, and in the porous disc beneath it, are now included in the measured head, but as this apparatus is only used for soils of very low permeability, the effect of these losses is insignificant.

Example 4.2

In a falling head permeability test, the water level in the standpipe was originally 1·584 m above the overflow, and dropped 1·0 m in 15·2 minutes. The sample was 0·1 m long and 0·1 m in diameter, and the area of the standpipe was 67 mm^2.

Let the level drop $-dh$ in time dt. Then the rate of flow

$$q = -a\frac{dh}{dt} = Ak\frac{h}{l}$$

Therefore

$$-\frac{dh}{h} = \frac{Ak}{al} \cdot dt$$

Then

$$\frac{Ak}{al} \cdot (t_2 - t_1) = -\log_e\frac{h_2}{h_1} = \log_e\frac{h_1}{h_2}$$

and

$$k = \frac{al}{A} \cdot \frac{\log_e (h_1/h_2)}{t}$$

where

$$t = (t_2 - t_1)$$

Then

$$k = \frac{0 \cdot 1 \times 67 \times 10^{-6}}{0 \cdot 1^2 \times (\pi/4)} \times \frac{2 \cdot 303}{15 \cdot 2 \times 60} \times \log_{10}\frac{1 \cdot 584}{0 \cdot 584}$$

$$= 9 \cdot 3 \times 10^{-6} \, \text{m/s}$$

4.11 Precautions required in laboratory tests: The principal causes of error in laboratory tests are listed below:

(a) *Air in the sample.* The hydraulic gradient is generally much greater in the laboratory sample than in the ground. The resulting pressure drop may cause any air which is dissolved in the soil water to be deposited in the pores. This would considerably reduce the measured permeability. To minimise this effect, all water used for testing is subjected to a vacuum to remove the air from solution.

(b) *Base exchange.* For the finer grained materials, the nature of the adsorbed ions affects the form of the adsorbed layers, and influences the permeability. In order to prevent exchange of the adsorbed ions, it would be best to test the sample, using the soil water as actually found on site. This is seldom possible in practice, and the water used is usually either distilled or chemically treated to reduce the ion content.

(c) *Unrepresentative samples.* Since the permeability may vary considerably over quite a small area, the results obtained from a few small samples may not give a very good picture of the average value of permeability over a large area.

(d) *Anisotropic permeability.* Many soils have been deposited in approximately horizontal layers of varying permeability. Also, there is often a preferred horizontal orientation of the particles within any layer. As a result, many soils have a much larger permeability horizontally than vertically. The ratio of horizontal to vertical permeability may be as much as 100:1. The normal form of laboratory test measures the vertical permeability, which may not have much relevance to cases where the seepage in the ground is predominantly horizontal.

(e) *Sample disturbance.* In practice, it is almost impossible to obtain undisturbed samples of clean coarse-grained soils. It is therefore usually necessary to carry out laboratory tests on samples which have been recompacted to the *in situ* density. However, it is impossible to reproduce the stratification and particle arrangement once these have been destroyed, and such tests may not therefore be very relevant to permeabilities in the ground.

IN SITU PERMEABILITY TESTS

4.12   Free surface and piezometric level: In Chapter 3, the water table was defined as an imaginary surface in the ground, joining points at which the pore water pressure is zero. Before considering seepage problems in the ground, it will be convenient to define two other terms, the *free surface* and the *piezometric level*.

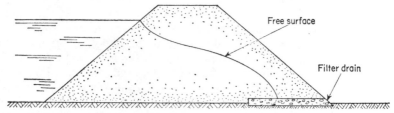

Fig. 4.5   Flow through a simplified model of an earth embankment—to illustrate unconfined flow.

Fig. 4.5 shows a section through an earth dam, through which water is seeping. In the lower part of the dam, the soil is saturated, and water flows under the influence of the hydraulic gradient. In the upper zone, the soil is only partially saturated, and such water as is retained in it is held in equilibrium with the air pressure by surface tension. There is negligible flow here. The boundary between the two zones is the free surface.

This surface is an air/water interface, along which it is reasonable to assume that the pore air pressure u_a is zero. The pore water pressure u_w is always less than u_a (see Section 3.4 above), and is therefore negative on this surface. The free surface, therefore, always lies above the water table (on which u_w is zero). Between them, there is a *capillary fringe* in which u_w is negative. However, in many cases, though not always, the width of the capillary fringe is small in relation to the dimensions of the structure. It is then sufficiently accurate to ignore the capillary suction $(u_a - u_w)$, and to assume that the free surface (on which u_a is zero) is coincident with the water table (on which u_w is zero).

The *piezometric level* for any point in the seepage zone is the level to which a column of water must rise to be in equilibrium with the pore water pressure at that point. The head h at any point is the height of the piezometric level above datum. If there is no vertical component of flow (*i.e.* no vertical hydraulic gradient), the piezometric level is everywhere coincident with the water table.

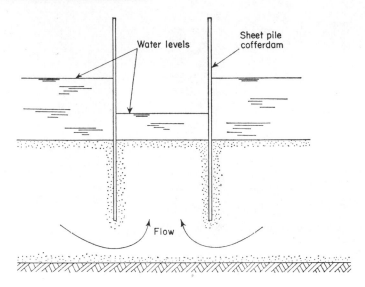

Fig. 4.6 Confined flow.

Where part of the boundary of the seepage zone is a free surface, as in Fig. 4.5, the flow is said to be *unconfined*. Fig. 4.6 shows an example of the alternative condition of *confined flow*, in which there is no free surface and no water table within the aquifer or water bearing stratum.

4.13 Pumping tests: In coarse-grained soils, for the reasons given above, *in situ* measurements of permeability are generally more reliable than laboratory tests. However, they are also much more expensive, and they are therefore only used in very large investigations.

The permeability of a stratum of soil is most commonly determined *in situ* by measuring the discharge from a well. The well is sunk into the stratum, and water is pumped from it at a constant rate. Pumping lowers the piezometric level in the vicinity of the well, and the resulting hydraulic gradient causes water to flow towards the well. The piezometric level may be charted by sinking observation holes at various distances from the well, and by measuring the height at which the water stands in each, once a steady state has been established.

4.14 Pumping tests—unconfined flow: Fig. 4.7 shows a well sunk through a permeable stratum to an impervious material beneath.

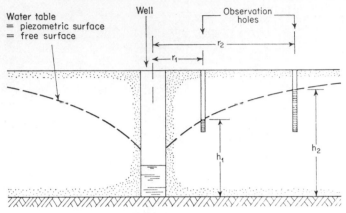

Fig. 4.7   Pumping test—unconfined flow.

In computing the permeability in this case, it will be assumed that

(a)   the flow is horizontal, so that the piezometric level is everywhere at the water table

(b)   the suction may be ignored, so that the free surface is coincident with the water table (and therefore with the piezometric level), and

(c)   the hydraulic gradient is constant at any given radius, and is equal to the rate of change of the piezometric level with radius $\left(\dfrac{dh}{dr}\right)$.

These assumptions were first suggested by Dupuit [4.5]. Although the assumptions are both incorrect and inconsistent, the method nevertheless yields values of permeability which usually agree closely with those obtained by more rigorous methods.

For steady conditions, the flow towards the well must be the same at all radii. Then, at radius r,

$$q = Aki = 2\pi rhk\,\frac{dh}{dr}$$

and

$$\frac{dr}{r} = \frac{2\pi k}{q}\,h\,dh$$

Integrating,

$$\log_e\frac{r_2}{r_1} = \frac{\pi k}{q}\,(h_2{}^2 - h_1{}^2)$$

$$k = \frac{q}{\pi}\left(\frac{\log_e(r_2/r_1)}{(h_2{}^2 - h_1{}^2)}\right)$$

Example 4.3 (see Fig. 4.8)

During preparations for a pumping test, a well was sunk through a stratum of dense sand 10 m deep, and into a clay of very low permeability beneath. Observation holes were drilled at 15 m and at 60 m from the well. The water in the well and in the observation holes stood originally at the same level, 2·35 m below the top of the well. After pumping until steady conditions had been achieved, the discharge

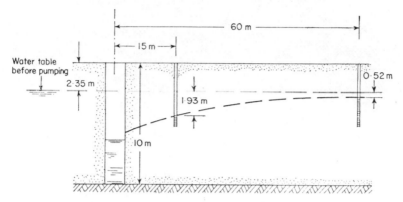

Fig. 4.8   Example 4.3.

was found to be 19·7 m³/hour. The water level in the observation hole 15 m from the well had dropped 1·93 m, and that in the hole 60 m away had dropped 0·52 m.

Then

$$r_1 = 15\cdot0 \text{ m} \quad \text{and} \quad h_1 = 10 - 2\cdot35 - 1\cdot93 = 5\cdot72 \text{ m}$$
$$r_2 = 60\cdot0 \text{ m} \quad \text{and} \quad h_2 = 10 - 2\cdot35 - 0\cdot52 = 7\cdot13 \text{ m}$$

$$k = \frac{19\cdot7}{\pi \times 3600} \times \frac{2\cdot303 \times \log_{10}(60/15)}{(7\cdot13^2 - 5\cdot72^2)}$$

$$= 1\cdot33 \times 10^{-4} \text{ m/s}$$

4.15   Pumping tests—confined flow: Fig. 4.9 shows a well sunk through an aquifer confined at both top and bottom by practically impervious strata. In this case, there is no water table and no free surface, and the piezometric level is everywhere above the top of the aquifer.

Then, for steady flow, at radius r

$$q = Aki = 2\pi r D k \frac{dh}{dr}$$

$$\frac{dr}{r} = \frac{2\pi D}{q} k \, dh$$

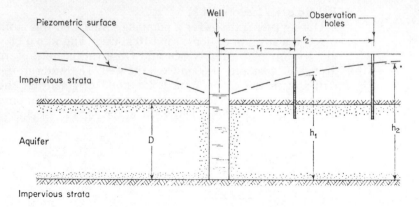

Fig. 4.9   Pumping test—confined flow.

Integrating,

$$\log_e \frac{r_2}{r_1} = \frac{2\pi D}{q} k(h_2 - h_1)$$

$$k = \frac{q}{2\pi D} \cdot \frac{\log_e (r_2/r_1)}{(h_2 - h_1)}$$

Example 4.4

Fig. 4.10 shows the same situation as was described in Example 4.3, except that the top 5 m of the soil is, in this case, of negligible permeability.

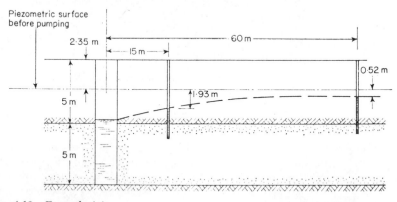

Fig. 4.10   Example 4.4.

Then,

$$k = \frac{19 \cdot 7}{2 \times \pi \times 5 \cdot 0 \times 3600} \times \frac{2 \cdot 303 \log_{10} (60/15)}{(7 \cdot 13 - 5 \cdot 72)}$$

$$= 1 \cdot 71 \times 10^{-4} \text{ m/s}$$

Tests of this type have the merit that the permeability of the soil is measured in the undisturbed state. For seepage problems, it is also an advantage that the measurement is made for practically horizontal flow. However, the measured permeability is really a mean value over quite a wide area, and considerable local variations may exist, which are not easy to detect from the test results.

It is possible that the free surface is not initially horizontal, since there is likely to be some natural seepage across the site. Errors resulting from this may be eliminated by measuring the height h from the mean of the levels in a number of observation holes arranged around the well.

It is important that the water from the well is not returned to the ground, where it can affect the free surface level, and it should generally be discharged into the surface water drainage system at some distance from the well.

Tests of this type are costly to perform. Not only do they require a great deal of expensive drilling, but pumping must continue for a long time before a steady state is reached. The results are also rather more difficult to interpret than the present somewhat simplified presentation might suggest.

4.16   Constant head in situ *permeability tests:*   When measuring the permeability of clays for settlement prediction (Section 12.20), it is convenient to use an *in situ* constant head test. A piezometer tip is installed in the ground, and the water pressure within it is maintained at a constant level above the ambient pore pressure in the ground.

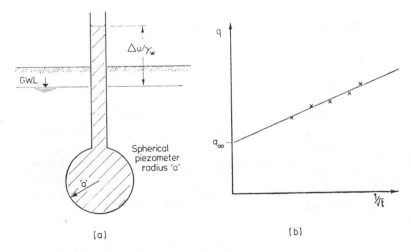

Fig. 4.11   Constant head in situ *permeability test.*

Gibson [4.12] has shown that the rate of flow of water from a spherical piezometer tip is (Fig. 4.11)

$$q_t = 4\pi a \frac{k}{\gamma_w} \Delta u \left(1 + \frac{1}{(\pi T)^{\frac{1}{2}}} \right)$$

where q_t is the rate of flow after time t

a is the radius of the tip

k is the permeability of the soil surrounding the tip

Δu is the pressure in the tip above the ambient pore pressure

T is the time factor $(= ct/a^2)$

c is the coefficient of swelling $(= k/m\gamma_w)$

After infinite time, a steady state would be reached, when

$$q_\infty = 4\pi a \frac{k}{\gamma_w} \Delta u$$

so that

$$k = \frac{q_\infty \gamma_w}{4\pi a \, \Delta u}$$

The value of q_∞ can be determined from the values of q_t at finite times, by plotting q_t against $1/(t)^{\frac{1}{2}}$, as shown in Fig. 4.11(b).

In deriving these expressions, it has been assumed that the soil behaves elastically, and that the permeability of the piezometer and any surrounding filter material is much larger than that of the soil. If either of these assumptions is untenable, some modification of the expressions is necessary (Gibson [4.13], [4.14]).

PERMEABILITY OF STRATIFIED SOILS

4.17 *Horizontal and vertical flow:* Where the soil profile consists of a number of strata having different permeabilities, the overall permeability of the soil is not the same in directions parallel to, and normal to, the strata.

Consider the soil profile shown in Fig. 4.12, consisting of two layers whose properties are indicated by the suffix 1 or 2 respectively.

For flow parallel to the strata, the hydraulic gradient in each layer is the same, the total flow is the sum of the flows in the two layers, and the total area of the section is the sum of the areas of the two layers.

$$i = i_1 = i_2$$
$$\bar{q} = q_1 + q_2$$
$$\bar{A} = A_1 + A_2$$

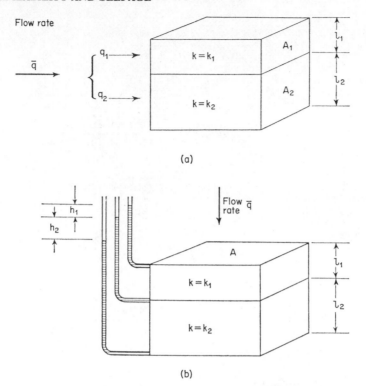

Fig. 4.12 *Permeability of stratified soils. (a) Horizontal flow. (b) Vertical flow.*

Then

$$\bar{k} = \frac{q}{\bar{A}i} = \frac{q_1 + q_2}{\bar{A}i}$$

$$= \frac{A_1 k_1 i_1 + A_2 k_2 i_2}{\bar{A}i}$$

$$= \frac{A_1 k_1 + A_2 k_2}{\bar{A}}$$

$$= \frac{l_1 k_1 + l_2 k_2}{l_1 + l_2}$$

For flow normal to the strata, the total flow is equal to the flow in each layer, the total area is equal to the area of each layer, and the head loss overall is equal to the sum of the losses in the two layers.

$$\bar{q} = q_1 = q_2$$
$$\bar{A} = A_1 = A_2$$
$$\bar{h} = h_1 + h_2$$

Then

$$\bar{k} = \frac{\bar{q}}{\bar{A}i} = \frac{\bar{q}}{\bar{A}} \cdot \frac{(l_1 + l_2)}{(h_1 + h_2)}$$

$$= \frac{\bar{q}}{\bar{A}} \cdot \frac{(l_1 + l_2)}{(i_1 l_1 + i_2 l_2)}$$

$$= \frac{\bar{q}}{\bar{A}} \cdot \frac{l_1 + l_2}{(q_1 l_1 / A_1 k_1) + (q_2 l_2 / A_2 k_2)}$$

$$= \frac{l_1 + l_2}{(l_1 / k_1) + (l_2 / k_2)}$$

In the general case, for n strata,

$$\bar{k} = \frac{l_1 k_1 + l_2 k_2 + \cdots + l_n k_n}{l_1 + l_2 + \cdots + l_n}$$

for flow parallel to the strata, and

$$\bar{k} = \frac{l_1 + l_2 + \cdots + l_n}{(l_1 / k_1) + (l_2 / k_2) + \cdots + (l_n / k_n)}$$

for flow normal to the strata.

Example 4.5

Consider a soil consisting of alternate horizontal layers of coarse and fine material. The thickness of the layers is the same for each material, but the permeability k_2 of the coarse layers is 100 times that of the fine layers (k_1). Then, for horizontal flow,

$$\bar{k}_H = \left(\frac{1 + 100}{2}\right) k_1 = 50 \cdot 5 k_1$$

and, for vertical flow,

$$\bar{k}_V = \left(\frac{2}{(1/1) + (1/100)}\right) k_1 = 1 \cdot 97 k_1$$

$$\frac{\bar{k}_H}{\bar{k}_V} = \frac{50 \cdot 5}{1 \cdot 97} = 25 \cdot 6$$

Such stratification is one of the reasons why a soil may have a horizontal permeability much greater than that in a vertical direction. In addition, individual strata may be anisotropic and have a greater permeability in a direction parallel to the bedding planes, which are often horizontal.

STEADY SEEPAGE

4.18 One-dimensional vertical flow: Consider the model shown in
Fig. 4.13.

Initially, when there is no flow, the total vertical normal
stress at the bottom of the tank is

$$\sigma = z\gamma_{\text{sat}} + d\gamma_w$$

The pore pressure at the same point is

$$u = (z + d)\gamma_w$$

The effective stress is therefore

$$\sigma' = z(\gamma_{\text{sat}} - \gamma_w)$$

Now let the head at the base of the tank increase by an
amount Δh.

This will cause an upward hydraulic gradient, and
upward flow.

$$i = \frac{\Delta h}{z}; \qquad q = Ak\frac{\Delta h}{z}$$

Since the total stress at the bottom of the tank is unaltered,
the effective stress is reduced, and

$$\sigma' = z(\gamma_{\text{sat}} - \gamma_w) - \Delta h\gamma_w$$

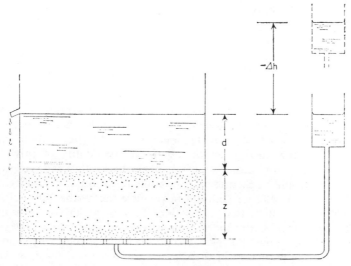

Fig. 4.13 One-dimensional vertical flow.

The shear strength of a clean sand is zero if the effective stress is zero. Thus the soil loses all strength if

$$z(\gamma_{sat} - \gamma_w) = \Delta h \gamma_w$$
$$= i_c z \gamma_w$$

where i_c is called the *critical hydraulic gradient*. Then

$$i_c = \frac{(\gamma_{sat} - \gamma_w)}{\gamma_w}$$

Since

$$\gamma_{sat} = \frac{G_s + e}{1 + e} \gamma_w$$

	by volume	by weight
Water	e	$e\gamma_w$
Solid	1	$G_s\gamma_w$

$$i_c = \frac{(G_s + e) - (1 + e)}{1 + e}$$

$$= \frac{G_s - 1}{1 + e}$$

Example 4.6

A soil has a porosity of 0·4, and the specific gravity of the particles is 2·70. Then

$$\gamma_{sat} = (0{\cdot}6 \times 2{\cdot}70 + 0{\cdot}4 \times 1{\cdot}0)9{\cdot}81 = 19{\cdot}7 \text{ kN/m}^3$$

and

$$i_c = \frac{19{\cdot}7 - 9{\cdot}81}{9{\cdot}81} = 1{\cdot}01$$

Alternatively

$$e = \frac{0{\cdot}4}{0{\cdot}6} = 0{\cdot}667$$

and

$$i_c = \frac{2{\cdot}70 - 1}{1 + 0{\cdot}667} = \frac{1{\cdot}7}{1{\cdot}667} = 1{\cdot}01$$

Cohesive soils usually have very low permeabilities, so that pore pressure changes are slow. Moreover, even when the effective stress has been reduced to zero, there may be some shear strength remaining. Instability of this kind is therefore not usually a problem in this type of soil. It would be theoretically possible to produce instability in any non-cohesive soil, but in practice the permeability of gravel is so high that a sufficient flow of water to maintain the critical hydraulic gradient is seldom available. In the case of fine sands, however, the permeability is relatively small, and the reduction in the shear strength caused by upward flow may be very dangerous. Thus, if an excavation is kept dry

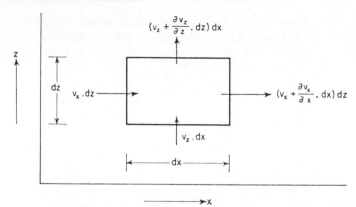

Fig. 4.14   General condition of continuity for two-dimensional fluid flow.

by pumping, too great a flow of water can cause the liquefaction of the soil at the bottom of the excavation. There will be a sudden inrush of water accompanied by the collapse of the supports, all in a very short space of time.

4.19   *General equations for two-dimensional flow:*   Consider an element $dx \times dz$, in an incompressible fluid having velocities v_x and v_z in the directions x and z respectively (Fig. 4.14). Then the flow of fluid into the element is

$$v_x\, dz + v_z\, dx$$

and the outflow from the element is

$$\left(v_x + \frac{\partial v_x}{\partial x} dx\right) dz + \left(v_z + \frac{\partial v_z}{\partial z} dz\right) dx$$

For steady conditions, the net inflow must be zero, and

$$\frac{\partial v_x}{\partial x}\, dx\, dz + \frac{\partial v_z}{\partial z}\, dz\, dx = 0$$

Therefore

$$\frac{\partial v_x}{\partial x} + \frac{\partial v_z}{\partial z} = 0$$

The next stage is to define a *potential function* (Φ), such that

$$v_x = \frac{\partial \Phi}{\partial x} \quad \text{and} \quad v_z = \frac{\partial \Phi}{\partial z}$$

Then

$$\frac{\partial^2 \Phi}{\partial x^2} + \frac{\partial^2 \Phi}{\partial z^2} = 0$$

This is the Laplace equation for the potential function (Φ).

It may be shown that the existence of a potential function requires *irrotational flow*, and that this in turn implies that

$$-\frac{\partial v_z}{\partial x} + \frac{\partial v_x}{\partial z} = 0$$

Then if a further function (Ψ) is defined such that

$$v_x = \frac{\partial \Psi}{\partial z} \quad \text{and} \quad v_z = -\frac{\partial \Psi}{\partial x}$$

it may be seen that

$$\frac{\partial^2 \Psi}{\partial x^2} + \frac{\partial^2 \Psi}{\partial z^2} = 0$$

Thus the function Ψ, which is called the *stream function*, is also governed by the Laplace equation.

Comparing the definitions of Φ and Ψ, it may be seen that, for the function Φ,

$$v_x = \frac{\partial \Phi}{\partial x} \quad \text{and} \quad v_z = \frac{\partial \Phi}{\partial z}$$

Since $\Phi = \Phi(x, z)$,

$$d\Phi = \frac{\partial \Phi}{\partial x}\,dx + \frac{\partial \Phi}{\partial z}\,dz$$

$$= v_x\,dx + v_z\,dz$$

Then, for constant Φ,

$$d\Phi = 0 \quad \text{and} \quad \frac{dz}{dx} = -\frac{v_x}{v_z}$$

For the function Ψ,

$$v_x = \frac{\partial \Psi}{\partial z} \quad \text{and} \quad v_z = -\frac{\partial \Psi}{\partial x}$$

Since $\Psi = \Psi(x, z)$,

$$d\Psi = \frac{\partial \Psi}{\partial x}\,dx + \frac{\partial \Psi}{\partial z}\,dz$$

$$= -v_z\,dx + v_x\,dz$$

Then, for constant Ψ,

$$d\Psi = 0 \quad \text{and} \quad \frac{dz}{dx} = \frac{v_z}{v_x}$$

Therefore the *equipotential lines* (lines of constant Φ) are normal to the *stream lines* (lines of constant Ψ), since the product of their gradients is -1. Then in any problem, the solutions of the two Laplace equations for the potential and stream functions take the form of two families of orthogonal curves, the equipotential lines on the one hand, and the stream lines on the other. The form of the curves is governed by the boundary conditions of the problem.

4.20   Application to ground water flow: The solutions obtained above may be applied to problems concerning the flow of water through the pores of a soil.

If Darcy's law is valid (and provided also that, if $k_x \neq k_z$, the axes of anisotropy are in the x and z directions),

$$v_x = -k_x \frac{\partial h}{\partial x} \quad \text{and} \quad v_z = -k_z \frac{\partial h}{\partial z}$$

where v_x and v_z are the *apparent velocities* in the x and z directions respectively. (The apparent velocity was defined in Section 4.2 above.) Then, for steady conditions in an incompressible fluid (see Section 4.19 above),

$$\frac{\partial v_x}{\partial x} + \frac{\partial v_z}{\partial z} = 0$$

and

$$\left(-k_x \frac{\partial^2 h}{\partial x^2} \right) + \left(-k_z \frac{\partial^2 h}{\partial z^2} \right) = 0$$

For the moment, it will be assumed that the soil is isotropic and that $k_x = k_z$. Then, if $\Phi = -kh$, the governing equation is, as before,

$$\frac{\partial^2 \Phi}{\partial x^2} + \frac{\partial^2 \Phi}{\partial z^2} = 0$$

4.21   The physical significance of Φ and Ψ: It can now be seen that the function Φ, which was originally defined in purely mathematical terms, has a physical significance. Equipotential lines are lines of constant head.

The significance of the stream function may be seen by considering the flow between any two points A and B, at which the values of the stream function are Ψ_A and Ψ_B respectively, as shown in Fig. 4.15. Consider unit length normal to the $x - z$ plane. The rate of flow across any element of length ds in the $x - z$ plane is,

$$dq = v_x \, dz - v_z \, dx$$

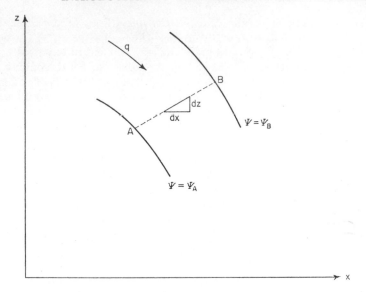

Fig. 4.15   Rate of flow in terms of the stream function.

Then, between points A and B,

$$q = \int_A^B dq = \int_A^B (v_x \, dz - v_z \, dx)$$

$$= \int_A^B \left(\frac{\partial \Psi}{\partial z} \, dz + \frac{\partial \Psi}{\partial x} \, dx \right)$$

$$= \Psi_B - \Psi_A$$

Thus, the rate of flow (per unit length normal to the $x - z$ plane) between any two stream lines is equal to the difference between the values of the stream function on those lines, and there is no flow across a stream line.

Seepage problems usually require the determination of

(a) the pore pressure at points within the seepage zone, and

(b) the rate of flow through the zone.

These can be rapidly evaluated, once we have constructed a *flow net* of equipotential and stream lines. The form of this flow net is governed by the internal and boundary conditions discussed below.

4.22 Flow nets—internal conditions: It has already been shown that the equipotential lines and the stream lines are orthogonal. There is one further condition.

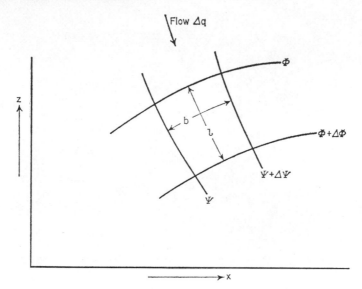

Fig. 4.16   Mesh dimensions in a flow net.

Consider a three-dimensional space of unit thickness, bounded by any two equipotential lines and any two stream lines in the $x - z$ plane, as shown in Fig. 4.16. Let the space have a mean length l in the direction of flow, and a mean breadth b. Then the rate of flow through this space is

$$\Delta q = \Delta A k i$$

Therefore

$$\Delta \Psi = \Delta q = -bk \frac{\Delta h}{l} = \frac{b}{l} \Delta \Phi$$

If we maintain $\Delta \Phi$ constant between adjacent equipotential lines, and $\Delta \Psi$ constant between adjacent stream lines, then b/l is constant. We may give b/l any value we choose, but it will be convenient to arrange that $b = l$, so that each of the quadrilaterals bounded by adjacent equipotential lines and stream lines has the same mean dimension in both directions, and is approximately 'square'. Then,

$$\Delta q = \Delta \Phi = \Delta \Psi$$

4.23   *Deflection of the flow net at a boundary between soils having different permeabilities:*   At a boundary between soils with different permeabilities, the stream lines are deflected, and a 'square' flow net cannot be drawn on both sides of the boundary. Consider the situation shown in Fig. 4.17. For a steady

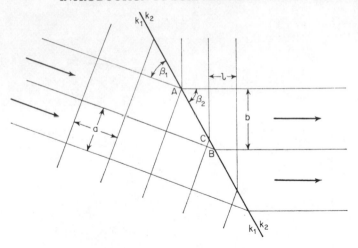

Fig. 4.17 Deflection of the flow net at a boundary between soils having different permeabilities.

state, the flow rate between any two stream lines must be the same on both sides of the boundary. Then

$$\Delta q = \frac{a}{a} k_1 \, \Delta h = \frac{b}{l} k_2 \Delta h$$

and

$$\frac{k_1}{k_2} = \frac{b}{l}$$

Also

$$AB = \frac{a}{\sin \beta_1} = \frac{b}{\sin \beta_2}$$

and

$$AC = \frac{a}{\cos \beta_1} = \frac{l}{\cos \beta_2}$$

so that

$$\frac{k_1}{k_2} = \frac{b}{l} = \frac{\tan \beta_2}{\tan \beta_1}$$

4.24 Flow nets—boundary conditions: Fig. 4.18 illustrates four possible types of boundary of a seepage zone. These are as follows:

(a) *A submerged permeable boundary (A–B and D–E).* The head h is constant on each of these surfaces, which are therefore equipotential lines.

(b) *An impermeable boundary (A–E).* Since there is no flow across this boundary, Ψ is constant and the boundary is a stream line.

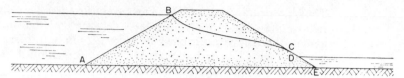

Fig. 4.18 *Flow through a simplified model of an earth embankment—to illustrate the boundary conditions in a seepage zone.*

(c) *A seepage surface (C–D).* Along this surface, the pore pressure is zero, so that

$$\Delta\Phi = -k\,\Delta h = -k\,\Delta z$$

If $\Delta\Phi$ is constant, the equipotential lines meet the seepage surface at constant vertical intervals.

(d) *A free surface (B–C).* There is an additional degree of freedom on this surface, as its position is not predetermined. An additional boundary condition is therefore required to make the flow net determinate.

The two boundary conditions are

(i) that $u = 0$, so that $\Delta\Phi = -k\,\Delta z$, as on the seepage surface, and

(ii) that Ψ is constant (*i.e.* the free surface is a stream line).

4.25 *Anisotropic soils:* In Section 4.20 it was shown that, provided that the axes of anisotropy are in the x and z directions, the governing equation for flow through a soil is

$$\left(-k_x\frac{\partial^2 h}{\partial x^2}\right) + \left(-k_z\frac{\partial^2 h}{\partial z^2}\right) = 0$$

If $k_x \neq k_z$, this equation may still be expressed in the Laplace form by writing

$$x' = \left(\frac{k_x}{k_z}\right)^{\frac{1}{2}} x$$

Then

$$\left(-k_z\frac{\partial^2 h}{\partial x'^2}\right) + \left(-k_z\frac{\partial^2 h}{\partial z^2}\right) = 0$$

Then, writing

$$\Phi = -k_z h$$

$$\frac{\partial^2\Phi}{\partial x'^2} + \frac{\partial^2\Phi}{\partial z^2} = 0$$

Thus, an orthogonal flow net can be constructed if we first redraw the boundaries of the seepage zone with the horizontal dimensions multiplied by $(k_z/k_x)^{\frac{1}{2}}$. On a true scale, the flow net is not orthogonal: the actual direction of flow is not normal to the equipotential lines.

4.26 *Construction of flow nets by sketching* [*Forchheimer, 4.6*]: Given these internal and boundary conditions, it is possible, with a little practice, to sketch in the flow net for confined flow with sufficient accuracy for most practical purposes. For unconfined flow, it is first necessary to assume an approximate form for the free surface. If this assumed form is incorrect, it will be found to be impossible to construct any flow net which satisfies both the boundary conditions for the free surface. It is then necessary to make a new approximation, and, by successive improvement, to arrive at the correct solution.

4.27 *The electrical analogue*: The equations governing the flow of electricity through a two-dimensional conducting medium are [Wyckoff and Reed, 4.9]:

$$I_x = -\frac{1}{R_x}\frac{\partial E}{\partial x} \quad \text{and} \quad I_z = -\frac{1}{R_z}\frac{\partial E}{\partial z}$$

where I_x, I_z is the current flowing in the x or z direction respectively.

R_x, R_z is the resistivity in the x or z direction respectively. E is the electrical potential.

This is exactly analogous to the flow of water through soil for which the equations are

$$v_x = -k_x\frac{\partial h}{\partial x} \quad \text{and} \quad v_z = -k_z\frac{\partial h}{\partial z}$$

Thus, the flow of electricity through a two-dimensional conducting medium may be used as a model for the flow of water through soil. The lines of constant potential on the electrical model represent the lines of constant head in the soil.

Two types of conducting medium are commonly used: conducting paper and a resistance network [see Herbert and Rushton, 4.10]. The graphite-impregnated paper provides a continuous conducting medium which can be easily cut to the shape of the seepage zone. A potential is applied to the equipotential boundaries, using silver paint to ensure contact with low resistance. The paper is cut through at all impervious boundaries. The electrical model provides no analogy for the free surface, which must be approximately located in advance. Using a resistance bridge, and a probe attached to a graphite-pointed pencil, the equipotential lines are located and drawn on the paper. The flow lines may then be sketched in, to complete the orthogonal net.

The paper has approximately the same resistivity longitudinally as transversely. Where the soil permeability is anisotropic, or where the resistivity of the paper is found to differ in the two directions, the flow net may be corrected in the way described in Section 4.25.

A network of resistances may be used in place of the conducting paper. Such a network is more laborious to set up, particularly where the shape of the boundaries is complicated. There may also be some distortion of the flow net, since the conducting medium is not continuous, and the potential can only be determined at the node points of the network. However, the resistance network allows much greater flexibility in dealing with problems where the soil permeability varies in different parts of the seepage zone. It is not easy to deal satisfactorily with such problems, using the conducting paper.

4.28 Methods of computation: Exact mathematical solutions to the Laplace equation have only been obtained in a few cases with very simple boundary conditions. Approximate solutions may be obtained in other cases using the finite difference method described by Southwell [4.11], the finite element method described by Zienkiewicz [4.15] and Verruijt [4.7], or the integral equation method [Symm and Pitfield, 4.8]. The finite difference method computes the potential at each of a number of node points arranged on a regular grid superimposed on the seepage zone. It is possible to vary the distance between node points in different parts of the grid, but only at the expense of considerable extra computation. For anisotropic soils, it is necessary that the direction of the greatest permeability should coincide with one of the co-ordinate axes of the grid. This makes it difficult to deal with contorted strata where the direction of the greatest permeability varies within the seepage zone. There are also difficulties in adjusting a regular grid to a seepage zone with an irregular boundary.

In the finite element method, the seepage zone is divided into elementary regions (the finite elements) and the potential is determined at node points on the boundaries of, and within, each element. Since the elements may be of any size, the spacing of the node points may be adjusted to suit

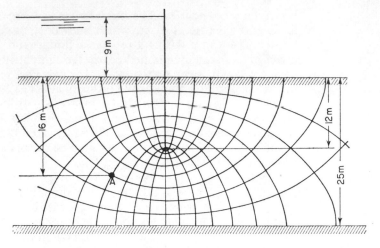

Fig. 4.19   Determination of pore pressure and rate of flow from a flow net.

the problem, and may easily be made closer in those parts of the seepage zone where hydraulic gradients are greatest, or where the solution is required with greatest precision. There is little difficulty in accommodating an irregular boundary, and the greatest and least permeabilities may be different in both magnitude and direction in each element. The finite element method is generally more flexible, therefore, in dealing with complicated problems, but solutions of acceptable accuracy may be obtained in suitable cases by either method.

4.29 *Determination of pore pressure from a flow net:*   The pore pressure at any point within the seepage zone may be determined from the flow net by computing the piezometric level for the equipotential line through that point. Thus, in Fig. 4.19,

$$\Delta h = 9/18 = 0.5 \text{ m}$$

and the piezometric level at point A is $5 \times 0.5 = 2.5$ m below the upper water surface. Then the pore pressure at A

$$u_A = (9 - 2.5 + 16)9.81 \text{ kN/m}^2$$
$$= 221 \text{ kN/m}^2$$

4.30 *Determination of the rate of flow from the flow net:*   In Section 4.22 we showed that, for a 'square' flow net, the flow rate per unit length between any two stream lines,

$$\Delta q = \Delta \Psi = \Delta \Phi$$

Then, if the number of spaces between the stream lines is n_Ψ, and the number of potential drops (Δh) between equipotential lines is n_Φ, the total rate of flow per unit length is

$$q = n_\Psi \, \Delta\Psi = n_\Psi \, \Delta\Phi$$

$$= \frac{n_\Psi}{n_\Phi} kH$$

In Fig. 4.19 there are 18 potential drops and 9 spaces between stream lines. If the permeability is 10^{-3} m/s, the rate of flow is

$$q = 9/18 \times 10^{-3} \times 9$$
$$= 4{\cdot}5 \times 10^{-3} \, \text{m}^3/\text{s/m}$$
$$= 16{\cdot}2 \, \text{m}^3/\text{hour/m}$$

REFERENCES

4.1 DARCY, H. 1856. *Les fontaines publique de la ville de Dijon.* Dalmont (Paris).

4.2 KOZENY, J. 1927. Uber kapillare Leitung des Wassers im Boden *Sitz. der Wien. Akad. der Wissenschaften,* **136.**

4.3 CARMAN, P. C. 1956. *Flow of gases through porous media.* Academic Press (New York).

4.4 AKROYD, T. 1957. *Laboratory testing in soil engineering.* Soil Mechanics Ltd.

4.5 DUPUIT, J. 1863. *Études théoriques et pratiques sur le mouvement des eaux à travers les terrains perméables.* Carilian-Goeury (Paris), 2nd Ed.

4.6 FORCHHEIMER, P. 1930. *Hydraulik.* Teubner (Leipzig), 2nd Edition.

4.7 VERRUIJT, A. 1970. *The theory of ground water flow.* Macmillan (London).

4.8 SYMM, G. T. and PITFIELD, R. A. 1974. *Solution of Laplace's equation in two dimensions.* National Physical Laboratory Report NAC 44.

4.9 WYCKOFF, R. D. and REED, D. W. 1935. Electrical conduction models for the solution of water seepage problems. *Physics,* **6.**

4.10 HERBERT, R. and RUSHTON, K. R. 1966. Ground water flow studies by resistance networks. *Géotechnique,* **16.**

4.11 SOUTHWELL, R. V. 1946. *Relaxation methods in theoretical physics.* Oxford University Press (London).

4.12 GIBSON, R. E. 1963. An analysis of system flexibility and its effect on time lag in pore water pressure measurements. *Géotechnique,* **13.**

4.13   GIBSON, R. E. 1966. A note on the constant head test to measure soil permeability *in situ*. *Géotechnique,* **16.**

4.14   GIBSON, R. E. 1970. An extension of the theory of the constant head *in situ* permeability test. *Géotechnique,* **20.**

4.15   ZIENKIEWICZ, O. C. 1971. *The finite element method in engineering science.* McGraw-Hill (London).

CHAPTER 5

A model for soil consolidation

MODELS FOR CONSOLIDATING SOILS

5.1 *Real soils and model soils:* In solving a problem in soil mechanics, the first step is to build up a *mathematical model* of the material. This model consists of a set of statements describing, in mathematical terms, the behaviour of the material when subjected to changes in external forces, pore pressures, displacements, etc. The statements are usually given in the form of general expressions describing the behaviour of some general class of materials, but they contain parameters to which numerical values may be assigned after tests on a particular material.

Any attempt to describe every feature of soil behaviour exactly would certainly be unproductive. In the first place, it would require a detailed knowledge of the soil which would be difficult and expensive, if not actually impossible to obtain. Secondly, the resulting mathematical expressions would be too complex to handle. Soil problems, being two- or three-dimensional, are always difficult to solve analytically: complex soil models usually make them nearly insoluble. Models must therefore be very simple if they are not to be almost useless.

A good soil model

   (a) describes the *significant* features of the behaviour with sufficient accuracy to solve the particular problem with which we are faced,

   (b) takes no account of insignificant features of the behaviour, and

   (c) is as simple as possible.

Features which are insignificant in some cases may be of major importance in others, so that a model which is appropriate to one problem may be quite unsuitable to solve another, although structure and material are identical. We should not be surprised, therefore, to find that several soil models are in use to solve different problems.

The next three chapters discuss some models used to describe the behaviour of soil subjected to stress changes. In this chapter we consider a model describing the time-dependent volume changes which result from changes in normal stress. This process of volume change is called *consolidation*.

5.2   The consolidation process: Consider a cylindrical specimen of saturated soil (Fig. 5.1) in a state of equilibrium under a uniform (or *spherical*) normal stress p_i which is being applied for the first time. We will assume that the boundaries of the specimen are permeable, and that the pore pressure on these

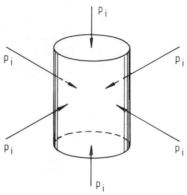

Fig. 5.1   Consolidation under a uniform (spherical) normal stress p_i.

boundaries is constant and equal to u_i. Then everywhere within the specimen the pore pressure is u_i and the effective stress p'_i is $p_i - u_i$.

Now suppose that the normal stress on the boundaries is suddenly increased to $p_i + \Delta p$. Since the soil is saturated, Skempton's pore pressure parameter B is nearly equal to $1 \cdot 0$, and the pore pressure within the specimen increases to $u_i + \Delta p$, so that the effective stress remains unchanged. Since both soil particles and water are nearly incompressible, the volume change at this stage is insignificant.

However, we have stipulated that the pore pressure on the boundary shall be maintained constant. There is therefore a hydraulic gradient between the inside of the specimen (where the pore pressure is $u_i + \Delta p$) and the boundary (where u_i is unchanged). Water leaves the element through the boundary, and the volume is reduced. Since the volume of the soil particles remains constant, this change in volume may be conveniently described as a change in void ratio. As the void ratio is reduced, the particles are packed more closely, and offer greater resistance to further reductions in volume. Experiments on real soils show that, for first-time loading of most soils, there is a single-valued relationship between void ratio and effective stress, which is almost independent of time. (Some cases where this is not true are discussed in Section 5.18 below.)

As the effective stress increases with decreasing void ratio, the pore pressure drops: the hydraulic gradient is correspondingly reduced, and the *rate* of volume change decreases. Eventually, after sufficient time, the specimen approaches a steady state in which the internal pore pressure is again equal to u_i. The effective stress everywhere within the specimen is then $p_i + \Delta p - u_i$ (Fig. 5.2).

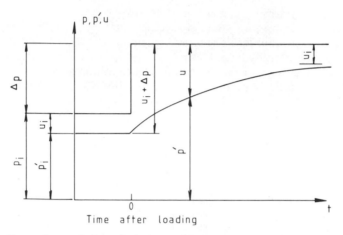

Fig. 5.2 *Stress changes during spherical consolidation.*

The *rate* at which the water is expelled from the specimen is mainly a function of the soil permeability. Coarse-grained soils, having large permeabilities, drain very rapidly. The excess pore pressure, generated by the total stress increment Δp, is almost totally dissipated nearly instantly, and at the same time the volume is reduced to the value corresponding to the ultimate effective stress $p_i + \Delta p - u_i$. In clay soils, however, the permeability is very much smaller. The volume change and dissipation of the excess pore pressure are very slow, and a long time may elapse before the specimen approaches the equilibrium condition. In considering the consolidation in soils of low permeability, therefore, we are concerned not only with the *magnitude* of the volume change and pore pressure change, but also with the *rate* at which these changes take place.

5.3 *One-dimensional consolidation:* In attempting to predict the long term settlement of structures on compressible soils, a good estimate can often be made by assuming that the consolidation of the soil is *one-dimensional*. By this we mean that

(a) all displacements are vertical, so that there are no lateral strains,

(b) all flow of water from the soil is in a vertical direction only, and

(c) the void ratio is a direct function of the *vertical* component of effective stress.

The condition of one-dimensional vertical consolidation generally corresponds to the state of the soil during deposition, when it is consolidating under the weight of overburden. The lateral extent of such a deposited stratum is generally large compared with its thickness, and the lateral strains can only be small. The condition does not correspond to the state under a load of finite lateral extent (such as a foundation), but the one-dimensional model may nevertheless give useful information. The reasons for this (and for the limitations on the usefulness of the model in some cases) are discussed in Chapter 12.

The one-dimensional model for consolidation is particularly attractive because

(a) it is mathematically simpler to handle than the three-dimensional equivalent, and

(b) the void ratio is assumed to be a function of the vertical component of effective stress (σ'_v) which is relatively easy to estimate.

In the remainder of this chapter, σ' will be assumed to refer to the *vertical* component of stress only. The suffix v will, for convenience, be omitted.

VOLUME CHANGE IN ONE-DIMENSIONAL CONSOLIDATION

5.4 *The consolidation test:* The rate and extent of the settlement, arising from volume changes in the soil, are generally predicted from the results of tests in an *oedometer* (see Fig. 5.3). A section is cut from an undisturbed sample of the soil, and is fitted into a steel ring. The ring effectively prevents lateral movement and lateral drainage, so that one-dimensional consolidation is enforced. Porous plates are placed at the top and bottom of the sample, which is therefore free to drain from both faces. The sample is loaded vertically, and a record is made of settlement against time. When the settlement is substantially complete, the

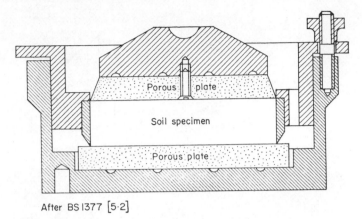

After BS 1377 [5·2]

Fig. 5.3   A typical oedometer cell.

load is increased and a further time/settlement record is prepared. This procedure is repeated, until the settlement under the whole of the relevant range of stress has been investigated.

For reasons which will be discussed later, the settlement is found to continue—although at an ever-decreasing rate —however long the sample is left under load. There are, therefore, differences in the results of tests in which different load increments or different periods of loading are employed. To ensure that the results of tests are comparable, it is usual in commercial practice to use a standard loading period of 24 hours, and a standard load increment ratio of 2:1 (that is, the load is doubled at the end of each stage). Different procedures may, however, be required in special cases.

From the results of oedometer tests, it is possible to derive

(a)   a relationship between the void ratio and the effective stress (measured at the end of each 24-hour period, when the pore pressure is assumed to be zero), and

(b)   a plot of settlement against time for each stage of the test. From the latter, the rate of settlement may be predicted.

5.5   *A mathematical model for one-dimensional consolidation:*   Figure 5.4(a) shows a typical relationship between the void ratio (e) and the vertical component of effective stress (σ'), obtained from tests on a clay specimen loaded in an oedometer. From this, we will construct a suitable mathematical model for one-dimensional consolidation (Fig. 5.4(b)).

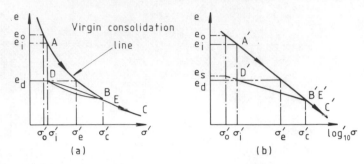

Fig. 5.4   One-dimensional consolidation: (a) Typical $e:\sigma'$ curves for a real soil. (b) Idealised $e:\log_{10}\sigma'$ curves for the model soil.

The line ABC in Fig. 5.4(a) shows the actual relationship between e and σ' for first-time loading. Tests show that, for any particular soil, there is generally a nearly linear relation between e and $\log_{10}\sigma'$. This relationship may therefore be idealised by the straight line $A'B'C'$ in Fig. 5.4(b), and may be expressed in the form

$$e = e_0 - C_c \log_{10} \frac{\sigma'}{\sigma'_0}$$

where e_0 is the void ratio *on the line $A'B'C'$* corresponding to any convenient value of the effective stress σ'_0.

C_c is a parameter called the *compression index*, equal to $-de/d(\log_{10}\sigma')$.

Now suppose that the effective stress σ' is increased from σ'_i (point A) to σ'_c (point B), and then reduced to σ'_i again. The swelling of the soil during unloading is shown by the line BD in Fig. 5.4(a), and the distance AD exhibits the extent of the irrecoverable plastic volumetric strain which has occurred during the first-time loading from A to B. The stress σ'_c, which is the greatest value of σ' applied so far, is called the *preconsolidation pressure*. The line BD in Fig. 5.4(a) may also (but rather less accurately) be idealised by a straight line $B'D'$ in Fig. 5.4(b), and may be expressed in the form

$$e = e_s - C_s \log_{10} \frac{\sigma'}{\sigma'_0}$$

where C_s, the *swelling index*, is a parameter similar to C_c, e_s is the void ratio *on the line $B'D'$* (*produced*) corresponding to the effective stress σ'_0.

Now let σ' be increased to σ'_c again. The resulting reconsolidation is shown by the line DE in Fig. 5.4(a).

Although not exactly reflecting the real soil behaviour (particularly when subjected to repeated loading), it is often convenient and sufficiently accurate to assume that, for values of σ' less than σ'_c, the soil behaves elastically, so that C_s is the same for loading and unloading. C_s may also be assumed to be the same for all values of σ'_c, so that all lines similar to $B'D'$ are parallel.

Reference to Fig. 5.4(b), therefore, shows that the line $A'B'C'$ is a boundary, separating possible combinations of e and σ' below the line from impossible states above. Any soil element whose state is described by a point below the line behaves elastically under small stress changes. Any element described by a point on the line reacts elastically to a reduction in stress, but a stress increment results in irrecoverable plastic volumetric strain. The material is said to *yield*, and the expression

$$e = e_0 - C_c \log_{10} \frac{\sigma'}{\sigma'_0}$$

is called the *yield condition*.

Notice that the yield under a given increment of effective stress is not unlimited. It only continues until the increment of effective stress is balanced by the increased resistance which results from the decrease in void ratio. A material whose resistance increases with plastic strain in this way is said to *strain harden* or *work harden*.

5.6 *Normally consolidated and overconsolidated soils:* A soil in which the present effective stress has never been exceeded (so that the state is defined by a point on the *virgin consolidation* line ABC) is said to be *normally consolidated*. Where, however, there has been previous consolidation to an effective stress σ'_c greater than that now applied (e.g. point D), the soil is said to be *overconsolidated*. Any point below the virgin consolidation line ABC defines an overconsolidated state.

For any void ratio e, we shall later find it convenient to define an *equivalent consolidation pressure* σ'_e which is the value of σ' on the virgin consolidation line corresponding to the void ratio e. Then

$$e = e_0 - C_c \log_{10} \frac{\sigma'_e}{\sigma'_0}$$

so that

$$\sigma'_e = \sigma'_0 \exp\left(\frac{2 \cdot 303(e_0 - e)}{C_c}\right)$$

For overconsolidated soils, we will define an *over-consolidation ratio R_σ* equal to σ'_e/σ'.* For normally consolidated soils, described by points on $A'B'C'$, $\sigma' = \sigma'_e$ and $R_\sigma = 1 \cdot 0$. Figure 5.4 shows the value of the equivalent consolidation pressure σ'_e corresponding to the void ratio e_d at point D, where the overconsolidation ratio R_σ is σ'_e/σ'_i.

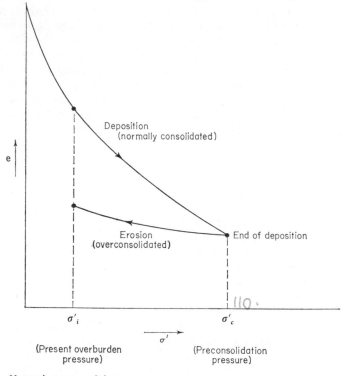

Fig. 5.5   Natural overconsolidation.

5.7   Overconsolidation in natural soils:

Many natural soils are over-consolidated, because they have been consolidated by the pressure of overburden which has since been removed by erosion (Fig. 5.5). The present overburden pressure is no indication of the preconsolidation pressure.

The fact that the soil carries a certain total overburden pressure is also no sure evidence that it has reached equilibrium under that pressure. Where the overburden is reduced, by excavation or by natural erosion, the subsequent

---

* Note.   The overconsolidation ratio is sometimes defined as $R_c = \sigma'_c/\sigma'$. In considering the critical state model (Section 7.9), however, it will be more convenient to use the definition given here.

swelling of the soil is a very slow process if the permeability is low. Equally, where the total stress is increased by loading, the resulting consolidation may not be complete for many years.

When a sample is taken from the ground, the total stress is reduced to zero. Provided that the sample is not disturbed during removal from the ground, and is immediately sealed to prevent absorption of water, its void ratio remains equal to that of the soil in the ground, and not to that corresponding to equilibrium under zero effective stress. (In practice, there is always some disturbance during sampling. The effect of this is discussed in Section 5.10 below.)

Desiccation of the soil near the surface in dry weather causes a large suction, and therefore a large positive effective stress, in a fine-grained soil. Desiccation, therefore, nearly always results in the upper one or two metres of such a soil being overconsolidated, even where there has been no erosion of the overburden. A fall and rise in the water table may cause some slight overconsolidation. This may result in a significant change in the behaviour under subsequent small stress changes.

5.8 *The compression index* (C_c): The compression index may be determined from the results of an oedometer test, by plotting the void ratio e against $\log_{10} \sigma'$, as shown in Fig. 5.6 and Example 5.1 below.

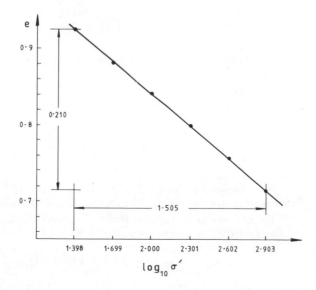

Fig. 5.6 Example 5.1.

Example 5.1 (Fig. 5.6)

During an oedometer test, the initial void ratio of the specimen was 0·945, and the thickness of the specimen after 24 hours at each load was found to be:

σ' (kN/m^2)	0	25	50	100	200	400	800
Thickness (mm)	19·05	18·85	18·44	18·03	17·63	17·21	16·80

Then

$\log_{10}\sigma'$	—	1·398	1·699	2·0	2·301	2·602	2·903
e	0·945	0·925	0·883	0·841	0·800	0·757	0·715

$$C_c = \frac{0·210}{1·505} = 0·138$$

$$e = 0·841 - 0·138 \log_{10}\frac{\sigma'}{100}$$

The volume change for virgin consolidation, as the effective stress changes from σ'_i to σ'_f, is

$$-dV = \left(\frac{e_i - e_f}{1 + e_i}\right)V_i$$

$$= \frac{1}{1 + e_i}V_i C_c \log_{10}\frac{\sigma'_f}{\sigma'_i}$$

where the suffixes i or f indicate the initial or final state respectively.

For one-dimensional consolidation, the change in thickness of a consolidating layer of initial thickness H_i is

$$-dH = \frac{1}{1 + e_i}H_i C_c \log_{10}\frac{\sigma'_f}{\sigma'_i}$$

Example 5.2

A soil stratum 3 m thick is normally consolidated under an overburden pressure of 50 kN/m^2 at a void ratio of 0·891. If the compression index is 0·138, the settlement, when the effective stress is increased to 100 kN/m^2, is

$$-dH = \left(\frac{1}{1 + 0·891}\right) \times 3 \times 0·138 \times \log_{10}\frac{100}{50}\,\text{m}$$

$$= 68\,\text{mm}$$

5.9 Coefficient of volume change (m_v): Since the compression index C_c only refers to the virgin part of the consolidation curve, and since many soils are overconsolidated over at least part of the relevant range of effective stress, it is often more convenient to express the e/σ' relationship in terms of the *coefficient of volume change*. This is defined as the change in volume, per unit volume, per unit change of effective stress:

$$m_v = -\frac{1}{V}\frac{dV}{d\sigma'}$$

For one-dimensional consolidation of a sample of thickness H,

$$m_v = -\frac{1}{H}\frac{dH}{d\sigma'}$$

Since the change in total volume equals the change in the volume of the voids,

$$\frac{dV}{V} = \frac{de}{1 + e}$$

Then

$$m_v = -\left(\frac{1}{1 + e}\right)\frac{de}{d\sigma'}$$

$$= -\left(\frac{1}{1 + e}\right) \times \text{(the slope of the } e/\sigma' \text{ curve)}$$

Example 5.3
Using the same test results as in Example 5.1

σ'(kN/m²)	0	25	50	100	200	400	800
Thickness (mm)	19·05	18·85	18·44	18·03	17·63	17·21	16·80

Then

$-dH$ (mm)	0·20	0·41	0·41	0·40	0·42	0·41
$\dfrac{-dH}{d\sigma'}$	0·0080	0·0164	0·0082	0·0040	0·0021	0·0010
m_v (m²/kN)	0·00042	0·00088	0·00045	0·00022	0·00012	0·00006

In a stratum 3 m thick, when the effective stress is increased from 50 kN/m² to 100 kN/m²,

$$m_v = 0\cdot000\,45\,\text{m}^2/\text{kN}$$

Then
$$-dH = 0{\cdot}00045 \times 3 \times 50\,\text{m}$$
$$= 68\,\text{mm}$$

5.10   The effect of sample disturbance on the e/σ' curves: The extent of the volume change under load is very sensitive to disturbance of the soil structure. The effect of this is two-fold:

(a)   When a soil is reloaded, the recompression curve turns gradually into the virgin consolidation curve, and does not follow the curve for unloading.

(b)   As the consolidation pressure is approached, the curve for a disturbed sample is steeper and at a lower value of the void ratio than that for the undisturbed soil.

The laboratory curve is inevitably obtained by reloading (from zero total stress) a sample which has been to some extent disturbed during removal from the ground. It is not always easy to reconstruct the early part of the *in situ* e/σ' curve, or to determine the preconsolidation pressure, from the laboratory test results.

Casagrande [5.3] has suggested the method shown in Fig. 5.7 for determining the preconsolidation pressure.

From the point of maximum curvature A, a line AB is drawn to bisect the angle between the horizontal and the tangent at A. The straight part of the $e/\log \sigma'$ curve is projected back to D. The intersection of AB and CD is taken to indicate the preconsolidation pressure (σ'_c).

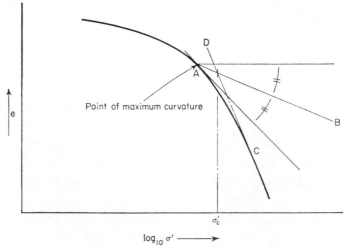

Fig. 5.7   Casagrande's construction for determining the preconsolidation pressure.

Leonards [5.4], adapting a method proposed by Schmertmann [5.5], has suggested the following construction for recovering the early part of the *in situ* curve (see Fig. 5.8). The oedometer test is continued until the virgin curve is reached (indicated by a straight line on the $e/\log \sigma'$ curve). The sample is then unloaded to the overburden pressure, and the mean slope of the unloading curve FG is determined.

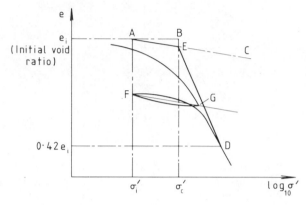

Fig. 5.8   *Correction for the effect of disturbance on the $e:\sigma'$ curves.*

The line AC is drawn parallel to the mean slope of FG, intersecting the line $\sigma' = \sigma'_c$ at E. ED is drawn to join the laboratory curve at $e = 0.42e_i$. AED is taken to represent the *in situ* curve.

RATE OF CONSOLIDATION

5.11   Terzaghi's theory of one-dimensional consolidation: In analysing the rate of one-dimensional consolidation, Terzaghi [5.1] made the following assumptions:

(a)   The soil is homogeneous and fully saturated.
(b)   The water and soil particles are incompressible.
(c)   Both the flow of water and the movement of the soil particles are in one dimension only.
(d)   Darcy's law is valid.
(e)   The permeability (k) is constant over the relevant range of effective stress.
(f)   The value of $de/d\sigma'$ is constant over the relevant range of effective stress.
(g)   The time lag in consolidation is entirely due to the low permeability of the soil.

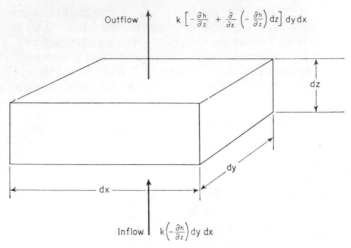

Fig. 5.9   Terzaghi's theory of one-dimensional consolidation.

Consider an element of soil of constant area $dx \times dy$, and of thickness dz, free to drain in the z direction only (see Fig. 5.9). Let the element be subjected to an increment of total stress $\Delta\sigma$, instantaneously applied and maintained constant.

If Darcy's law is valid, the rate of flow of water into the element is

$$v_z \, dx \, dy = -k \frac{\partial h}{\partial z} \, dx \, dy$$

Then the net rate of flow of water from the element is

$$\frac{\partial}{\partial z}\left(-k \frac{\partial h}{\partial z} dx \, dy\right) dz = -k \, dx \, dy \, dz \frac{\partial^2 h}{\partial z^2}$$

since k, dx, and dy are assumed to be constant.
The volume of the voids is

$$\frac{e}{1 + e} \, dx \, dy \, dz$$

Then the rate of change of volume of the voids is

$$\frac{\partial}{\partial t}\left(\frac{e}{1 + e} dx \, dy \, dz\right) = \left(\frac{dx \, dy \, dz}{1 + e}\right) \frac{\partial e}{\partial t}$$

since

$$\left(\frac{dx \, dy \, dz}{1 + e}\right)$$

is the volume of the soil particles, which is assumed to be constant.

Since the soil is assumed to be saturated, and since the water and soil particles are assumed to be incompressible, the rate of reduction of the volume of the voids must equal the net rate of flow of water from the element.

Therefore

$$k \frac{\partial^2 h}{\partial z^2} = \frac{1}{1 + e} \frac{\partial e}{\partial t}$$

But

$$h = z + u/\gamma_w$$

Therefore

$$\frac{k}{\gamma_w} \frac{\partial^2 u}{\partial z^2} = \frac{1}{1 + e} \frac{\partial e}{\partial t}$$

since γ_w is constant.

But

$$m_v = -\frac{1}{1 + e} \frac{de}{d\sigma'}$$

$$= -\frac{1}{1 + e} \frac{\partial e/\partial t}{\partial \sigma'/\partial t}$$

so that

$$\frac{k}{\gamma_w} \frac{\partial^2 u}{\partial z^2} = -m_v \frac{\partial \sigma'}{\partial t}$$

Since σ remains constant,

$$\frac{\partial \sigma'}{\partial t} = -\frac{\partial u}{\partial t}$$

Therefore

$$\frac{k}{\gamma_w m_v} \frac{\partial^2 u}{\partial z^2} = \frac{\partial u}{\partial t}$$

Writing

$$c_v = \frac{k}{\gamma_w m_v}$$

$$c_v \frac{\partial^2 u}{\partial z^2} = \frac{\partial u}{\partial t}$$

where c_v is called the coefficient of consolidation.

This may be written in non-dimensional terms,

$$\frac{\partial^2 U_v}{\partial Z^2} = \frac{\partial U_v}{\partial T_v}$$

where $U_v = (u_i - u)/(u_i - u_f)$, and is called the *degree of consolidation.*

$Z = z/d$

$T_v = c_v t/d^2$, and is called the *time factor.*

u_i = the initial pore pressure.

u = the pore pressure after time t.

u_f = the final pore pressure after infinite time, when equilibrium has been reached.

d = the length of the maximum drainage path (Fig. 5.10).

Since $de/d\sigma'$ and σ are both assumed constant,

$$d\sigma' = -du \quad \text{and} \quad \frac{\partial e}{\partial u} \text{ is constant}$$

Then

$$U_v = \frac{u_i - u}{u_i - u_f} = \frac{e_i - e}{e_i - e_f}$$

where e_i and e_f are the initial and final values of the void ratio respectively

e is the void ratio after time t.

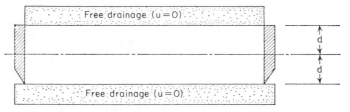

Fig. 5.10   Drainage conditions in the oedometer test.

Thus, the term U_v represents both the proportion of the pore pressure dissipated after time t, and also the proportion of the total volume change which has taken place at that time.

For the sample in the oedometer test, the boundary conditions are (see Fig. 5.10) as follows:

For $0 \leq T_v \leq \infty$ and $Z = 0$ $\qquad u = 0$
For $0 \leq T_v \leq \infty$ and $Z = 2$ $\qquad u = 0$
For $T_v = 0$ $\qquad$ and $0 < Z < 2$ $\qquad u = \Delta\sigma$

For these boundary conditions, Terzaghi, using the method of separation of variables, derived the solution

$$U_v = 1 - \sum_{m=0}^{m=\infty} \frac{2}{M} \sin(MZ) \exp(-M^2 T_v)$$

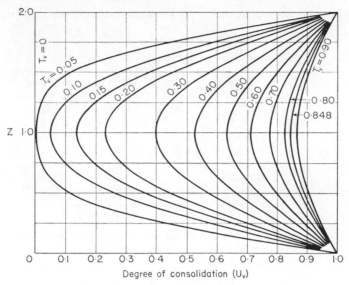

Taylor [5.6]

Fig. 5.11   Degree of consolidation (U_v) as a function of Z and T_v in the oedometer test.

where
$$M = \tfrac{1}{2}\pi(2m + 1)$$
$$m = 0, 1, 2, \ldots$$

Figure 5.11 shows the values of U_v plotted against Z and T_v.

Let $\bar{U}$ be the mean value of U_v over the thickness $2d$. This represents the proportion of the total volume change in the whole sample which has taken place at time t. Then it may be shown [Taylor, 5.6] that

$$\bar{U} = 1 - \sum_{m=0}^{m=\infty} \frac{2}{M^2} \cdot \exp(-M^2 T_v)$$

The values of $\bar{U}$ derived from this expression are given in Table 5.1.

5.12   *Determination of c_v in the oedometer test:*   The parameters $\bar{U}$ and T_v are non-dimensional, and the solution quoted above is therefore applicable to all cases having the same boundary conditions, provided only that the assumptions made in deriving the solution are valid. If, therefore, the theoretical relation between $\bar{U}$ and T_v is plotted alongside the experimental curve for settlement against time, corresponding

TABLE 5.1   *Variation of T_v with $\bar{U}$ for the boundary conditions of the oedometer test*

$\bar{U}$	T_v	$\bar{U}$	T_v
0·10	0·008	0·55	0·238
0·15	0·019	0·60	0·287
0·20	0·031	0·65	0·342
0·25	0·050	0·70	0·403
0·30	0·071	0·75	0·490
0·35	0·097	0·80	0·567
0·40	0·126	0·85	0·700
0·45	0·160	0·90	0·848
0·50	0·196	0·95	1·125

values of T_v and t may be determined from a comparison of the curves.

5.13   *Taylor's (time)$^{\frac{1}{2}}$ curve fitting method:*   For $\bar{U} \not> 0·6$,

$$\bar{U} \simeq \frac{2}{(\pi)^{\frac{1}{2}}} (T_v)^{\frac{1}{2}}$$

with an error of less than 1%. If $\bar{U}$ is plotted against $(T_v)^{\frac{1}{2}}$, the upper part of the curve is an almost straight line whose slope is $2/(\pi)^{\frac{1}{2}}$. This slope may be compared with the equivalent slope of the curve obtained in the laboratory test (see Fig. 5.12). If the straight section of the curve is projected to meet the line $\bar{U} = 1·0$, then

$$t = t_1 \quad \text{and} \quad T_v = \frac{\pi}{4}$$

Therefore

$$\frac{\pi}{4} = T_v = \frac{c_v t_1}{d^2}$$

$$c_v = \frac{\pi d^2}{4t_1}$$

5.14   *Casagrande's log (time) curve fitting method:*   It is sometimes more convenient to plot the settlement against the logarithm of time, as shown in Fig. 5.13. From Table 5.1 it may be seen that when

$$\bar{U} = 0.5 \qquad T_v = 0·196$$

Then

$$0·196 = \frac{c_v t_{50}}{d^2}$$

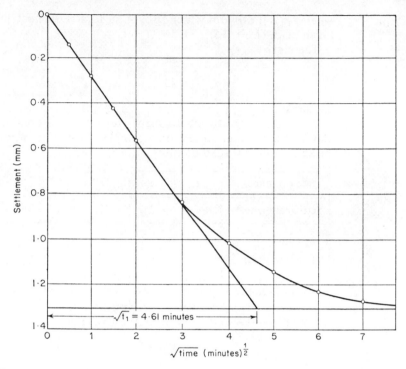

Fig. 5.12   Example 5.4(a).

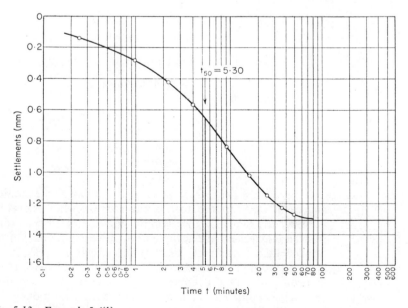

Fig. 5.13   Example 5.4(b)

where t_{50} is the time at which 50% of the total settlement has taken place ($\bar{U} = 0.5$).

Therefore

$$c_v = \frac{d^2}{t_{50}} \times 0.196$$

Example 5.4

The following time/settlement record was obtained during a consolidation test on a sample of soft clay, whose mean thickness was 19·05 mm.

Time (minutes)	$\frac{1}{4}$	1	$2\frac{1}{4}$	4	9
Settlement (mm)	0·140	0·282	0·425	0·565	0·840

Time (minutes)	16	25	36	49	24 hours
Settlement (mm)	1·020	1·148	1·231	1·275	1·308

(a) From the settlement/(time)$^{\frac{1}{2}}$ curve (Fig. 5.12),

$$(t_1)^{\frac{1}{2}} = 4.61 \text{ (minutes)}^{\frac{1}{2}}$$

Therefore

$$c_v = \frac{\pi(19\cdot05/2)^2}{4 \times 4\cdot61^2}$$
$$= 3\cdot35 \text{ mm}^2/\text{minute}$$
$$= 1\cdot76 \text{ m}^2/\text{year}$$

(b) From the settlement/log (time) curve (Fig. 5.13),

$$t_{50} = 5\cdot30 \text{ minutes}$$

Therefore

$$c_v = \frac{(19\cdot05/2)^2 \times 0\cdot196}{5\cdot30}$$
$$= 3\cdot35 \text{ mm}^2/\text{minute}$$
$$= 1\cdot76 \text{ m}^2/\text{year}$$

5.15 *Sources of error in the oedometer test:* The principal mechanical source of error in the test is the friction which develops between the sides of the sample and the steel ring. This friction may relieve the sample of an appreciable part of the applied load increment, so that the effective stress in the sample is over-estimated. Errors in the effective stress exceeding 6% have been recorded. A greased Teflon liner or a polished chrome surface inside the ring can reduce this error considerably, but it can never be entirely eliminated.

A further disadvantage of this apparatus is that there is no provision for controlling the pore pressure in the sample, which is allowed to drain freely until u is zero. This may allow air to come out of solution, blocking the pores of the soil with small bubbles, and altering the permeability of the soil. The small size of the sample (generally only 19·05 mm thick) is also a disadvantage when testing 'varved' or stratified soils.

Apart from this, the main discrepancies between the results of laboratory tests and *in situ* observations are the consequences of the assumptions made in deriving Terzaghi's theory. These assumptions will now be examined. The assumption that the soil particles and pore fluid are incompressible is close to the truth for any saturated soil. One-dimensional consolidation is enforced in the oedometer, although it is very different from the form of consolidation in the ground under a load of limited lateral extent. There is some evidence for deviations from Darcy's law if the hydraulic gradient is very low, as it may be in the ground. However, this is unlikely to be a major source of the observed discrepancies, since the latter are, in general, the reverse of those to which this error would lead.

Several attempts have been made to improve on Terzaghi's assumptions of constant permeability and constant value of $de/d\sigma'$. For a normally consolidated clay, the coefficient of volume change (m_v) is almost inversely proportional to the effective stress (σ'), while the coefficient of consolidation (c_v) is fairly constant over quite a wide range of stress. Based on these two conditions, Davis and Raymond [5.7] obtained the same solution as Terzaghi for the rate of settlement. However, they predicted a rate of pore pressure dissipation appreciably slower than that given by Terzaghi's theory, and they were able to demonstrate this difference in tests. At large increment ratios, this discrepancy could be important, but at the usual ratio of 2:1 the effect is marginal, and is masked by the contrary effect of viscous forces between the soil particles. The latter effect is discussed in Section 5.18 below.

5.16 *Comparison of theoretical and experimental settlement/time curves:* In practice, there are considerable differences between the shapes of the theoretical and experimental settlement/time curves, both at the beginning and at the end of consolidation. These differences are the results of

(a) 'immediate' settlement, and
(b) secondary compression.

5.17   *'Immediate' settlement:*   At the beginning of the test, there is often a measurable amount of rapid settlement, mainly as a result of the compression of small quantities of air in the soil. Although this settlement is not quite instantaneous, it generally occurs so quickly that we may ignore it when considering the rate of consolidation in the long term. On

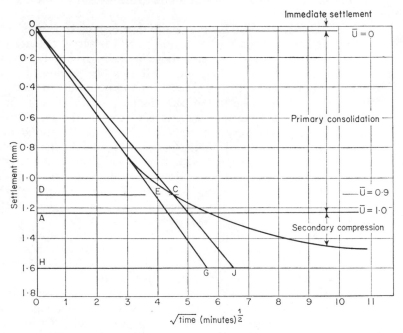

Fig. 5.14 'Immediate' settlement and secondary compression (Taylor's (time)$^{\frac{1}{2}}$ curve fitting method).

the settlement/(time)$^{\frac{1}{2}}$ curve, the point at which the true or primary consolidation begins may be estimated by projecting back the straight part of the curve to meet the line $t = 0$, as shown in Fig. 5.14.

On Casagrande's settlement/log (time) curve, the beginning of the test (when $t = 0$) cannot be recorded, but the starting point of the primary consolidation settlement may be found by a simple construction, illustrated in Fig. 5.15. For values of $\bar{U}$ less than 0·6, T_v is nearly proportional to $\bar{U}^2$. If we select any convenient time t_1, the consolidation settlement between $t = 0$ and $t = t_1$ is equal to the settlement between $t = t_1$ and $t = 4t_1$.

5.18   *Secondary compression:*   It is often found that the settlement in the oedometer test continues at a considerable, although ever-decreasing, rate after the excess pore pressure has practically vanished. This settlement is called *secondary compression*, and is mainly the consequence of the viscous resistance to relative movement of the soil particles, caused by the high viscosity of the adsorbed water layers. Although Terzaghi was well aware of this effect, he specifically excluded it from

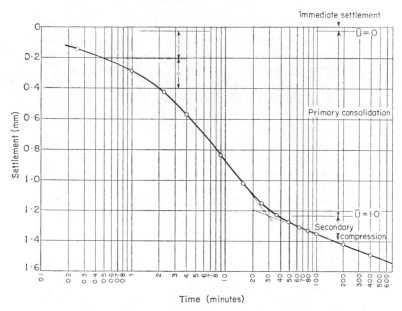

Fig. 5.15 *'Immediate' settlement and secondary compression (Casagrande's log (time) curve fitting method).*

consideration in the development of his theory, which is based on the implicit assumption that the void ratio is a unique function of effective stress, independent of time.

On a settlement/log (time) curve, the secondary compression generally appears as a straight line—at least in the early stages. The point at which u vanishes is given approximately by the intersection of the tangent through the point of contraflexure with the backward projection of this straight section, as illustrated in Fig. 5.15.

On a settlement/(time)$^{\frac{1}{2}}$ curve (Fig. 5.14), the corresponding point may be determined by the construction described below. It is found in practice that the experimental curve follows the form of Terzaghi's solution until $\bar{U} \simeq 0.9$. At this point,

$$(T_v)^{\frac{1}{2}} = (0.848)^{\frac{1}{2}} = DC$$

But

$$DE = \frac{(\pi)^{\frac{1}{2}}}{2} \times 0.9$$

$$\frac{DC}{DE} = \frac{2 \times (0.848)^{\frac{1}{2}}}{(\pi)^{\frac{1}{2}} \times 0.9}$$

$$= 1.15$$

The straight part of the curve is therefore projected to cut any convenient base line at G. HJ is set off equal to 1.15 HG. The point of intersection of OJ with the experimental curve is the point C, corresponding to $\bar{U} = 0.9$. If OA is set off equal to $OD/0.9$, the point A corresponds to $\bar{U} = 1.0$. All settlement below this level is assumed to be secondary compression.

5.19   *Interpretation of laboratory test curves showing secondary compression:* Even where the viscous resistance is large, the early part of the experimental curve closely follows the *form* of Terzaghi's solution. There is, therefore, a temptation to assume that we may apply Terzaghi's theory directly to that part of the curve representing primary consolidation, and in this way estimate the initial rate of settlement in a soil mass, regarding secondary compression as an aberration affecting the later stages only. This implies that the viscous forces have no effect during primary consolidation, whereas, in fact, they act from the instant of loading and affect the shape of the whole curve. Moreover, the laboratory test is greatly accelerated, because of the thinness of the sample. The viscous forces in the laboratory sample are much greater than they would be during the slower consolidation of a large soil mass. Direct extrapolation may therefore lead to considerable error.

Taylor and Merchant [5.8] first analysed the problem of primary and secondary compression, assuming a linear relation between the viscous forces and the rate of volume change. It is unlikely that the real relationship is as simple as this, and a number of more complex cases have since been examined [see, for example, Barden, 5.9]. These solutions provide a means of estimating the rate of settlement in the ground, when secondary compression is too large to be ignored. The details of these solutions are beyond the scope of this book, but we can draw certain general conclusions from them. Firstly, a small load increment ratio increases the proportion of secondary compression. The load increment ratio is generally standardised at 2:1 in the laboratory test, but may differ considerably from this in the ground.

Secondly, increasing the sample thickness greatly reduces the proportion of secondary compression. Settlement records of structures often show little or no secondary compression, even though the corresponding laboratory tests show it to a marked degree. Direct extrapolation of the laboratory tests leads to an under-estimate of the initial rate of settlement, and an over-estimate of the rate at which the pore pressure is dissipated.

REFERENCES

5.1 TERZAGHI, K. 1943. *Theoretical soil mechanics*. Wiley (New York).

5.2 British Standards Institution. Methods of testing soils for civil engineering purposes. BS:1377: 1975.

5.3 CASAGRANDE, A. 1936. The determination of the pre-consolidation load and its practical significance. *Proc. 1st Int. Conf. Soil Mech. and Found. Eng.*, **3.**

5.4 LEONARDS, G. A. 1962. *Foundation Engineering*. McGraw-Hill (New York).

5.5 SCHMERTMANN, J. M. 1955. The undisturbed consolidation of clay. *Trans. American Soc. Civil Eng.*, **120.**

5.6 TAYLOR, D. W. 1948. *Fundamentals of soil mechanics*. Wiley (New York).

5.7 DAVIS, E. H. and RAYMOND, G. P. 1965. A non-linear theory of consolidation. *Géotechnique*, **15.**

5.8 TAYLOR, D. W. and MERCHANT, W. 1940. A theory of clay consolidation accounting for secondary compressions. *Journal of Math. Phys.*, **19.**

5.9 BARDEN, L. 1955. Consolidation of clay with non-linear viscosity. *Géotechnique*, **15.**

CHAPTER 6

A model for shear strength of soil

THE MOHR–COULOMB SOIL MODEL

6.1 Shear strength, yield, and failure: The last chapter discussed soil
models describing behaviour caused by one-dimensional
changes in normal stress. In this chapter, we will consider a
very simple soil model describing the behaviour of soil
subjected to shear. Before examining this model, however, it
will be useful to have a general understanding of the way in
which real soils are found to behave.

Consider a small element of soil, subjected to a constant
and uniform normal stress σ, and a variable shear stress τ, as
shown in Fig. 6.1(a). Fig. 6.1(b) shows a typical relationship
between shear stress and shear strain for such an element of
real soil. For very small values of the shear stress, the
corresponding strains may be nearly linear and elastic. As the
shear stress increases, a point is reached where significant
plastic shear strains starts to develop. At this point, the
material is said to *yield*. The resistance of the soil to

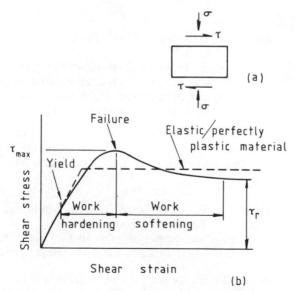

Fig. 6.1   (a) *A soil element in simple shear.* (b) *Stress/strain relationships: typical real soil*
(*solid line*), *and elastic/perfectly plastic model soil* (*broken line*).

119

increasing plastic shear strains is called the *shear strength*. At first, the plastic strains are strictly limited, because they result in an increased resistance to further deformation (that is, an increased shear strength). The material is said to *work harden* or *strain harden*. In discussing compressibility (Section 5.5), we showed that soils subjected to increasing normal stress are theoretically capable of unlimited strain hardening. The yield is said to be *stable*, and unlimited volumetric strain cannot occur. However, strain hardening can only increase the resistance to shear stress to a strictly limited extent. Once the applied stress exceeds the limiting value τ_{max}, shear. strains increase continuously for as long as the shear stress is maintained, because the external work done exceeds the internal dissipation of energy in overcoming the resistance of the soil. The yield is said to be *unstable* (Drucker, [6.1]) and the soil is said to *fail*. In some soils, and under some loading conditions, the shearing resistance decreases after failure. This decrease is described as *strain softening* or *work softening*. Eventually, after quite large strains have occurred, a nearly constant residual strength (τ_r) is developed, and this is nearly independent of further strains.

6.2 *Coulomb's soil model:* Coulomb [6.2] suggested that the shear strength of a soil might be expressed in the form (see Fig. 6.2(a))

$$\tau_f = c + \sigma_n \tan \varphi$$

where τ_f is the absolute value of the shear stress in the soil at failure,

σ_n is the normal stress on the surface of failure, compressive stress being taken as positive,

c, φ are parameters which are approximately constant for any particular soil.

Experience has since shown that τ_f cannot generally be expressed in terms of total stress (σ_n). Coulomb's model must therefore be redefined in the form (Fig. 6.2(b))

$$\tau_f = c' + \sigma'_n \tan \varphi'$$

where σ'_n is the effective normal stress, as defined in Section 3.2 above,

c', φ' are approximately constant parameters expressing the shear strength in terms of effective stress.

With sufficient accuracy for all practical purposes, it is found that, for saturated soils,

$$\sigma'_n = \sigma_n - u$$

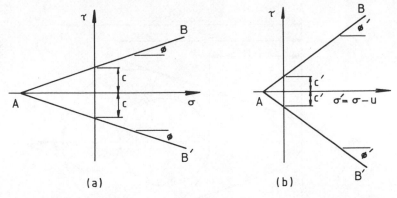

Fig. 6.2   Coulomb's failure condition. (a) In terms of total stress. (b) In terms of effective stress.

where u is the pore water pressure. This relationship, first explicitly stated by Terzaghi, is found to be approximately valid for a wide variety of materials and conditions of loading.

The shear strength parameters (c', φ'), defined in this way, should not be thought of as having a simple physical meaning. The angle φ' is not a true angle of friction: it is the slope of the line representing shear strength in terms of effective normal stress on the surface of failure, and is called the *angle of shearing resistance*. The parameter c' represents that part of the shear strength which is independent of the effective normal stress, and is called the *apparent cohesion*.

The defects of Coulomb's model as a means of describing real soil behaviour will be discussed in Section 6.18 below. Two points must however be made before proceeding further. Firstly, the model makes no statement about strains, and can therefore provide no information about displacements resulting from the application of any load. Secondly, the model implies that the shear strength parameters are constants, so that the shearing resistance is unaffected by work softening after failure. It may also sometimes be useful to assume that the model soil does not strain harden before failure. For such a material, *yield* and *failure* are simultaneous. A material which neither hardens nor softens in this way is said to be *perfectly plastic*. If the model soil is assumed to be linearly elastic for all values of the shear stress less than τ_f, the stress/strain relationship is of the form shown by the broken line in Fig. 6.1. Such a material is said to be *elastic/perfectly plastic*. This is not, however, a necessary feature of Coulomb's model, which essentially makes no statement whatever about stress states below failure.

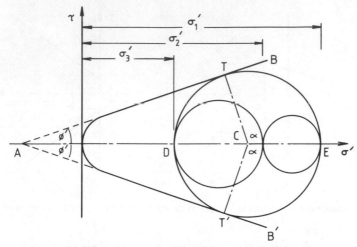

Fig. 6.3   Coulomb's failure condition and the Mohr stress circles.

The lines AB and AB' in Fig. 6.2 are boundaries, dividing possible stress states on and between the lines, from impossible stress states outside them. For all stress states described by points on the lines, the model soil is in a state of failure. The expression

$$\tau_f = c' + (\sigma'_n - u)\tan\varphi'$$

is called the *failure* (and also the *yield*) *condition*. The lines AB and AB' are called *failure* (and also *yield*) *loci*. For states described by points between the lines, all we can say is that the strains are of finite magnitude.

6.3   *Coulomb's failure condition and the Mohr stress circles:*   Figure 6.3 shows the Mohr circles for effective stress at some point within a soil mass, superimposed on the yield loci. The three principal components of effective stress are represented by σ'_1, σ'_2, and σ'_3: in this chapter we will observe the usual convention and assume that $\sigma'_1 \geq \sigma'_2 \geq \sigma'_3$. If the soil is to be stable, all three circles must lie wholly between the lines AB and AB'. Thus, another way of stating Coulomb's failure condition is to say that failure occurs if the largest Mohr stress circle (defined by the greatest and least principal stress components σ'_1 and σ'_3), is tangential to Coulomb's failure loci. This is generally known as the *Mohr–Coulomb failure condition*. The failure loci are also known as the *Mohr envelopes*.

Note that the shear strength is independent of the intermediate principal stress component σ'_2. This is a feature of the Mohr–Coulomb model which will be examined in the next chapter.

Cohesive soils are capable of supporting some tension, at least temporarily. However, cracks may develop in the ground, and tensile stress cannot be sustained across these cracks. It is therefore usual to ignore the tensile strength of all soils, and to assume that they are capable of sustaining compressive (that is, positive) normal effective stress only. If this is the case, the possible stress states are limited by the Mohr stress circle for which σ'_3 is zero, as shown in Fig. 6.3.

6.4   *Slip lines and slip-line fields:*   The angles ECT and ECT' in Fig. 6.3 are equal to $\alpha = \pm(\pi/2 + \varphi')$. The tangent points T and T' on the Mohr stress circle $ETDT'$ therefore define the stress components on two planes inclined at $\pm\frac{1}{2}(\pi/2 + \varphi')$ to the plane on which the greatest principal stress component σ'_1 acts. These two planes are therefore inclined at $\pm(\pi/4 - \varphi'/2)$ to the direction of σ'_1. The soil has reached the limit of its strength on these two planes, which therefore form *slip lines* along which movement will occur at failure. Notice that these slip lines are a feature of the Mohr–Coulomb model soil. This does not necessarily mean that real soils will be found to behave in this way.

As a soil structure approaches collapse, a growing part of it reaches the limit of its shear strength. Once this *plastic zone* is sufficiently extensive to form an unstable mechanism, unlimited plastic shear strains occur. The pattern of slip lines within the plastic zone is called the *slip-line field*.

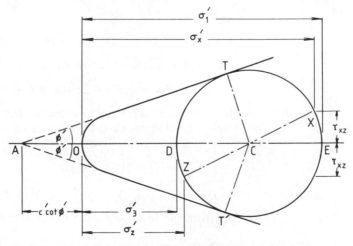

Fig. 6.4   Geometry of the Mohr stress circle at failure.

6.5 *Other forms of the Mohr–Coulomb failure condition:* From the geometry of Fig. 6.4 we can derive other forms of the Mohr–Coulomb failure condition, which will be useful later.

(a) $AC = \frac{1}{2}(\sigma'_1 + \sigma'_3) + c' \cot \varphi'$

$TC = CE = \frac{1}{2}(\sigma'_1 - \sigma'_3)$

But $TC = AC \sin \varphi'$

so that

$$(\sigma'_1 - \sigma'_3) = (\sigma'_1 + \sigma'_3) \sin \varphi' + 2c' \cos \varphi'$$

Alternatively

$$\sigma'_1(1 - \sin \varphi') - \sigma'_3(1 + \sin \varphi') - 2c' \cos \varphi' = 0$$

(b) Taking the previous result as our starting point,

$$\sigma'_3 = \sigma'_1 \left(\frac{1 - \sin \varphi'}{1 + \sin \varphi'} \right) - \left(\frac{2c' \cos \varphi'}{1 + \sin \varphi'} \right)$$

$$= \sigma'_1 \left(\frac{1 - \sin \varphi'}{1 + \sin \varphi'} \right) - 2c' \left(\frac{(1 - \sin^2 \varphi')}{(1 + \sin \varphi')^2} \right)^{\frac{1}{2}}$$

$$= \sigma'_1 \left(\frac{1 - \sin \varphi'}{1 + \sin \varphi'} \right) - 2c' \left(\frac{1 - \sin \varphi'}{1 + \sin \varphi'} \right)^{\frac{1}{2}}$$

By simple trigonometry, it can be shown that

$$\left(\frac{1 - \sin \varphi'}{1 + \sin \varphi'} \right) = \tan^2 (\pi/4 - \varphi'/2)$$

so that

$$\sigma'_3 = \sigma'_1 \tan^2 (\pi/4 - \varphi'/2) - 2c' \tan (\pi/4 - \varphi'/2)$$

Since $\tan (\pi/4 - \varphi'/2) = 1/\tan (\pi/4 + \varphi'/2)$

$$\sigma'_1 = \sigma'_3 \tan^2 (\pi/4 + \varphi'/2) + 2c' \tan (\pi/4 + \varphi'/2)$$

(c) Let points X and Z represent the stress components on planes normal to two co-ordinate axes x and z in the plane of σ'_1 and σ'_3. Then

$$OC = \frac{1}{2}(\sigma'_1 + \sigma'_3) = \frac{1}{2}(\sigma'_x + \sigma'_z)$$

$$XZ = DE = (\sigma'_1 - \sigma'_3) = ((\sigma'_x - \sigma'_z)^2 + 4\tau^2_{xz})^{\frac{1}{2}}$$

so that, from the result obtained in (a) above,

$$((\sigma'_x - \sigma'_z)^2 + 4\tau^2_{xz})^{\frac{1}{2}} = (\sigma'_x + \sigma'_z) \sin \varphi' + 2c' \cos \varphi'$$

6.6   *Total stress and effective stress analyses:*   In Section 6.2, we said that
the shear strength cannot generally be stated in terms of total
stress. Calculations must generally be made in terms of
effective stress, and this requires a knowledge of the pore
pressure. There is one important case, however, where this is
not necessary. In a saturated soil, Skempton's pore pressure
parameter B is nearly 1·0 (see Section 3.6). Any increase in
the mean normal stress results in an almost equal rise in the
pore pressure, so that the mean normal effective stress
remains constant. Thus the shear strength is unaffected by
changes in total normal stress. Provided that the soil is
saturated, and is undrained (so that pore pressures generated
by the total stress changes cannot dissipate), Coulomb's
failure condition may be stated in terms of total stress in the
form

$$\tau_f = c_u \qquad \text{and} \qquad \varphi_u = 0$$

where c_u and φ_u are parameters defining the shear strength in
terms of total stress for the undrained condition. It will be
shown later (Section 7.14) that c_u is a function of the void
ratio e.

TESTS FOR SHEAR STRENGTH OF SOIL

6.7   *The direct shear test:*   In this test (see Fig. 6.5) a sample of the soil
is placed in a square box, which is split in half horizontally.
In the usual form of the apparatus, the box is 60 mm
square, but a larger box is sometimes used for testing coarse-
grained soils or fissured clays. A dead weight is applied to

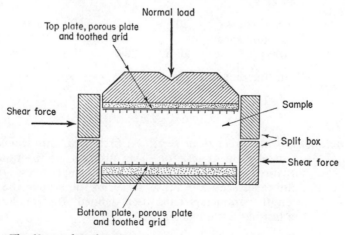

Fig. 6.5   The 60 mm shear box.

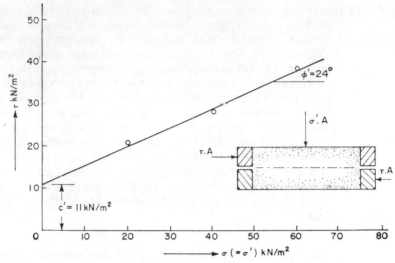

Fig. 6.6   Example 6.1.

the top of the sample, and a shear stress is applied by forcing the two halves of the box in opposite directions until failure occurs on the horizontal plane between them. The test is repeated with different normal stresses, and a direct plot is obtained of shear strength against normal stress on the horizontal failure surface.

Example 6.1
    The following results were obtained during slow drained tests in a 60 mm shear box on undisturbed samples of a silty clay:

Normal stress	20	40	60	(kN/m^2)
Maximum shear stress	20·6	28·0	38·1	(kN/m^2)

From the diagram (Fig. 6.6),

$$c' = 11 \ kN/m^2$$
$$\varphi' = 24°$$

6.8   *Limitations of the direct shear test:*   At first sight, this seems to be an ideal form of test for shear strength, as the quantities measured—the shear stress and normal stress on the failure surface—are those in terms of which the shear strength parameters have been defined. In fact, however, the test has a number of defects:

    (a)   *Non-uniform stress distribution in the sample.* The

distribution of stress is far from uniform, and the whole stress pattern is far more complex than the simple interpretation of Fig. 6.6 would suggest.

(b) *Indeterminate pore pressures.* There is no provision for controlling the drainage or for measuring the pore pressure in the sample, and the test can therefore only be used for total stress measurements. Provided that the test is carried out rapidly, the undrained strength of saturated clays may be determined. Alternatively, drained tests, carried out so slowly that the pore pressure is virtually zero throughout, may be used to determine c' and φ', the shear strength parameters with respect to effective stress.

The test has advantages over the triaxial compression test (see below) in three cases:

(a) *For cohesionless soils.* Samples of clean sands or gravels may be set up in the triaxial compression test apparatus, but the technique is not easy, particularly if the sample is to be compacted to a specified density. Since these soils have a high permeability, fully drained shear box tests may be carried out in a reasonably short time.

(b) *For tests at large deformations.* For reasons which will be discussed later, the residual strength of some clays, after large displacements along the failure surfaces, is less than the maximum. The strain in the sample in the triaxial compression test is limited to about 20%. Although the maximum value of the shear stress is generally reached within this limit, quite large deformations are required before the residual value is fully established, and, with samples of conventional size, these deformations are not generally obtainable within the limits of the test. Tests requiring large deformations in clay soils have been carried out [Skempton, 6.4] using a modified shear box apparatus. The sample is loaded in the normal way, but, when the box has reached the end of its travel, the motion is reversed. The deformation then continues, but in the opposite direction. Except for a slight distortion at the moment of reversal, this procedure seems to have little effect on the shape of the shear stress/deformation curve, and the process may be continued until the residual strength is fully established. Fig. 6.7 shows a typical curve obtained in this way. Other methods of determining the residual strength are discussed in Section 7.25 below.

(c) *For large samples.* Triaxial compression tests on small

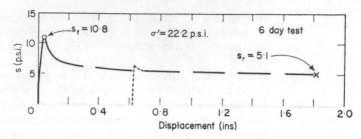

Fig. 6.7   Large displacement shear box test on a clay soil (after Skempton [6.4]).

samples of fissured clay may give misleadingly high values for the shear strength, since the only samples which can be prepared for testing come inevitably from the intact material between the fissures. Larger samples, which contain both fissures and intact material, will give measured values of the shear strength closer to the real strength of the bulk material. These samples, where they can be obtained, may be tested in the large shear box.

6.9   *The triaxial compression test apparatus:*   The soil samples used in this test are cylindrical, and are usually 38·1 mm (1½ in) in diameter and 76·2 mm (3 in) long, although larger samples may be used.

The sample is fitted between rigid end caps, and covered with a latex sheath, and is then placed inside a Perspex cell which is filled with water (see Fig. 6.8).

Pressure is applied to the water in the cell, which thus applies a uniform compressive stress σ_3 to the sample. An additional stress $(\sigma_1 - \sigma_3)$ is then applied in a vertical direction, by loading the sample through a ram bearing on the top end cap. The ram is forced down at a constant rate and the load on the ram is recorded.

A number of pieces of peripheral equipment have been developed, which allow considerable flexibility in the choice of test procedure. The most important of these are as follows:

(a)   *The compression testing machine.* The vertical load is applied to the ram in a compression testing machine, operating at a constant rate of displacement. The load is measured with a proving ring. Since the strain is controlled, the load on the ram may be measured at all stages of the test, even after the peak strength has passed.

(b)   *Constant pressure apparatus.* For tests of short duration, the cell pressure is generally maintained by applying

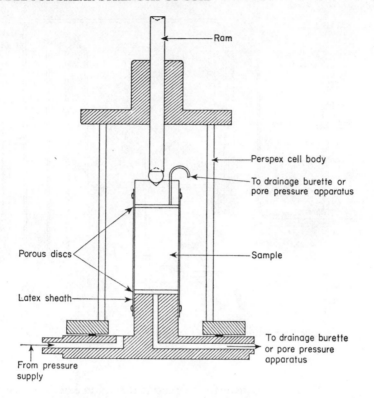

Fig. 6.8   The triaxial compression test cell.

an air pressure to the surface of water in a reservoir attached to the cell. For long term tests, the pressure may be maintained by a mercury column, or by a piston supporting a dead weight. In either case, there must be some provision for compensating for small leaks in the system over a long period.

(c) *Apparatus for pore pressure measurement.* Porous discs may be fitted to the top or bottom of the sample, or to both top and bottom, and these may be connected to apparatus for measuring the pore water pressure. The pressure measurement is usually made with a mercury manometer or a Bourdon gauge. If a true measurement of pressure is to be made, there must be no movement of water between the sample and the measuring apparatus, but in each of these instruments a considerable volume change accompanies any change in pressure. To overcome this difficulty, a small capillary tube filled with mercury is fitted between the sample and the pressure gauge or manometer (see Fig. 6.9).

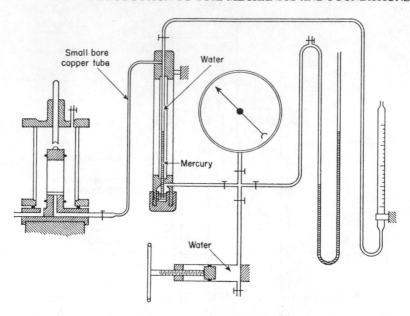

Fig. 6.9   Apparatus for pore pressure measurement (after Bishop [6.5]).

Provision is made for correcting the volume of the measuring system, so that the mercury column remains stationary. For tests of short duration, this volume correction may be effected by operating a piston by hand. For longer tests, the piston may be replaced by an oil reservoir containing a heating element. The volume of the oil is controlled by the heater, which is switched on and off by small movements of the mercury column. Alternatively, the pressure may be measured using a transducer of the diaphragm type, in which the volume change during operation is negligible.

(d)   *The drainage burette.* The porous discs may be connected to a burette, to measure the quantity of water expelled during consolidation. For saturated samples, this gives a direct measure of the volume change. In some samples, the complete dissipation of pore water pressure within the sample allows air to come out of solution, so that the sample is no longer fully saturated. This may be prevented by applying a pressure to the water in the burette, and at the same time increasing the cell pressure (σ_3) by the same amount. Since both the total stress and the pore pressure are equally increased, the effective stress remains unchanged.

6.10 *Conventional triaxial compression tests:* The principal types of triaxial compression test in common use are as follows [Bishop and Henkel, 6.3]:

(a) Undrained tests on samples of saturated clay, to determine the shear strength parameters with respect to total stress.

(b) Undrained tests, with pore pressure measurement, to determine the shear strength parameters with respect to effective stress.

(c) Drained tests and consolidated-undrained tests, to determine the shear strength parameters with respect to effective stress.

In addition, certain other tests are carried out in the triaxial apparatus. In particular,

(a) dissipation tests and permeability tests, and

(b) tests to determine the pore pressure parameters A and B.

6.11 *Undrained tests on saturated clays:* The sample is fitted between solid end caps, so that no change in water content is possible, and is loaded to failure as already described. Usually, not less than three samples are tested, each test being at a different value of σ_3. Mohr circles are drawn for the total stresses at failure in each test, and the Mohr envelope is drawn tangential to the circles.

Provided that the soil is fully saturated, the effective stresses at failure are the same for each test (see Section 6.6 above). Then the Mohr envelope, plotted with respect to total stress, is horizontal, φ_u is zero, and the shear strength is constant and equal to c_u.

Example 6.2

At a certain point in a saturated clay, the total vertical stress is 200 kN/m^2 and the horizontal stress is 150 kN/m^2. The pore pressure at this point is 50 kN/m^2. The pore pressure parameter A (which, for the purposes of this example, will be assumed to be constant) is 0·7.

If a sample of this soil is removed from the ground, the total stress is reduced to zero. Then

$$\Delta u = B[\Delta\sigma_3 + A(\Delta\sigma_1 - \Delta\sigma_3)]$$
$$= 1{\cdot}0[-150 + 0{\cdot}7(-200 + 150)] = -185 \text{ kN/m}^2$$
$$u = -135 \text{ kN/m}^2$$
$$\sigma'_1 = \sigma'_3 = +135 \text{ kN/m}^2$$

An undrained triaxial test is now carried out on part of

this sample, at a cell pressure of 100 kN/m^2. When the cell pressure is applied,

$$\Delta u = B \Delta \sigma_3 = +100 \text{ kN/m}^2$$
$$u = -135 + 100 = -35 \text{ kN/m}^2$$
$$\sigma'_3 = 100 + 35 = +135 \text{ kN/m}^2$$

Thus, the effective stress is unchanged.
When the vertical stress is applied,

$$\Delta u = B \times A(\Delta \sigma_1 - \Delta \sigma_3)$$
$$= 0 \cdot 7(\Delta \sigma_1 - \Delta \sigma_3) = 0 \cdot 7(\sigma_1 - \sigma_3)$$
$$\Delta \sigma'_1 = \Delta \sigma_1 - \Delta u = (1 \cdot 0 - 0 \cdot 7)(\sigma_1 - \sigma_3)$$
$$\Delta \sigma'_3 = -\Delta u = -0 \cdot 7(\sigma_1 - \sigma_3)$$

If the shear strength parameters with respect to effective stress are

$$c' = 20 \text{ kN/m}^2 \quad \text{and} \quad \varphi' = 25°$$

then from the diagram (Fig. 6.10), it may be seen that, at failure,

$$(\sigma'_1 - \sigma'_3) = 184 \text{ kN/m}^2 = (\sigma_1 - \sigma_3)$$

Tests at cell pressures of 200 kN/m^2 and 300 kN/m^2 will give the same result. Then, when these results are plotted in terms of total stress (the solid lines in Fig. 6.10), we find that

$$c_u = \frac{184}{2} = 62 \text{ kN/m}^2$$

$$\varphi_u = 0$$

6.12   *The unconfined compression test:*   If the soil can be assumed to be fully saturated—and this is commonly the case with clay soils, except in an arid climate—φ_u is zero, and only one parameter (c_u) remains to be determined. Only one test is therefore required, and this may be most easily carried out in unconfined compression (that is, with $\sigma_3 = 0$). This test

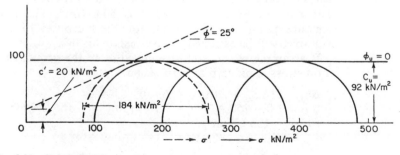

Fig. 6.10   Example 6.2.

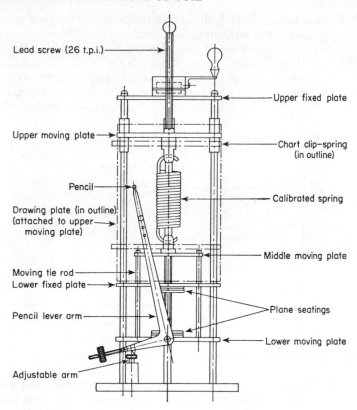

Lead screw (26 t.p.i.)

Upper fixed plate

Upper moving plate

Chart clip–spring
(in outline)

Pencil

Calibrated spring

Drawing plate (in outline)
(attached to upper
moving plate)

Middle moving plate

Moving tie rod
Lower fixed plate

Pencil lever arm

Plane seatings

Lower moving plate

Adjustable arm

Fig. 6.11   Unconfined compression test apparatus (after BS 1377:1975 [6.6]).

may be made either in the triaxial apparatus with zero cell pressure, or in a small portable apparatus designed for the purpose. This is shown in Fig. 6.11.

6.13   *Undrained tests with pore pressure measurement, on unsaturated samples:* For unsaturated soils, the pore pressure parameter B is less than $1 \cdot 0$, and an increase in the total stress causes some increase in effective stress. The effective stress may be determined by deducting the measured value of the pore pressure from the total stress. The sample is set up between porous discs which are connected to apparatus for measuring pore pressure. At least three samples are tested at different cell pressures, and the effective stresses at failure are computed. The Mohr circles are drawn for effective stresses at failure, and the parameters c' and φ' are determined from the Mohr envelope.

6.14 Drained tests: In fully drained samples, where the pore pressure is zero, the total and effective stresses are the same.

In this form of test, the sample is fitted between porous discs, through which it may drain freely. The sample is set up in the cell, a uniform total stress is applied, and time is allowed for the resulting pore pressure to dissipate. The cell is then transferred to a compression testing machine,

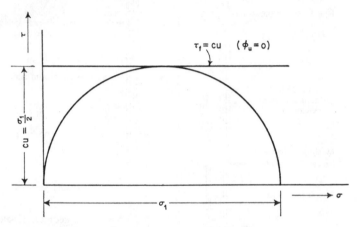

Fig. 6.12 Conventional interpretation of the unconfined compression test.

and the sample is loaded vertically to failure, but at such a slow rate that no appreciable pore pressure can build up. This stage of the test may take up to 48 hours for a small sample, and much longer for a larger one. At least three samples are tested at different cell pressures. Mohr circles are drawn for effective stresses at failure, and the parameters c' and φ' are determined from the Mohr envelope.

6.15 Consolidated-undrained tests: A drained test on a soil with low permeability has to be very slow. The parameters c' and φ' may be determined more rapidly by measuring the pore pressure and calculating the effective stress as the difference between the total stress and the pore pressure.

If the pore pressure is measured during an undrained test, the effective stress may be calculated. However, for saturated soils, the effective stress circle is the same for each sample, and the Mohr envelope cannot be drawn. This difficulty can be avoided if each sample is first consolidated, so that the effective stresses are not the same.

Three samples are set up to allow drainage from one end, and measurement of pore pressure at the other, and each sample is consolidated under a different stress (σ_3), as in the first stage of the drained test. At the end of this stage, the effective stress in each sample is practically equal to the cell pressure, which is different for each of the samples. Each cell is transferred in turn to the compression testing machine, the drainage tap is closed, and each sample is loaded to failure without further drainage. The pore pressure at failure is measured, Mohr circles are drawn for the effective stresses at failure, and the parameters c' and φ' are determined from the Mohr envelope.

Example 6.3

The following results were obtained during a consolidated-undrained triaxial test on a sample of overconsolidated clay.

Cell pressure (σ_3)	100	200	300	kN/m^2
Principal stress difference at failure ($\sigma_1 - \sigma_3$)	146	191	239	kN/m^2
Pore pressure at failure (u_f)	56	126	176	kN/m^2

Then, at failure,

$(\sigma'_1 - \sigma'_3) = (\sigma_1 - \sigma_3)$	146	191	239	kN/m^2
$\sigma'_3 = (\sigma_3 - u_f)$	44	74	124	kN/m^2

From the diagram (Fig. 6.13),

$$c' = 32 \text{ kN/m}^2$$
$$\varphi' = 22°$$

6.16 Dissipation tests: The coefficient of consolidation (c_v), required for settlement analysis, is generally determined from measurements of volume change in an oedometer test.

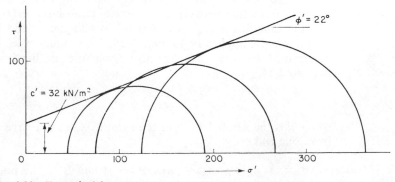

Fig. 6.13 Example 6.3.

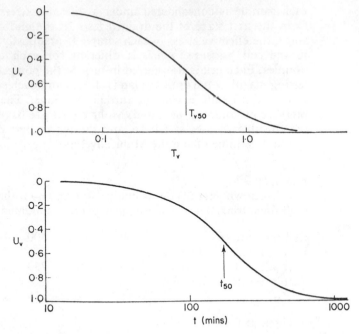

Fig. 6.14   The dissipation test.

However, where c_v is required for predicting the rate of dissipation of pore pressure, there are some advantages in measuring the pore pressure changes directly in a dissipation test, rather than deducing them indirectly from the volume changes. This is particularly true in the case of partially saturated soils, and soils with very low compressibilities.

For a dissipation test, the sample is set up in a triaxial cell, with provision for measurement of pore pressure at one end of the sample, and for drainage from the other. The theoretical relation between the degree of consolidation (U_v) and the time factor (T_v) was derived in Chapter 5, and is shown diagrammatically in Fig. 6.14. The value of c_v is determined by the curve fitting procedure outlined in Section 5.14.

$$c_v = \frac{T_{v50}}{t_{50}} \cdot d^2$$

where d is the whole length of the sample, since drainage is from one end only.

6.17   *Measurement of the pore pressure parameters A and B:* The pore pressure parameters may be measured in a consolidated-undrained test. The value of B may be determined from the initial rise in pore pressure which results from the application of the cell pressure σ_3. The value of B is almost $1\cdot0$ for saturated soils, but varies considerably with the degree of saturation. A typical relationship between B and S_r is shown in Fig. 6.15. The parameter B is also considerably reduced if the pore pressure resulting from a previous stress change has only partially dissipated. This is important in the analysis of pore pressures in dams, where the pore pressures

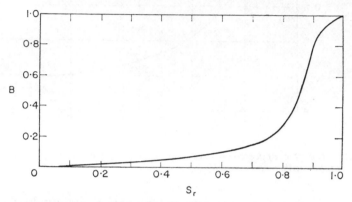

Fig. 6.15   Typical relationship between the pore pressure parameter B and the degree of saturation S_r.

caused by the weight of the structure are partially dissipated during the construction period.

The pore pressure parameter A may be determined during the second stage of the consolidated-undrained test, when the pore pressure changes result only from the change in the principal stress difference $(\Delta\sigma_1 - \Delta\sigma_3)$. The value of A is not constant, and the factors affecting it were discussed in Chapter 3. Dense sands and overconsolidated clays increase in volume when sheared, if they are free to do so. Under undrained conditions, the volume change is restricted. As a result, the pore pressure is reduced and the effective stress is increased. These soils have low, even negative, values of A, while loose sands and normally consolidated clays generally have values of A which, at failure, approach $1\cdot0$. Figure 6.16 shows a typical relationship between A_f (the value of A at failure) and the overconsolidation ratio, obtained from tests on a remoulded London clay.

6.18   Limitations of the Mohr–Coulomb model: As a method of describing real soil behaviour, the Mohr–Coulomb model has three obvious defects.

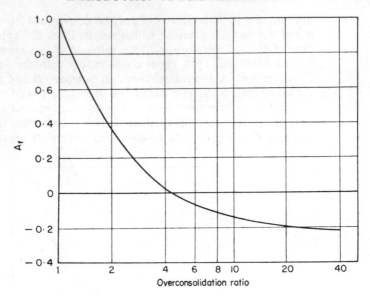

Fig. 6.16 The effect of overconsolidation on the value of the pore pressure parameter A at failure (after Bishop and Henkel [6.3]).

(a) Excessive ground movements are usually the first sign of an unsatisfactory soil structure. The model, however, makes no statement about strains, and cannot be used to predict displacements. In practice, we usually avoid this problem by

(i) using simple models (such as linear elasticity and one-dimensional consolidation) to predict displacements under service loads, and

(ii) using the Mohr–Coulomb model to check that there is an adequate factor of safety against complete structural collapse due to soil failure.

Handled with care, this method is usually adequate in practice. However, it is essentially illogical to separate considerations of displacement and strength in this way, and it may lead to serious misconceptions about structural behaviour. Moreover, in some structures, local failure of part of the soil mass occurs at much smaller loads than those required to cause structural collapse. Under these conditions, the simple elastic models may give a very poor estimate of the displacement.

(b) The volume changes described in Chapter 5 and the shear failure described in this chapter are only different views of the same soil behaviour. The Mohr–Coulomb model, however, implies that the volume changes do not affect the shear strength, which is certainly untrue. If, therefore, the parameters c' and φ' are to be treated as soil constants, we have to adopt a very restrictive definition of 'soil'. This has to mean not just 'a particular material', but 'a particular material at a particular void ratio and under a particular loading regime'. Clearly we need a model which will describe the behaviour of a particular material under *any* conditions.

(c) The model implies (see Section 6.3) that the intermediate principal stress σ'_2 does not affect the shear strength. This is unlikely to be absolutely true for any real material, and we should certainly not accept it in our soil models without further investigation.

In the next chapter, we will examine some more general soil models which seek to avoid these difficulties. What we gain in generality, however, we lose in increased complexity of the models and in the much greater computational effort required to handle them. In spite of these defects, therefore, the models of Chapters 5 and 6 are still used, and will continue to be used, to solve most routine practical problems.

REFERENCES

6.1 DRUCKER, D. C. 1959. A definition of stable inelastic material. *Trans. ASCE*, **81**.

6.2 COULOMB, C. A. 1776. Essai sur une application des règles des maximis et minimis à quelques problèmes de statique relatif à l'architecture. *Mem. Acad. Roy. Pres. à Div. Sav. Etr.*

6.3 BISHOP, A. W. and HENKEL, D. J. 1962. *The measurement of soil properties in the triaxial test*. Arnold (2nd Ed.).

6.4 SKEMPTON, A. W. 1964. Long term stability of clay slopes. *Géotechnique*, **14**.

6.5 BISHOP, A. W. 1961. The measurement of pore pressure in the triaxial test. In *Pore pressure and suction in soils*. Butterworths.

6.6 British Standards Institution. Methods of test for soils for civil engineering purposes. BS 1377: 1975.

CHAPTER 7

General soil models

STRESS PATHS AND INVARIANTS

7.1 The implications of the Mohr–Coulomb model: It was shown in Section 6.18 above that the classical Mohr–Coulomb soil model has two important implications. These are:

> (a) that the volume changes which commonly accompany shear strains have no effect on the shear strength, and
>
> (b) that the shear strength is independent of the intermediate principal stress component.

In this chapter, we will examine the suitability of the Mohr–Coulomb model as a means of describing the behaviour of real soil, and will consider some other models which might be expected to describe soil behaviour rather better. In the light of this discussion, we will then consider how we should interpret the results of the routine tests described in Chapter 6. First, however, we must construct a reference frame within which our comparisons may be made.

7.2 Principal stress space: The behaviour of a three-dimensional soil element subjected to general stress changes can best be examined by plotting the stresses and yield loci in a three-dimensional *principal stress space* (Fig. 7.1). Any point in

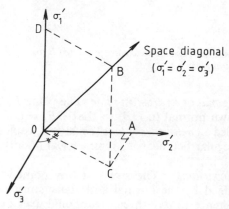

Fig. 7.1 Principal stress space.

141

this space defines the magnitudes (but not the directions) of the three principal effective stress components σ'_1, σ'_2 and σ'_3 with reference to three orthogonal co-ordinate axes. Since we generally assume that soil cannot resist effective tensile stress, we are only concerned with that part of the space in which all three principal stress components are positive. In this chapter, we shall abandon the convention that $\sigma'_1 \geq \sigma'_2 \geq \sigma'_3$: unless specifically stated to the contrary, the greatest (or least) principal stress component may be any one of the three.

Then point A (Fig. 7.1) represents a stress state in which $\sigma'_2 = OA$ and $\sigma'_1 = \sigma'_3 = 0$. The plane $OCBD$ (produced if necessary) represents all stress states in which $\sigma'_2 = \sigma'_3$ (the condition of axial symmetry in the triaxial test apparatus). $OC = \sqrt{2}\sigma'_2 = \sqrt{2}\sigma'_3$. The line OB (produced if necessary) is the locus of all points for which $\sigma'_1 = \sigma'_2 = \sigma'_3$ (spherical normal stress—see Fig. 5.1). This line is called the *space*

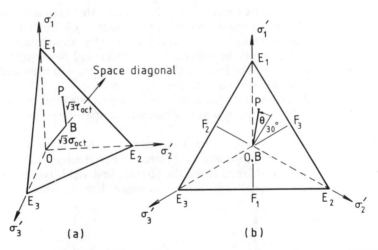

Fig. 7.2 *The stress invariants and the octahedral plane.* (a) *In principal stress space.* (b) *View normal to the octahedral plane.*

diagonal or *hydrostatic line*. The plane $E_1E_2E_3$ (Fig. 7.2) is drawn normal to OB, so that $OE_1 = OE_2 = OE_3$. This is called an *octahedral plane* (because it would form one face of a regular octahedron, symmetrical about O).

7.3 *The stress invariants:* The stress at any point in a body is fully defined by the normal and shear stress components with reference to *any* three orthogonal axes (x, y, and z). For isotropic materials, therefore, it must be possible to state the yield and failure conditions in terms of functions of the effective stress components which do not change as the axes are rotated. These functions are called *stress invariants*. It may be shown that there are three such functions which are independent. They may be expressed in a number of ways, of which the following are the most convenient for our purposes:

(a) The first stress invariant is $\frac{1}{3}(\sigma'_x + \sigma'_y + \sigma'_z)$, and is called the *octahedral normal stress* (σ'_{oct}). Since the axes x, y, and z may be rotated to coincide with the principal stress directions, it follows that $\sigma'_{oct} = \frac{1}{3}(\sigma'_1 + \sigma'_2 + \sigma'_3)$. The reader may readily verify that, for all points on the octahedral plane $E_1 E_2 E_3$ (Fig. 7.2), σ'_{oct} is constant and $OB = \sqrt{3}\sigma'_{oct}$.

(b) The second stress invariant is

$$\frac{1}{3}[(\sigma'_x - \sigma'_y)^2 + (\sigma'_y - \sigma'_z)^2 + (\sigma'_z - \sigma'_x)^2 + 6(\tau_{xy}^2 + \tau_{yz}^2 + \tau_{zx}^2)]^{\frac{1}{2}}$$

or

$$\frac{1}{3}[(\sigma'_1 - \sigma'_2)^2 + (\sigma'_2 - \sigma'_3)^2 + (\sigma'_3 - \sigma'_1)^2]^{\frac{1}{2}}$$

This is called the *octahedral shear stress* (τ_{oct}). For any point P in the plane $E_1 E_2 E_3$, $BP = \sqrt{3}\tau_{oct}$.

(c) The third stress invariant defines the magnitude of the intermediate principal stress component with respect to the other two, and may be expressed as the angle θ (Fig. 7.2(b)). It may be shown that

$$\tan \theta = \frac{\sigma'_1 - 2\sigma'_2 + \sigma'_3}{\sqrt{3}(\sigma'_1 - \sigma'_3)}$$

(d) Similar invariants may also be written in terms of the total stress components. Thus

$$\begin{aligned} \sigma_{oct} &= \tfrac{1}{3}(\sigma_1 + \sigma_2 + \sigma_3) \\ &= \tfrac{1}{3}(\sigma'_1 + \sigma'_2 + \sigma'_3 + 3u) \\ &= \sigma'_{oct} + u \end{aligned}$$

In a similar way, the reader may readily verify that τ_{oct} and θ are identical whether stated in terms of total or effective stress.

7.4 *Stress paths:* For an elastic material, the deformations resulting from an application of load are fully determined if the initial and final stress states are known, regardless of the way in

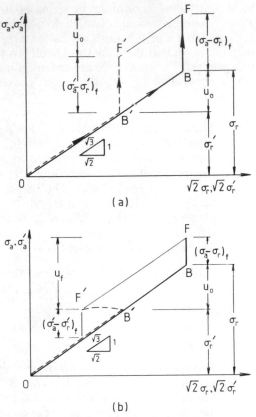

Fig. 7.3 Stress paths for triaxial compression tests in principal stress space. (a) A drained test. (b) A consolidated undrained test. —— Total stress paths. - - - Effective stress paths.

which the final state is reached. In examining the behaviour of an element of plastic material, however, it is necessary to describe the whole sequence of stress changes which take place as the material passes from initial yield towards failure. This may be done by plotting a *stress path*, which is a locus of points defining the successive stress states through which the element passes. For example, in principal stress space, we may plot stress paths for triaxial compression tests on the plane *OCBD* in Fig. 7.1 (assuming that σ'_1 is the axial stress component σ'_a, and $\sigma'_2 \ (=\sigma'_3)$ is the radial stress component σ'_r). In this case, $\sigma'_{oct} = \frac{1}{3}(\sigma'_a + 2\sigma'_r)$ and $\tau_{oct} = ((\sqrt{2})/3)(\sigma'_a - \sigma'_r)$. It is often useful to superimpose total stress and effective stress paths on the same diagram. This is illustrated by the following example:

Consider a specimen of saturated clay in a triaxial cell, initially under zero cell pressure and with zero internal pore

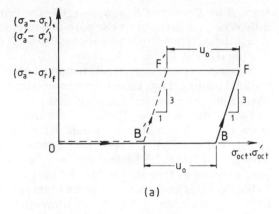

(a)

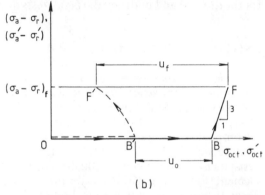

(b)

Fig. 7.4   Stress paths for triaxial compression tests in $(\sigma'_a - \sigma'_r):\sigma'_{oct}$ space and $(\sigma_a - \sigma_r):\sigma_{oct}$ space. (a) A drained test. (b) A consolidated undrained test. —— Total stress paths. - - - Effective stress paths.

pressure. Both total and effective stress states are defined by point O in Fig. 7.3(a). Let the cell pressure be increased to σ_r and then be maintained constant. The total stress plot moves along the space diagonal from O to B, where $OB = \sqrt{3}\sigma_r$. Since the soil is saturated, so that Skempton's pore pressure parameter B is nearly 1·0, the increase in pore pressure is nearly equal to σ_r, so that the effective stress remains almost zero.

Now let the specimen drain against a back pressure u_0 (see Section 6.9(d)). The total stress plot remains unchanged, but the effective stress plot moves from O to B', where $BB' = \sqrt{3}\sigma_{oct} - \sqrt{3}\sigma'_{oct} = \sqrt{3}u_0$.

Now suppose that the specimen is loaded to failure by increasing the axial total stress to σ_a, while allowing the specimen to drain against the back pressure, so that the pore pressure remains equal to u_0. The total stress plot moves from B to F where $BF = (\sigma_a - \sigma_r)_f$—the principal stress difference at failure. Since the pore pressure remains equal to u_0, the effective stress plot moves from B' to F', where $B'F' = BF$ and $FF' = \sqrt{3}u_0$.

If, however, the specimen had remained undrained during the final loading stage, there would have been a considerable change in pore pressure. For a normally consolidated clay, this would have been nearly equal to the change in axial total stress $(\sigma_a - \sigma_r)_f$, so that σ'_a would have remained nearly constant. The effective stress plot, therefore, would move from B' to F' (Fig. 7.3(b)) and $FF' = \sqrt{3}u_f$, where u_f is the pore pressure at failure. The line FF' is parallel to OB.

We shall also find it useful to trace stress paths on a plot of $(\sigma'_a - \sigma'_r)$ against σ'_{oct}. Figure 7.4 shows these stress paths for the drained and undrained triaxial tests described above.

SHEAR STRENGTH AND VOLUME CHANGE

7.5   *The volume change component of shear strength:* In a true continuum, shear strain does not normally cause significant volume change. Soil, however, is a particulate material, and is not a true continuum. Shear strain is commonly accompanied by significant volume change, which results from repacking the grains. The effect of this on the shearing resistance is conveniently illustrated by considering an element (Fig. 7.5(a)) within a layer of clean sand subjected to shear between two rough rigid bodies. The bodies will be assumed to be of very large lateral extent, so that volume change within the sand layer can only occur as a result of a change in this thickness. The pore pressure within the layer will be assumed to be zero.

Consider two grains within the element (Fig. 7.5(b)) sliding on a surface of contact inclined at an angle β to the base of the element. Let the intergranular force be resolved into components N normal to, and T parallel to, the base of the element.

Then

$$T/N = \tan(\beta + \varphi_\mu)$$

where φ_μ is the angle of friction at the intergranular contact.

$$T/N = \frac{\tan\varphi_\mu + \tan\beta}{1 - \tan\varphi_\mu\tan\beta}$$

and

$$T = N\tan\varphi_\mu + N\tan\beta + T\tan\varphi_\mu\tan\beta$$

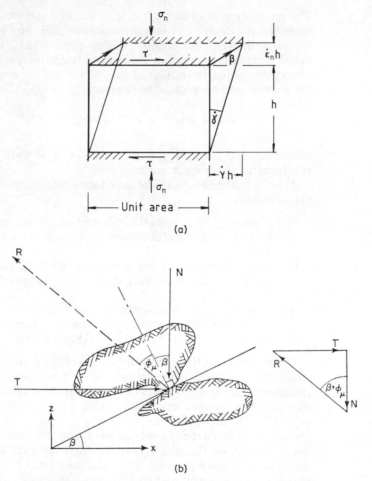

Fig. 7.5   *An element of sand in simple shear. (a) Rates of strain and displacement. (b) Forces at the contact between two grains.*

If β were constant for all sliding surfaces, we could integrate the intergranular forces over unit horizontal area:

$$\Sigma T = \tau \qquad \text{and} \qquad \Sigma N = \sigma'_n$$

Let the shear strain γ increase at a rate $\dot{\gamma}$. Also, let the shear strain be accompanied by a normal strain ε_n increasing at a rate $\dot{\varepsilon}_n$. Then the ratio of the rates of strain increment is $\dot{\varepsilon}_n/\dot{\gamma}$. If we assume all interparticle sliding surfaces to be inclined at a constant angle β, then (see Fig. 7.5(a)) $\tan \beta = \dot{\varepsilon}_n/\dot{\gamma}$. Then

$$\tau = \sigma'_n \tan \varphi_\mu + \sigma'_n \frac{\dot{\varepsilon}_n}{\dot{\gamma}} + \tau \tan \varphi_\mu \frac{\dot{\varepsilon}_n}{\dot{\gamma}}$$

The grains do not, in fact, all slide on parallel surfaces, and there is some degree of random movement. Let the angle β represent the *average* direction of sliding, equal to $\tan^{-1}(\dot{\varepsilon}_n/\dot{\gamma})$. Since increasingly random movements may increase the energy dissipated in friction at the intergranular contacts, the expression above must be rewritten

$$\tau = \sigma'_n \tan \varphi'_{cv} + \sigma'_n \frac{\dot{\varepsilon}_n}{\dot{\gamma}} + \tau \tan \varphi'_{cv} \frac{\dot{\varepsilon}_n}{\dot{\gamma}}$$

where $\varphi'_{cv} \geq \varphi_\mu$, and φ'_{cv} is the value of φ' defining the resistance to shearing at constant volume.

The physical meaning of the three terms in this expression are as follows:

(a) $\sigma'_n \tan \varphi'_{cv}$ represents the resistance of the element, had it been sheared at constant volume (when $\dot{\varepsilon}_n/\dot{\gamma}$ is zero).

(b) $\sigma'_n(\dot{\varepsilon}_n/\dot{\gamma})$ represents the resistance resulting from work done in expanding the element against the normal stress σ'_n.

(c) $\tau \tan \varphi'_{cv}(\dot{\varepsilon}_n/\dot{\gamma})$ represents a small amount of additional frictional resistance resulting from the inclination (β) of the average sliding surface.

Figure 7.6 shows the relationship between shear stress and strain for two specimens, initially in a dense and loose condition respectively. In each case, as shear strains develop, the void ratio changes, eventually reaching a constant value (e_c), called the *critical void ratio* by Casagrande [7.1]. This critical void ratio depends only on the shape and grading of the grains and on the effective stress, and is independent of the original state of packing. Once this void ratio has been reached, the soil continues to deform as a frictional fluid with constant strength, and at constant volume. The soil is then said to be in the *critical state*, and the shear strength has the residual value

$$\tau_r = \sigma'_n \tan \varphi'_r$$

where φ'_r is the value of φ' defining the residual strength. Notice that at the critical state $\varphi'_{cv} = \varphi'_r$, but φ'_{cv} is not necessarily constant at earlier stages of the test, since it may depend on the void ratio, which is changing.

The increase in the volume of the densely packed specimen results in a considerable contribution to the shear strength from components (b) and (c) above. The shear strength reaches its peak value approximately when $\dot{\varepsilon}_n/\dot{\gamma}$ is a maximum. The magnitude of τ_{max} depends on the volume

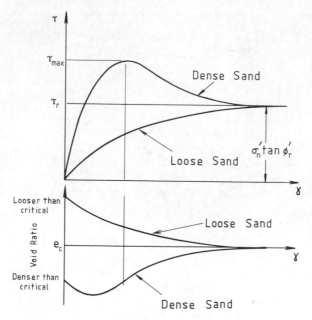

Fig. 7.6 .Stress/strain behaviour of sand in simple shear.

change, which in turn depends on the original density of packing.

In the loosely packed specimen, in which the initial void ratio is greater than the critical value, shearing causes a reduction in volume and a collapse of the soil structure, so that the components (b) and (c) are negative. The maximum strength is therefore at the critical state, and is the same as the residual strength for the densely packed specimen.

We may draw two general conclusions from the argument above:

(a) The maximum shear strength depends upon the initial void ratio. It may exceed the strength at the critical state, if the initial void ratio is less than e_c.

(b) At the critical state, both shear strength and void ratio are functions of the effective normal stress, and are independent of the initial void ratio.

7.6 *Rendulic's plot of principal stress and void ratio:* Rendulic [7.2] examined the results of a number of triaxial tests by plotting failure loci and lines of constant void ratio on the plane OCBD of Fig. 7.1. For our purposes, Rendulic's method is most conveniently illustrated by a similar plot prepared by Henkel [7.3]. Henkel compared the results of a series of very

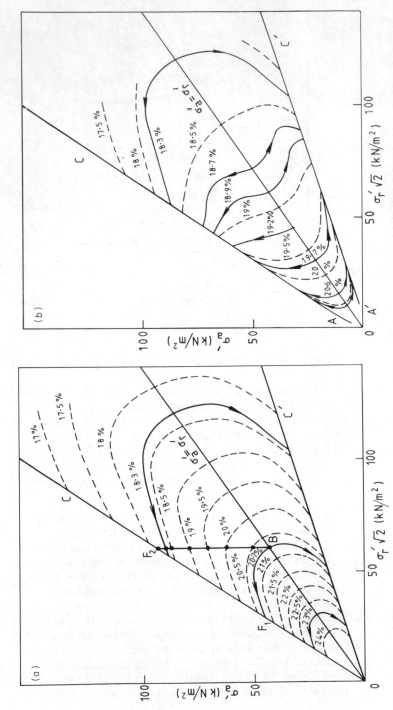

Fig. 7.7 *Rendulic plots, showing results of triaxial compression and extension tests. (a) Tests on normally consolidated specimens. (b) Tests on overconsolidated specimens having a common preconsolidation pressure. (After Henkel [7.3].)*

careful triaxial tests on specimens of saturated Weald clay, remoulded at a water content near the liquid limit, and then normally consolidated under spherical normal stress.* The line OC in Fig. 7.7(a) shows the observed failure locus for triaxial compression tests in which $\sigma'_a > \sigma'_r$, while the line OC' is the corresponding failure locus for triaxial *extension* tests, in which $\sigma'_a < \sigma'_r$. In each case, Henkel found that there was a unique failure locus for all tests, both drained and undrained.

The broken lines in Fig. 7.7(a) are contours of constant water content (and therefore of constant void ratio) derived from observations made during drained tests. The solid lines are stress paths observed during undrained tests. Since the volume change of a saturated undrained specimen is insignificant, each of these stress paths is also a line of constant void ratio. It is clear from the figure that both sets of lines form part of a common pattern, and that there is a unique relationship between the principal effective stress components (σ'_a, σ'_r) and the normally consolidated void ratio, which is independent of the drainage conditions.

Henkel also tested overconsolidated specimens. Figure 7.7(b) shows the results of a series of tests on specimens having the same preconsolidation pressure, but tested at different cell pressures, and therefore at different over-consolidation ratios. Henkel's results for failure lie approximately along the lines AC and $A'C'$, which are always outside the lines OC and OC'. In this set of results, also, there is clearly a unique relationship between the principal effective stress components (σ'_a, σ'_r) and the void ratio, independent of the drainage conditions in the tests.

Notice that all the stress paths for normally consolidated and lightly overconsolidated specimens turn towards the origin at failure. On the other hand, those with the highest overconsolidation ratios, initially nearest to the origin, turn sharply away from it. This prompts us to ask whether, had Henkel been able to observe the behaviour after failure, he would have found that all specimens reached a common critical state, independent of the original condition. We shall consider this question in Section 7.12 below.

7.7 *The unified Rendulic plot:* Now consider the effective stress paths of Fig. 7.3 superimposed on Henkel's plot (Fig. 7.7(a)). The undrained stress path for a normally consolidated specimen follows a line of constant void ratio BF_1, and there is no

* The remoulding process destroys permanent interparticle bonds, and creates an artificial material whose behaviour is more constant, but differs somewhat from that of the natural soil.

volume change during the test. The drained test path BF_2, on the other hand, crosses a series of lines of constant void ratio, as plastic volumetric strain develops. In Section 5.5 above, we identified the development of plastic strain with *yield*. The lines of constant void ratio therefore form a family of *yield loci*, through which the stress path passes as the material strain hardens with decreasing volume. The process ends when the stress path reaches the *failure locus OC*, after which unlimited (unstable) shear strain develops at constant volume.

Henkel's specimens were prepared by consolidation under spherical stress. For any void ratio e, therefore, we will define an *equivalent (spherical) consolidation pressure p'_e*, which is the *spherical* effective stress ($\sigma'_a = \sigma'_r$) which will result in normal consolidation to the given void ratio e. This is similar in principle to the equivalent one-dimensional consolidation pressure σ'_e which was defined in Section 5.6 above. For any void ratio e, p'_e is the value of σ'_r ($= \sigma'_a$) at the intersection of the corresponding line of constant void ratio with the space diagonal OB (Fig. 7.7(a)).

It appears from Fig. 7.7 that all the lines of constant void ratio for normally consolidated specimens are of similar shape. We can therefore greatly simplify the diagram by plotting the results in terms of the non-dimensional quantities σ'_a/p'_e and σ'_r/p'_e. This *unified* plot is shown in Fig. 7.8. The family of contours of constant void ratio (Fig. 7.7(a)) is now represented by the single line CBC', and this forms the stress path for all normally consolidated specimens, whether they are tested drained or undrained. All such specimens fail when the path reaches C (in triaxial compression tests) or C' (in triaxial extension tests).

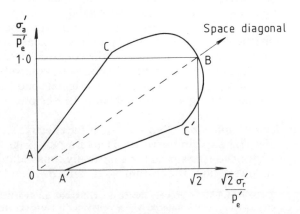

Fig. 7.8   A unified Rendulic plot for triaxial compression and extension tests.

In Fig. 7.8, the failure states for overconsolidated specimens lie along the lines AC and $A'C'$ (lying always outside OC and OC'). The line $OACBC'A'O$ is therefore the boundary separating possible stress states within from impossible stress states outside. States on the boundary CBC' result in stable yield. States on the lines AC and $A'C'$ result in failure (i.e. unstable yield). OA and OA' express the condition that the effective stress components must always be non-negative.

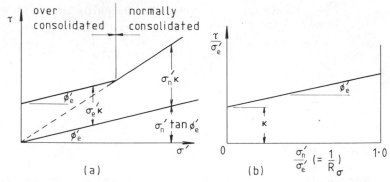

Fig. 7.9 *Hvorslev's effective cohesion and effective angle of internal friction. (a) In terms of τ_f and σ'_n. (b) In terms of τ_f/σ'_e and σ'_n/σ'_e.*

7.8 Hvorslev's analysis of shear strength and void ratio in clays:

Hvorslev [7.4, 7.5] has discussed the results of a series of direct shear tests on specimens of saturated remoulded clay, prepared by one-dimensional consolidation. He showed (Fig. 7.9(a)) that the measured shear strength might be divided into two components, one being proportional to the effective normal stress σ'_n, and the other being a function of the void ratio. He also showed that the latter was nearly proportional to the equivalent consolidation pressure (as defined in Section 5.6 above), so that

$$\tau_f = c'_e + \sigma'_n \tan \varphi'_e$$
$$= \sigma'_e K + \sigma'_n \tan \varphi'_e$$

where

$$\sigma'_e = \sigma'_0 \exp\left[\frac{2 \cdot 303(e_0 - e)}{C_c}\right]$$

Hvorslev called the parameters c'_e and φ'_e the *effective cohesion* and *effective angle of internal friction* respectively. For normally consolidated clay, $\sigma'_e = \sigma'_n$ so that

$$\tau_f = \sigma'_n(K + \tan \varphi'_e)$$

and the shear stress is proportional to the effective normal stress, so that Coulomb's apparent cohesion c' is zero. For overconsolidated clay, the overconsolidation ratio

$$R_\sigma \left(= \frac{\sigma'_e}{\sigma'_n} \right) > 1 \cdot 0$$

Then

$$\tau_f = \sigma'_n (R_\sigma K + \tan \varphi'_e)$$

The two components may be more clearly separated if the equation is expressed in non-dimensional form by dividing τ_f and σ'_n by σ'_e. Then

$$\frac{\tau_f}{\sigma'_e} = K + \frac{\sigma'_n}{\sigma'_e} \tan \varphi'_e$$

$$= K + \tan \varphi'_e / R_\sigma$$

This is shown graphically in Fig. 7.9(b).

THE CRITICAL STATE MODEL

7.9 *The critical state model in $p':q:v$ space:* We will now consider a soil model which seeks to draw together the relationships between volume and shear strength which have been discussed in the preceding sections. This is the critical state model developed by Roscoe and his colleagues [7.6, 7.7, 7.8]. In describing the model, the definitions and symbols of refs. [7.7 and 7.8] will be used as far as possible, as this should make it easier for the reader to study the literature further. Only the salient features of the model will be described here. For a fuller treatment, the reader should consult ref. [7.8]. We will first consider the model with reference to the results of drained and undrained triaxial tests on remoulded specimens of saturated clay.

In its original form, the critical state model is best described by reference to a three-dimensional space (Fig. 7.10(a)) whose three axes define the magnitudes of the variables p', q, and v, where

$$p' = \sigma'_{\text{oct}}$$

$$q = \frac{3}{\sqrt{2}} \tau_{\text{oct}}$$

$$v = 1 + e$$

The specific volume v is the total volume of soil containing unit volume of solid particles.

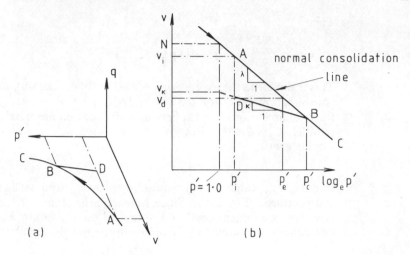

Fig. 7.10 Consolidation and swelling under spherical stress. (a) In $p':q:v$ space. (b) On a $v:\log_e p'$ plane.

Notice that the stress is here defined by reference to only two of the three stress invariants, and that the third invariant θ must be assumed to have no effect. This is perfectly admissible *so long as we are only concerned with analysing the effects of triaxial compression tests.* In this case, we are only concerned with the upper part of the plane $OCBD$ of Fig. 7.1, above the space diagonal OB. This intersects the octahedral plane in the line BE_1 (Fig. 7.2), on which the angle $\theta = 30°$. When, later on, we wish to extend the use of the model to cases where the two lesser principal stresses are not equal, we shall have to investigate the effect of different values of θ. In describing the model, however, we will consider only the effects of triaxial compression tests, in which $\sigma_a > \sigma_r$. In these cases,

$$q = \frac{3}{\sqrt{2}}\tau_{\text{oct}} = (\sigma'_a - \sigma'_r), \quad \text{and} \quad p' = \tfrac{1}{3}(\sigma'_a + 2\sigma'_r).$$

7.10 The normal consolidation line: First consider a specimen, normally consolidated under an increasing spherical effective stress ($\sigma'_a = \sigma'_r$). As in the case of one-dimensional consolidation (Section 5.5) there is a nearly linear relationship between void ratio and the logarithm of effective stress which may be expressed (Fig. 7.10(b)) as

$$v = N - \lambda \log_e p'$$

where N is the specific volume corresponding to $p' = 1\cdot0$.

Then

$$p' = \exp\left(\frac{N - v}{\lambda}\right)$$

This is similar to the expression derived for one-dimensional consolidation in Section 5.5 above. Indeed, since the curves for one-dimensional and spherical consolidation are nearly parallel, $C_c = 2 \cdot 303\lambda$. For overconsolidated soil, we may similarly write

$$v = v_\kappa - \kappa \log_e p'$$

where v_κ is the value of v corresponding to $p' = 1 \cdot 0$ on the line BD produced (Fig. 7.10). Since however, the slopes of the lines for one-dimensional and spherical consolidation are not exactly the same, C_s is only approximately equal to $2 \cdot 303\kappa$.

In Section 7.7 above, we defined the equivalent consolidation pressure (p'_e) as the value of p' on the normal consolidation line corresponding to any value of e. Then for any value of v,

$$v = N - \lambda \log_e p'_e$$

so that

$$p'_e = \exp\left(\frac{N - v}{\lambda}\right)$$

For overconsolidated soils, where $p' < p'_e$, the overconsolidation ratio (for spherical consolidation) will be defined as

$$R_p = \frac{p'_e}{p'}$$

7.11 *The critical state line and the state boundary surface:* Now consider a series of specimens, normally consolidated as described above, and then loaded to failure by increasing the axial total stress σ_a. Henkel's results show that there is a single failure locus for all such tests, which is independent of the drainage conditions. The projection of this failure locus on the $p':q$ plane is a straight line through the origin (Fig. 7.11(a)), while the projection on the $v:p'$ plane is a curve which, on the semi-log plot in Fig. 7.11(b), is nearly parallel to the normal consolidation line. The equations of this failure locus may therefore be written

$$q_{(cs)} = Mp'_{(cs)}$$
$$v_{(cs)} = \Gamma - \lambda \log_e p'_{(cs)}$$

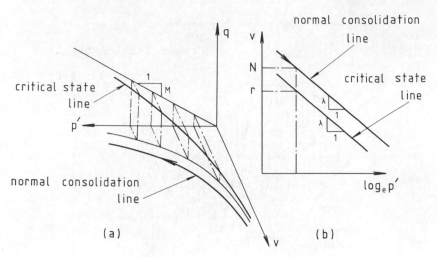

Fig. 7.11   The critical state line. (a) In p':q:v space. (b) Projected on the v:log_e p' plane.

where M, Γ and λ are constants for the particular soil. Once the test path reaches this line, unlimited shear strains occur at constant volume. The soil is said to be in the *critical state*, and the failure locus is called the critical state line.

On the critical state line,

$$p'_{(cs)} = \exp\left(\frac{\Gamma - v_{(cs)}}{\lambda}\right)$$

so that

$$R_{p(cs)} = p'_e/p'_{(cs)}$$

$$= \exp\left(\frac{N - v_{(cs)}}{\lambda}\right)\bigg/\exp\left(\frac{\Gamma - v_{(cs)}}{\lambda}\right)$$

$$= \exp\left(\frac{N - \Gamma}{\lambda}\right)$$

Fig. 7.7(a) shows that there is a unique relationship between stress and void ratio for all normally consolidated specimens, independent of the drainage conditions. It follows that there is a unique surface between the normal consolidation and critical state lines in which all paths for normally consolidated specimens must lie. Overconsolidated states are described by points below this surface, but no point above it is possible. The surface therefore forms a boundary, dividing possible states on or below it from impossible states above, and is called the *state boundary surface* (see Fig. 7.12).

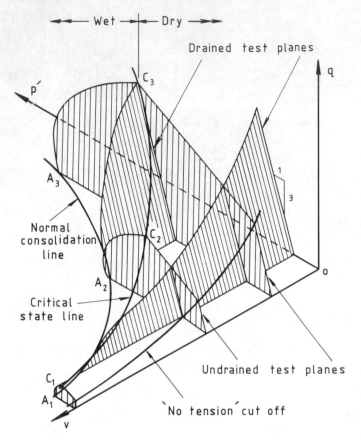

Fig. 7.12   *The state boundary surface. (a) In $p':q:v$ space. (After Roscoe et al. [7.6].)*

7.12   *The state boundary surface in the 'dry' region:*   The state boundary surface for a normally consolidated soil lies wholly in the region outside (that is, further from the origin than) the projection of the critical state line on the $v:p'$ plane. Drained tests result in reductions in volume in this region, and the soil is described as 'wet' (that is, having a higher water content than at the critical state). Heavily overconsolidated specimens, on the other hand, have initial states on the opposite, or 'dry' side of the critical state line. We must now consider

     (a)   whether overconsolidated specimens eventually reach · the same critical state as normally consolidated specimens, and

     (b)   whether we can extend the state boundary surface into the dry region.

These questions are not easy to answer experimentally, for three reasons [7.5]

(a) In drained tests, heavily overconsolidated soils soften after failure, so that the critical state strength is less than that at failure.

(b) Large strains are required to establish the critical state, and these cannot be developed in the triaxial apparatus without excessive changes in the specimen geometry.

(c) As the strength reduces after failure, the strains are concentrated in narrow bands of weakened soil, and the specimen ceases to be homogeneous. It is then difficult to determine the state of this weakened soil by means of measurements made at the specimen boundaries.

However, Parry [7.9] has shown that test paths for overconsolidated specimens tend towards the critical state as strains increase after failure, and it seems reasonable to assume for our model soil that the critical state is the same for all specimens, regardless of the initial state. Also, although the evidence for this is not conclusive, Roscoe et al. [7.6] suggest that it is possible to assume with sufficient accuracy for our purposes that the state boundary surface in the 'dry' region may be expressed in a form similar to Hvorslev's failure condition, as follows:

$$q = p'\mathrm{M}_0 + p'_e I$$

where M_0 and I are constants.
Then

$$q = p'(\mathrm{M}_0 + R_p I)$$

so that

$$\frac{q}{p'} = \mathrm{M}_0 + R_p I$$

where $R_p\,(=p'_e/p')$ is the overconsolidation ratio, defined in Section 7.10. Where this state boundary surface intersects the critical state line,

$$\frac{q_{(cs)}}{p'_{(cs)}} = \mathrm{M} \quad \text{and} \quad R_{p(cs)} = \exp\left(\frac{\mathrm{N} - \Gamma}{\lambda}\right)$$

(see Section 7.11). Then

$$\mathrm{M} = \mathrm{M}_0 + R_{p(cs)} I$$

$$I = \frac{\mathrm{M} - \mathrm{M}_0}{R_{p(cs)}} = (\mathrm{M} - \mathrm{M}_0)\exp\left(\frac{\Gamma - \mathrm{N}}{\lambda}\right)$$

Then the equation of the state boundary surface is

$$q = p'\mathrm{M}_0 + p'_e(\mathrm{M} - \mathrm{M}_0)\exp\left(\frac{\Gamma - \mathrm{N}}{\lambda}\right)$$

$$= p'\mathrm{M}_0 + (\mathrm{M} - \mathrm{M}_0)\exp\left(\frac{\Gamma - \mathrm{N}}{\lambda}\right)\exp\left(\frac{\mathrm{N} - v}{\lambda}\right)$$

$$= p'\mathrm{M}_0 + (\mathrm{M} - \mathrm{M}_0)\exp\left(\frac{\Gamma - v}{\lambda}\right)$$

The first part of this expression for q depends on the mean effective stress p', and the second part is a function of the specific volume v.

Since we usually assume that soil is incapable of sustaining effective tensile stress, the state boundary surface is limited by the condition that σ'_r cannot be less than zero. When σ'_r is zero, $q = \sigma'_a$ and $p' = \frac{1}{3}\sigma'_a$, so that $q/p' = 3$. The state boundary surface is therefore limited by a plane inclined at 3 to 1 to the p' axis, as shown in Fig. 7.12.

7.13 *Drained test paths in p':q:v space:* In a drained test, where σ_r is maintained constant, an increase $\Delta\sigma'_a$ in the effective axial stress results in an increase $\Delta\sigma'_a$ in q but only $\frac{1}{3}\Delta\sigma'_a$ in p. A drained test path must therefore lie in a plane inclined at 3 to 1 to the plane $q = 0$ (Fig. 7.13(a)). For normally consolidated specimens, the path passes from A (Fig. 7.13(b)) on the normal consolidation line to C on the critical state line, and lies wholly on the state boundary surface. For overconsolidated soils, the initial state point lies below the state boundary surface. It will be convenient to assume that the model soil behaves elastically for all states below the state boundary surface, although this is not quite true for real

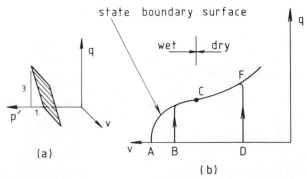

Fig. 7.13   *Drained test paths in p':q:v space. (a) A drained test plane. (b) Test paths projected on the q:v plane.*

soils. For the model soil, therefore, the state boundary surface defines the yield condition. For overconsolidated specimens, the test path moves up the drained test plane to the state boundary surface, where yield commences. It then travels along the boundary surface to the critical state line.

For normally and lightly overconsolidated specimens (*A* and *B* Fig. 7.13(b)), the stress path reaches the state boundary surface on the 'wet' side of the critical state line. The volume reduces and the soil strain hardens as the specimen approaches the critical state, which is also the failure state. In heavily overconsolidated soils (*D* in Fig. 7.13(b)), on the other hand, the stress path reaches the boundary surface at a point (*F*) on the 'dry' side of the critical state. At this point, *q* is a maximum, and the soil fails. Subsequently, as the stress path extends towards the critical state line, the specific volume increases and *q* decreases (i.e. the soil softens) so that the strength at the critical state is less than at failure.

7.14 *Undrained test paths in p':q:v space:* Since volume change is insignificant in undrained tests, paths for such tests must lie in planes of constant *v* (Fig. 7.14). For a normally consolidated specimen, the path passes from *A* on the normal consolidation line to *C* on the critical state line, and lies wholly on the state boundary surface. Assuming again that the soil behaves elastically for all points below the boundary surface, all paths for overconsolidated specimens pass vertically to the state boundary surface and then travel over it to the critical state line. Lightly and heavily overconsolidated soils approach the critical state from opposite sides of the line, but in every case failure (defined by the maximum value of *q*) occurs at the critical state.

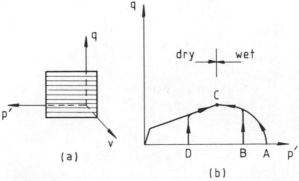

Fig. 7.14 *Undrained test paths in p':q:v space. (a) An undrained test plane (constant v). (b) Test paths for undrained tests.*

7.15 Pore pressure changes in undrained tests: The pore pressure changes during undrained tests may be estimated by comparing the total and effective stress paths, as described in Section 7.4 above. Consider a specimen, tested at a cell pressure p_0, and loaded to failure in undrained triaxial compression. Any increment of axial total stress $\Delta\sigma_a$ results in an increase of $\Delta\sigma_a$ in q and $\frac{1}{3}\Delta\sigma_a$ in p. The total stress path is therefore inclined at 3 to 1 to the p axis.

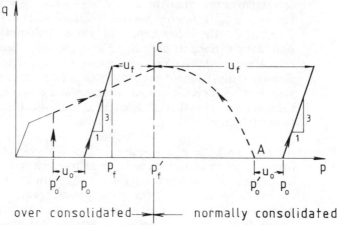

over consolidated → ← normally consolidated

Fig. 7.15   Pore pressure changes in undrained tests. ——— *Total stress paths.* --- *Effective stress paths. (After Atkinson and Bransby* [7.8].)

Figure 7.15 shows the stress paths for two specimens, one normally consolidated and the other heavily over-consolidated. In each case, failure occurs at the critical state, so that

$$p_f = p_0 + \tfrac{1}{3}q_f$$
$$= p_0 + \tfrac{1}{3}Mp'_f$$

and

$$v = \Gamma - \lambda \log_e p'_f$$

so that

$$p'_f = \exp\left(\frac{\Gamma - v}{\lambda}\right)$$

Then

$$u_f = p_f - p'_f$$
$$= p_0 - (1 - \tfrac{1}{3}M)p'_f$$
$$= p_0 - (1 - \tfrac{1}{3}M)\exp\left(\frac{\Gamma - v}{\lambda}\right)$$

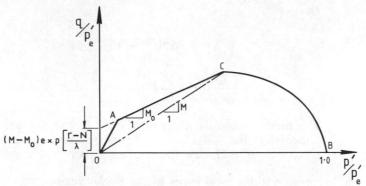

Fig. 7.16 Unified plot of the state boundary surface.

For normally consolidated and lightly overconsolidated specimens, the pore pressure increases during the test, but in heavily overconsolidated specimens $u_f - u_0$ is negative.

7.16 *A unified plot of the state boundary surface:* Figure 7.12 shows sections on planes of constant v through the state boundary surface. All such sections are of the same shape, although of different sizes. The whole state boundary surface may therefore be displayed by a single unified plot of p'/p'_e and q/p'_e where $p'_e = \exp[(N - v)/\lambda]$. This unified plot is shown in Fig. 7.16. The normal consolidation line is represented by the single point B where $p'/p'_e = 1\cdot0$ and $q/p'_e = 0$. The critical state is represented by the single point C, and the equation of the state boundary surface in the dry region is

$$\frac{q}{p'_e} = \frac{p'M_0}{p'_e} + (M - M_0)\exp\left(\frac{\Gamma - v}{\lambda}\right)\Big/p'_e$$

But

$$p'_e = \exp\left(\frac{N - v}{\lambda}\right) \quad \text{and} \quad p'_e/p' = R_p$$

Hence

$$\frac{q}{p'_e} = \frac{M_0}{R_p} + (M - M_0)\exp\left(\frac{\Gamma - N}{\lambda}\right)$$

In the 'wet' region, a number of shapes for the state boundary surface have been suggested. Of these, the most generally useful seems to be the elliptical surface used in the 'modified cam clay' model (Roscoe and Burland, [7.14]). The state boundary surface is defined by the expression

$$p'_e = p'\left[1 + \frac{(q/p'^2)}{M^2}\right]$$

so that

$$\left(\frac{q}{p'}\right)^2 = M^2\left(\frac{p'_e}{p'} - 1\right)$$

$$\left(\frac{q}{p'_e}\right)^2 = \left(\frac{M}{R_p}\right)^2 (R_p - 1)$$

The unified Rendulic diagram (Fig. 7.8) and the unified critical state diagram (Fig. 7.16) describe essentially the same behaviour. In each case the diagram shows

(a) a stable yield locus BC in the wet region,
(b) an unstable yield (failure) locus AB in the dry region, and
(c) a no-tension condition OA.

Additional features of the critical state model are

(a) the concept of a critical state which lightly consolidated soils reach at failure, and to which heavily overconsolidated soils revert after failure, and
(b) mathematical expressions defining the yield conditions and the critical state.

7.17 *Yield, failure and the critical state in principal stress space:* We showed in Section 7.9 that we are only entitled to ignore the effects of the third stress invariant θ for so long as we limit our investigation to triaxial compression tests, in which $\sigma'_a \geq \sigma'_r$. We must now see if the model can be extended to cover cases in which the two lesser principal stress components are not equal. We will start by transferring the unified state boundary surface from Fig. 7.16 to the unified Rendulic diagram (Fig. 7.17(a)). For the conditions of the triaxial test, $\theta = 30°$, and the plot lies in the upper part of the plane $OCBD$, above the space diagonal. For other values of θ, the yield condition, which forms the state boundary surface in the 'dry' region, is represented by a cone, not necessarily circular in section, around the space diagonal. The figure is closed by a cap, again not necessarily circular in section, representing the yield condition in the 'wet' region. The general forms of the two yield conditions, and the critical state may be studied by examining their intersections with octahedral planes such as $E_1E_2E_3$ (Fig. 7.2). Notice that, since we have not stipulated which principal stress component is greatest, the sections must be symmetrical about E_1F_1, E_2F_2, and E_3F_3. They are therefore fully

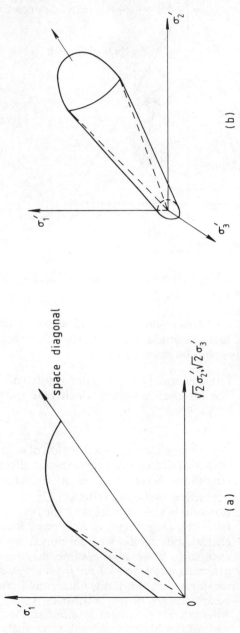

Fig. 7.17 Yield and critical state conditions in principal stress space. (a) For triaxial compression tests ($\sigma'_a \geq \sigma'_r$). (b) For general stress states.

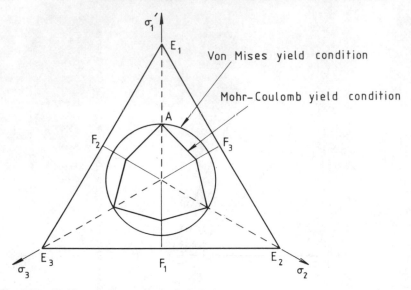

Fig. 7.18   Yield conditions in the octahedral plane.

described by values of θ between $30°$ and $-30°$ (i.e. by the sector E_1BF_3).

7.18   The extended von Mises and generalised Mohr–Coulomb models:
Of the many general soil models which have been suggested, only two will be examined here.

(a)   Drucker *et al.* [7.10] suggested an extended form of von Mises criterion, in which yield is assumed to occur when

$$\tau_{\text{oct}} = f(\sigma'_{\text{oct}})$$

This implies that τ_{oct} is a function of σ'_{oct} only, and that the third invariant θ does not affect the yield condition. Since σ'_{oct} is a constant on any octahedral plane, and $BP = \sqrt{3}\tau_{\text{oct}}$, the intersection of this yield surface with the octahedral plane is a circle (Fig. 7.18). In Sections 7.9 to 7.16, the critical state model was developed without reference to θ. If we now assume that the model is generally applicable *in the form developed above*, we imply that the yield, failure, and critical state conditions are of the von Mises type.

(b)   Alternatively, we might assume that the conditions were of the Mohr–Coulomb type, such that

$$(\sigma'_{\text{max}} - \sigma'_{\text{min}}) = f(\sigma'_{\text{max}} + \sigma'_{\text{min}})$$

This implies that the yield condition is not affected by the intermediate principal stress component. It may be shown that the surface defined by such a condition intersects the octahedral plane in an irregular hexagon, as shown in Fig. 7.18.

7.19 General yield, failure and critical state conditions for real soils: There are considerable practical difficulties in the way of conducting tests in which all three principal stress components are to be controlled independently, and the evidence about the behaviour of real soils subjected to general stress changes is therefore far from conclusive. However, the following points are fairly well established (see for example references [7.11] to [7.15]).

(a) The shear strength of real soils is generally predicted from the results of triaxial compression tests. Figure 7.18 shows the failure conditions which would be predicted from the result of such a test (represented by point A), assuming

   (i)   a von Mises, and
   (ii) a Mohr–Coulomb failure condition.

The evidence seems to show that the failure condition for most real soils lies somewhere between the two.

(b) A triaxial compression test interpreted in terms of the Mohr–Coulomb failure condition will slightly under-estimate the real strength in plane strain, but may predict the strength in triaxial extension almost exactly.

(c) A triaxial compression test interpreted in terms of the von Mises failure condition may seriously over-estimate the strength in plane strain.

(d) When the intermediate principal stress component is small (i.e. when θ is nearly $30°$), the real failure condition may be nearer to the von Mises than to the Mohr–Coulomb model.

(e) Generally similar behaviour is observed at the critical state.

(f) Near the tip of the cap (where the principal stress components are nearly equal), the yield condition may be nearer to the von Mises model.

Although there is nothing unacceptable in the critical state theory as such, the model is clearly not generally applicable to all possible stress states in the form in which it was derived in Sections 7.9 to 7.15 above. Of course, there is no reason why we should not construct a critical state model in which the yield conditions are defined in terms of all three stress invariants, and such models have been proposed [7.15]. However, the complexity of such models and the difficulty of evaluating the parameters, often make this approach unproductive. It is usually simpler and more efficient to use either Mohr–Coulomb conditions (which always err on the safe side) or von Mises conditions with the parameters suitably reduced. The actual choice will depend on the purpose for which the model is required. The choice of suitable models is discussed by Zienkiewicz and Pande [7.16] and Naylor [7.17].

7.20   *Applications of the critical state model:* Critical state theory provides an excellent *qualitative* model of soil behaviour, by bringing together a number of apparently unconnected concepts, such as those of Casagrande, Rendulic, Coulomb, and Hvorslev, and the critical state concept itself. In addition, some simple quantitative problems may be solved directly. For example, the prediction of pore pressure changes is often better done by the stress path techniques described in Section 7.15 than by the use of Skempton's pore pressure parameters, which are very sensitive to stress levels [7.18].

The most promising quantitative application of the model is as a basis for elasto-plastic stress–strain laws for use with the finite element method. At the moment, the computational effort required for this technique restricts its use to the research field, but the method appears to have considerable potential [7.17].

For ease and simplicity, most routine problems are still solved by using simple elastic theory and consolidation theory to predict displacements, and the classical Mohr–Coulomb model to predict collapse loads. Nevertheless, it is important to realise the limitations of these models when applying them to predict the behaviour of real soils.

SHEAR STRENGTH OF REAL SOIL

7.21   *The maximum strength of natural cohesive soil:* In developing the soil models, we considered only remoulded soils. The maximum strength of a cohesive soil in its natural undisturbed state is commonly significantly greater than that of the same soil after remoulding. This is because certain features which contribute to the shear strength of the soil are destroyed during remoulding. These features are as follows:

(a) *Orientation of the soil particles.* Clay particles are flat, and their orientation and mode of assembly affects their behaviour under load. Anisotropic consolidation tends to produce a preferred orientation of the particles normal to the direction of the greatest component of principal stress. Large strains, however, cause the particles to be rearranged parallel to the direction of sliding. This results in a significant drop in strength.

(b) *Physico-chemical bonds.* These develop gradually in the adsorbed layers surrounding the particles in contact. If the bonds are broken by deformation of the soil, they do not reform immediately, and the shear strength is correspondingly reduced.

(c) *Intergranular cement.* In some soils (see Section 1.6), the particles are cemented with iron oxide, calcite, etc. deposited at and around the particle contacts. If this cement is broken up by relative movement of the particles, the shear strength is permanently reduced.

7.22 *Undrained strength of saturated cohesive soils:* The undrained strength of a saturated soil (Section 7.14) is unaffected by changes in the total normal stress. Then

$$\tau_f = c_u \quad \text{and} \quad \varphi_u = 0$$

where c_u and φ_u are Coulomb's shear strength parameters for the undrained condition in terms of total stress.

In overconsolidated soils, however, the pore pressure drops sharply as the specimen approaches the critical state. In this condition, the strength is slightly less than at failure, as a result of sample disturbance. Local soil failure then occurs on quite narrow bands of softened soil, in which the pore pressure is less than in the adjacent material. This causes fairly rapid local migration of water towards these bands, leading to a rise in pore pressure and a loss of strength. Such soils, therefore, may not remain truly undrained for long [7.19].

7.23 *Variation of the undrained shear strength with depth:* The undrained shear strength (c_u) of a normally consolidated clay is almost directly proportional to the consolidation pressure (σ'_c). Skempton [7.20] has suggested the following empirical relationship between the index of plasticity (PI) and the ratio c_u/σ'. (See Fig. 7.19):

$$\frac{c_u}{\sigma'} = 0 \cdot 11 + 0 \cdot 003\ 7\ (\text{PI})$$

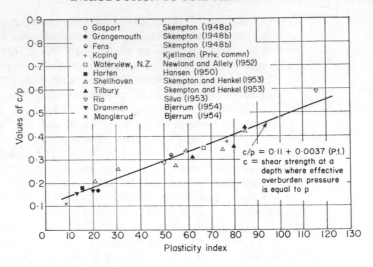

Fig. 7.19   Skempton's relationship between undrained shear strength, consolidation pressure and the index of plasticity.[7.20].

If the ground water level is at the ground surface, σ' (and therefore c_u) is often directly proportional to depth, as shown by the solid line in Fig. 7.20(a).

Figure 7.20(b) shows a typical relationship between shear strength for a heavily overconsolidated clay, for which the preconsolidation pressure σ'_c is much greater than the present effective overburden pressure. The effects of weathering and seasonal water content changes near the surface have curved both lines. Close to the surface of an otherwise normally consolidated clay, desiccation in dry weather may cause a large suction (that is, a large negative pressure). This results in an increase in effective stress, and produces the effect of overconsolidation, even though the present overburden pressure has never been exceeded. This is reflected in the increased strength of the drying crust near the surface in Fig. 7.20(a). Below this crust, there is often a considerable layer in which the strength has been increased by physico-chemical effects.

7.24   *Drained strength of clay soils:*   The maximum drained strength of a normally consolidated soil may generally be expressed in terms of effective stress in the form

$$\tau_f = \sigma'_n \tan \varphi' \quad \text{so that} \quad c' = 0$$

For a remoulded and normally consolidated soil, the strength at failure and at the critical state are almost

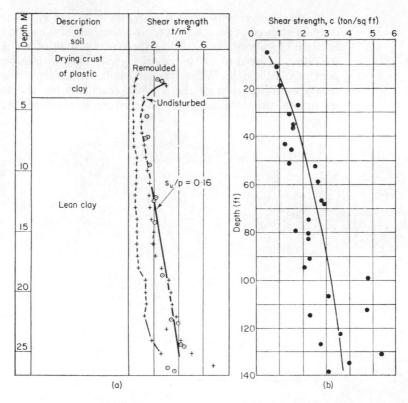

Fig. 7.20 *Typical strength/depth profiles for* (a) *normally consolidated clay* [7.21], *and* (b) *over consolidated clay* [7.22].

identical. In an undisturbed natural soil, however, there is usually some loss of strength after failure due to disruption of physico-chemical bonds and interparticular cement. The strength of a natural normally consolidated soil at the critical state is therefore generally somewhat below the maximum.

In overconsolidated soils, the lower void ratio is reflected in a higher strength at failure, and

$$\tau_f = c' + \sigma'_n \tan \varphi'$$

where φ' is less than for the same soil when normally consolidated, but c' is greater than zero and is a function of the preconsolidation pressure. A typical relationship between the maximum strengths of a soil in the normally and overconsolidated conditions is shown in Fig. 7.21. A comparison with Fig. 7.9(a) shows that this is in good agreement with the behaviour of Hvorslev's model.

In heavily overconsolidated soils, there is a substantial drop in strength between failure and the critical state (see Section 7.13 above). In assessing the collapse load of a soil structure, we need to know the *average* strength within the zone of plastic yield. Where the strains before collapse are nearly uniform, most of the soil will reach the maximum strength at the same moment. The average strength will be

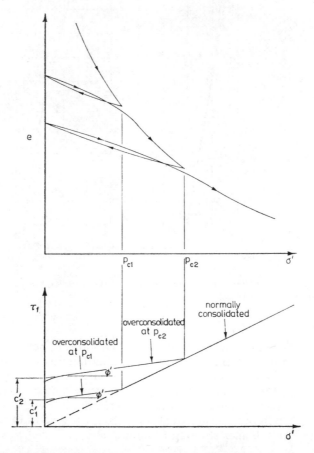

Fig. 7.21   Shear strength of clay soils in terms of effective stress.

nearly equal to the maximum. In some structures, however, quite large strains occur locally in the soil mass before total collapse. In these parts, the soil strength will have been reduced to the critical state value. The average strength then lies between the maximum and the critical state value, and may be much nearer to the latter.

7.25 *Shear strength of clay soils after large deformations:* In some overconsolidated clays (particularly those without fissures), the strength computed from observations of the collapse of natural and excavated slopes has been near the measured peak strength of the undisturbed material. However, in some overconsolidated fissured clays (notably the London clay) there has evidently been a gradual loss of strength before collapse, extending over many years, and tending towards the critical state value. This is presumed to be the effect of the fissures causing progressive failure. In such soils, the critical state strength is clearly the appropriate value to use in design. As explained above, this critical state strength is somewhat less than the maximum strength of the normally consolidated soil. Skempton [7.24] suggests that, as a conservative estimate, we may take the peak value of φ', assuming c' to be zero.

In previous sections, it has been implied that the critical state and residual strengths are the same, but in real soils this is not entirely true. In overconsolidated soil, collapse usually results in movements of substantial bodies of soil, by sliding along relatively thin deforming layers within the soil mass. Displacements of the order of 100 mm along such layers are generally sufficient to reduce most clays to the critical state, after which deformation continues at nearly constant volume. However, further displacements—of the order of 1 m or more—may reduce the strength still further. Skempton [7.23] originally made tests on London clay using a reversing shear box (see Section 6.8) and obtained values of φ'_r of about 16°. Similar values have been obtained using cut-plane samples in the triaxial apparatus. These values may be compared with the value of $22\frac{1}{2}°$ for the critical state strength predicted by Schofield and Wroth [7.7]. More recently, Bishop and others, using the ring shear apparatus, which facilitates much larger displacements, have obtained much lower values [7.25]. Evidence from field investigations indicates that this further reduction in strength reflects the formation of large and continuous slip surfaces within the soil mass. The displacements necessary to develop these slip surfaces are generally much greater than are usually observed in cuttings or natural slopes which have not previously slipped. However, where there is a risk of reactivating an old slip surface, the residual shear strength which must be used in design may be much less than the critical state strength.

7.26 *Slip lines and rupture surfaces:* There is a fundamental difference between the drained behaviour of a soil initially on the 'wet' or loose side of the critical state, and one initially on the 'dry' or dense side. In a normally consolidated or lightly overconsolidated soil, having an initial void ratio on the loose side of the critical state line, the approach to failure is accompanied by continuous progressive strain hardening and a general collapse of the soil structure. Dense soil, on the other hand, dilates during shear and its strength decreases. As a result, the strains are concentrated in a few narrow bands of weakened material which appear as slip lines. It is important to realise that when, in later chapters, reference is made to slip lines and rupture surfaces, it is in the context of the behaviour of a perfectly plastic model soil. In real loosely packed soil, no rupture surfaces will occur under shear loading, at least until after failure. Even in heavily overconsolidated soils, the rupture surfaces which do develop may not be very close to those implied by the Mohr–Coulomb model.

REFERENCES

7.1 TAYLOR, D. W. 1948. *Fundamentals of soil mechanics.* Wiley (New York).

7.2 RENDULIC, L. 1936. Porenziffer und Porenwasserdruck in Tonen. *Der Bauingenieur,* **17.**

7.3 HENKEL, D. J. 1960. The shear strength of saturated remoulded clays. *Research Conf. on Shear Strength of Cohesive Soils,* Boulder, Colorado (ASCE).

7.4 HVORSLEV, M. J. 1937. Uber die Festigkeitseigenshaften gestorter bindiger Boden. *Danmarks Naturvidenskabelige Samfund.* Copenhagen.

7.5 HVORSLEV, M. J. 1960. Physical components of the shear strength of saturated clays. *Research Conf. on Shear Strength of Cohesive Soils,* Boulder, Colorado (ASCE).

7.6 ROSCOE, K. H., SCHOFIELD, A. N. and WROTH, C. P. 1958. On the yielding of soils. *Géotechnique,* **8.**

7.7 SCHOFIELD, A. N. and WROTH, C. P. 1968. *Critical state soil mechanics.* McGraw-Hill (London).

7.8 ATKINSON, J. H. and BRANSBY, P. L. 1978. *The mechanics of soils.* McGraw-Hill (London).

7.9 PARRY, R. G. H. 1958. On the yielding of soils (Discussion of ref. 7.6). *Géotechnique,* **8.**

7.10 DRUCKER, D. C., GIBSON, R. E. and HENKEL, D. J. 1957. Soil mechanics and work hardening theories of plasticity. *Trans. ASCE,* **122.**

7.11 BISHOP, A. W. and ELDIN, A. K. G. 1953. The effect of stress history on the relation between φ and porosity in sand. *Proc. 3rd Int. Conf. Soil Mech. and Found. Eng.*

7.12 KIRKPATRICK, W. M. 1957. The condition of failure for sands. *Proc. 4th Int. Conf. Soil Mech. and Found. Eng.*

7.13 BISHOP, A. W. 1966. The shear strength of soils as engineering materials. *Géotechnique,* **16.**

7.14   ROSCOE, K. H. and BURLAND, J. B. 1968. On the generalised stress–strain behaviour of wet clay. In *Engineering Plasticity*. (Ed. HEYMAN, J. and LECKIE, F.) Cambridge Univ. Press (London).

7.15   VERMEER, P. A. 1978. A double hardening model for sand. *Géotechnique*, **28**.

7.16   ZIENKIEWICZ, O. C. and PANDE, G. N. 1976. Some useful forms of isotropic yield surfaces for soil and rock mechanics. *Numerical methods in soil and rock mechanics*. (Ed. BORN, G. and MEISSNER, H.), **2**. Institut für Bodenmechanik und Felsmechanik, Karlsruhe.

7.17   NAYLOR, D. J. 1978. Stress–strain laws for soil. In *Developments in soil mechanics*. (Ed. SCOTT, C. R.) Applied Science Publ. (London).

7.18   BURLAND, J. B. 1971. A method of estimating the pore pressures and displacements beneath embankments on soft natural clay deposits. *Proc. Roscoe Memorial Symp., Cambridge*.

7.19   SKEMPTON, A. W. and LAROCHELLE, P. 1965. The Bradwell Slip: a short term failure in the London clay. *Géotechnique*, **15**.

7.20   SKEMPTON, A. W. 1957. Discussion on 'The planning and design of the new Hong Kong airport'. (GRACE, H. and HENRY, J. K. M.) *Proc. Inst. Civil Eng.*, **7**.

7.21   BJERRUM, L. 1967. Engineering geology of Norwegian normally consolidated clays as related to settlements of buildings. *Géotechnique*, **16**.

7.22   SKEMPTON, A. W. and HENKEL, D. J. 1960. Tests on London clay from deep borings at Paddington, Victoria and the South Bank. *Proc. 4th Int. Conf. Soil Mech. and Found. Eng.*

7.23   SKEMPTON, A. W. 1964. Long term stability of clay slopes. *Géotechnique*, **14**.

7.24   SKEMPTON, A. W. 1970. First-time slides in over-consolidated clays. *Géotechnique*, **20**.

7.25   BISHOP, A. W., GREEN, G. E., GARGA, V. K., ANDRESEN, A. and BROWN, J. D. 1971. A new ring shear apparatus and its application to the measurement of residual strength. *Géotechnique*, **21**.

CHAPTER 8

Plasticity, limit analysis, and limiting equilibrium

PLASTICITY AND COLLAPSE

8.1 Deformation, yield, and collapse: A system of surface and body forces applied to a body will cause changes in the stresses and strains within it. The changes in strain will usually be partly elastic and immediately recoverable, but may also result in part from irrecoverable plastic yield. As the forces increase, a growing part of the body may reach the limit of its shear strength. Once this part is sufficiently extensive to form an unstable mechanism, unlimited yield will occur. This will continue as long as the forces can be maintained in the face of the changing geometry of the body. This unlimited yield is called *collapse*, and the system of forces which will produce it is called the *collapse load*.

In principle, the conditions of equilibrium and of strain compatibility, together with a knowledge of the stress/strain behaviour of the material, are sufficient to determine the distribution of stress and displacement up to the moment of collapse. In practice, however, the stress/strain behaviour of soil is so complex, and so dependent on the loading history, that analysis of soil structures must be drastically simplified. It is not that we are evading the mathematical complexities, great though these are; but a full solution would require a knowledge of the behaviour of the soil, and of the past and future loads on it, which in practice we cannot hope to obtain.

Fortunately, most practical problems may be solved by answering the following two questions:

(a) What are the structural displacements under the working loads?

(b) Is the working load substantially less than the collapse load?

Provided that the applied load is substantially less than the collapse load, the first question can often be answered with sufficient accuracy by assuming that the stresses (and sometimes the displacements also) are those which would occur in a linearly elastic body. This was discussed in Chapter

177

5 and will be considered further in Chapter 12. Only where there are significant plastic strains under the working loads is it necessary to make a more elaborate analysis of displacement.

This chapter is concerned with the answer to the second question—that is, with the attempt to predict the collapse load.

8.2   *The condition of plane strain:*   The application of a system of external forces to the body would, in general, cause changes in both stress and strain components in three dimensions. However, many of the structures with which we are concerned—such as embankments, retaining walls and ground beams— are very long in comparison with their transverse dimensions. The longitudinal strain changes in and under such structures can only be very small, and no serious error is introduced by assuming the condition of plane strain. In this chapter, the collapse loads are only derived for the plane strain state, but the principles adduced apply equally to analysis of collapse in three dimensions.

8.3   *Work softening and perfect plasticity:*   Real soils commonly have stress/strain curves of the type shown by the solid line in Fig. 8.1. With increasing shear strain, the shearing resistance increases to a peak value and then drops to a nearly constant residual value. Between the peak and residual values, the material is said to *work soften*.

All the methods of analysis discussed in this chapter assume that the yield condition is independent of strain, so that the stress/strain curve is of the form shown by the broken line in Fig. 8.1. A material with this form of yield condition is said to be *perfectly plastic*. Yield and failure conditions in this case are identical (see Section 6.3 above).

If the predicted collapse load is to be reliable, it is important that the average resistance of the real soil should match that of the model soil within the plastic zone. Clearly,

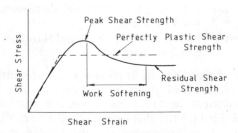

Fig. 8.1   Stress/strain relationships for work softening and perfectly plastic materials.

if the strain in the real soil is nearly uniform at the collapse load, much of this soil will develop its peak resistance at the same moment. The greatest average resistance will be near the peak value. In some structures, however, much of the soil undergoes substantial plastic strain before collapse. In such cases, the average resistance in the plastic zone is much nearer to the residual value. Some care is therefore needed in choosing the values to be assigned to parameters defining the failure condition of the model soil. These values will depend on the collapse mechanism of the structure, as well as on the properties of the real soil.

8.4   *Analysis of the collapse state:*   When a structure formed of perfectly plastic materials is on the point of collapse, a zone of material must exist which has reached the limit of its shear strength. This zone must be sufficiently extensive to form an unstable mechanism. Within this plastic zone, stress and strain components must satisfy

   (a)   the condition of equilibrium
   (b)   the yield condition, and
   (c)   a *flow rule*, governing the relation between the components of yield stress and *strain rate*.
         (The significance of these terms is explained in Sections 8.8 and 8.9 below.)

EQUILIBRIUM IN PLANE STRAIN

8.5   *Stress components in plane strain:*   The stress at any point in a three-dimensional continuum may be expressed in terms of the normal and tangential components of stress on the surfaces of a rectangular element, as shown in Fig. 8.2. In this chapter, the collapse loads are only determined for the state of plane strain. If the co-ordinate axes are chosen so that the strains are limited to the plane containing two of them (say x and z), and are zero in the third direction (y), the number of independent stress components is reduced from nine to four. Four of the shear stress components (τ_{xy}, τ_{yx}, τ_{yz} and τ_{zy}) and the longitudinal normal stress (σ_y) are no longer independent of the others.

   The remaining independent stress components are shown in Fig. 8.3. Since normal stress in soil is predominantly compressive, we elect to treat such stress as positive. Shear stress is considered positive when acting outwards on the faces of the element nearest to the origin. This sign convention is convenient for our purposes, and is usual in

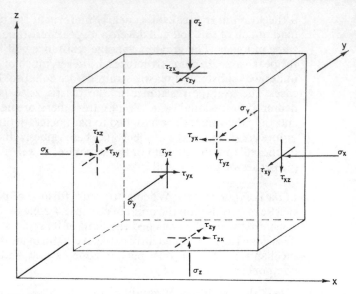

Fig. 8.2   Stress components in cartesian co-ordinates.

soil mechanics, but the reader should note that it is the reverse of the convention commonly used in texts on strength of materials. One consequence of this convention is that positive normal strains imply *shortening* of the longitudinal dimension (Fig. 8.3(a)). Unfortunately, this convention for stress is not suitable for the Mohr circle diagrams. For these diagrams *only*, shear stress τ is considered positive if exerting an anti-clockwise moment on the element (Fig. 8.3(b)).

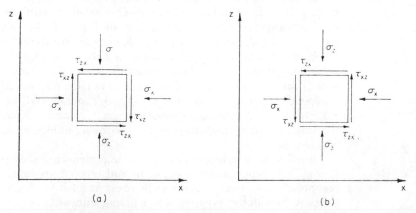

Fig. 8.3   Positive directions of stress components (a) except for Mohr circle diagrams, and (b) for Mohr circle diagrams only.

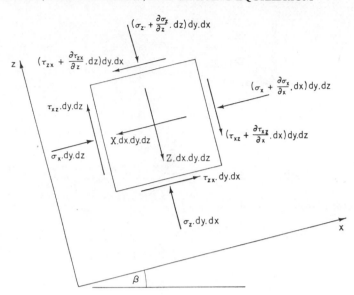

Fig. 8.4   Forces on an element in plane strain.

8.6   *The condition of equilibrium in plane strain:*   Now consider a body in which the stress components vary with x and z, and in which there are body forces X and Z per unit volume of the material, acting in the x and z directions respectively. Let the x axis be inclined at an angle β to the horizontal. Body forces are taken to be positive when acting in the negative directions of x and z. The forces now acting on the element are shown in Fig. 8.4.

Each element within the body must be in equilibrium under the effect of the stresses on the boundaries of the element and of the body forces acting within it. For the case of plane strain in the x–z plane, this condition is satisfied if

(a)   there is no moment about an axis normal to the x–z plane, and
(b)   there is no resultant force on the element in either of the x or z directions.

The first of these requires that $\tau_{xz} = \tau_{zx}$. The second shows that

$$\left(\frac{\partial\sigma_x}{\partial x}\,dx\right)dy\,dz + \left(\frac{\partial\tau_{xz}}{\partial z}\,dz\right)dx\,dy + X\,dx\,dy\,dz = 0$$

$$\left(\frac{\partial\sigma_z}{\partial z}\,dz\right)dx\,dy + \left(\frac{\partial\tau_{xz}}{\partial x}\,dx\right)dy\,dz + Z\,dx\,dy\,dz = 0$$

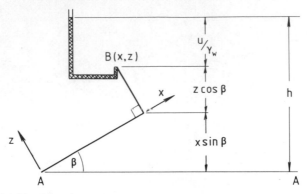

Fig. 8.5   Total head h and pore pressure u.

Therefore

$$\frac{\partial \sigma_x}{\partial x} + \frac{\partial \tau_{xz}}{\partial z} + X = 0$$

$$\frac{\partial \sigma_z}{\partial z} + \frac{\partial \tau_{xz}}{\partial x} + Z = 0$$

If the weight of the material is the only body force, and if the x axis is inclined at β to the horizontal,

$$\frac{\partial \sigma_x}{\partial x} + \frac{\partial \tau_{xz}}{\partial z} + \gamma \sin \beta = 0$$

$$\frac{\partial \sigma_z}{\partial z} + \frac{\partial \tau_{xz}}{\partial x} + \gamma \cos \beta = 0$$

These two equations define the equilibrium condition in terms of total stress. The condition in terms of effective stress may be derived directly as follows.

Writing $\sigma'_x + u$ for σ_x in the appropriate equilibrium equation gives

$$\frac{\partial \sigma'_x}{\partial x} + \frac{\partial \tau_{xz}}{\partial z} + \gamma \sin \beta + \frac{\partial u}{\partial x} = 0$$

But (see Fig. 8.5), the total head

$$h = x \sin \beta + z \cos \beta + u/\gamma_w$$

so that

$$\frac{\partial u}{\partial x} = \gamma_w \left(\frac{\partial h}{\partial x} - \sin \beta \right)$$

Then

$$\frac{\partial \sigma'_x}{\partial x} + \frac{\partial \tau_{xz}}{\partial z} + (\gamma - \gamma_w) \sin \beta + \frac{\partial h}{\partial x} \gamma_w = 0$$

Similarly

$$\frac{\partial \sigma'_z}{\partial z} + \frac{\partial \tau_{xz}}{\partial x} + (\gamma - \gamma_w) \cos \beta + \frac{\partial h}{\partial z} \gamma_w = 0$$

Writing

$$i_x = -\frac{\partial h}{\partial x} \quad \text{and} \quad i_z = -\frac{\partial h}{\partial z}$$

$$\frac{\partial \sigma'_x}{\partial x} + \frac{\partial \tau_{xz}}{\partial z} + (\gamma - \gamma_w) \sin \beta - i_x \gamma_w = 0$$

$$\frac{\partial \sigma'_z}{\partial z} + \frac{\partial \tau_{xz}}{\partial x} + (\gamma - \gamma_w) \cos \beta - i_z \gamma_w = 0$$

Thus, the equilibrium condition may be stated in terms of effective stress, provided that the body forces are taken to be the sum of

(a) the relevant component of the submerged unit weight, and

(b) the *seepage pressure*, defined as the negative product of the hydraulic gradient and the unit weight of water.

THE YIELD CONDITION AND THE FLOW RULE

8.7   The Mohr–Coulomb yield condition: The methods described in this chapter may be applied to any *perfectly plastic* model soil. For such models, the yield and failure conditions are identical. The methods cannot however be applied to those soil models, such as the critical state model, which take account of strain hardening or strain softening. For the reasons discussed in Section 6.18 above, the Mohr–Coulomb model is most commonly used. The yield (and failure) condition may be stated in the form

$$\tau_f = c + \sigma_n \tan \varphi$$

in terms of total stress, or

$$\tau_f = c' + (\sigma_n - u) \tan \varphi'$$

in terms of effective stress.

Other forms of this condition were discussed in Section 6.5 above.

Both the equilibrium and yield conditions may be expressed in similar forms in terms of either total or effective stress. In the remainder of this chapter, the primes denoting effective stress have, for convenience, been omitted, but the methods used and the expressions derived are equally applicable to analysis in terms of total or effective stress.

8.8   *Strain rate and velocity:*   A consequence of our assumption of perfect plasticity is that any stress state satisfying the yield condition will, if maintained, cause unlimited plastic strain. There is, therefore, no direct relationship between yield stress and plastic strain. We therefore need to define, not the strain, but the *strain rate*—that is, the rate at which the strain is increasing with respect to time. The absolute value of the strain rate is not determinate, since, in designing the soil models, we have not specified any property (such as viscosity) which would control it. This turns out to be not very important as we are concerned only with the *relative* magnitudes of the strain rate components. These define the directions of the strain rate vectors, and the shape of the deformed body.

Knowing the strains everywhere within a body, we may determine the relative displacements of different points within it. In a similar way, knowing the strain rates, we may determine the *velocities*—that is, the rates of displacement. As in the case of the strain rates, the absolute magnitudes of the velocities are not determinate. Our concern is with the *relative* magnitudes of the velocity components, since these define the directions of the velocity vectors and hence the directions of motion. A pattern of velocity vectors, defining the motion everywhere within the plastic zone is called a *velocity field.*

8.9   *The flow rule and the plastic potential:**   In a linearly elastic body, Hooke's law defines the relation between the components of stress and strain. Similarly, in a perfectly plastic body, the *flow rule* defines the relation between the components of the yield stress (for example σ_n, τ_f) and the corresponding plastic strain rates (e.g. $\dot{\varepsilon}_n^p$, $\dot{\gamma}^p$).

Von Mises [8.1] suggested that the flow rule might be expressed in terms of a *plastic potential function* (f) which may be defined, for a Mohr–Coulomb material, by the equation

$$\frac{\dot{\gamma}^p}{\dot{\varepsilon}_n^p} = \frac{\partial f/\partial \tau}{\partial f/\partial \sigma_n}$$

---

* In the discussion which follows ε = longitudinal strain with a subscript to denote direction, γ = shear strain and ^{p} = *plastic* strain. The strain rate is denoted by a dot over the symbol for the corresponding strain.

Von Mises also suggested that it may often be useful to assume that the potential function is identical with the yield condition. For a Mohr–Coulomb material, where yield and failure conditions are identical,

$$f = \tau - c - \sigma_n \tan \varphi \leq 0$$

For all stress states on the yield locus, $\tau = \tau_f$ and $f = 0$. A flow rule defined in this way by the yield condition is said to be *associated*.

An associated flow rule makes sense in studies of metal plasticity, since it implies that

(a) for isotropic materials, the directions of principal stress and principal strain rate coincide, and

(b) for frictionless materials, there is no volumetric strain during plastic yield.

Certain problems arise if an associated flow rule is applied to a frictional material. These are discussed in Section 8.16 below.

8.10 *Normality of the strain rate vector:*   The Mohr–Coulomb yield condition may be expressed as

$$\tau_f = c + \sigma_n \tan \varphi$$

The associated flow rule is defined by

$$\frac{\dot{\gamma}^p}{\dot{\varepsilon}_n^p} = \frac{\partial f/\partial \tau}{\partial f/\partial \sigma_n}$$

where

$$f = \tau - c - \sigma_n' \tan \varphi$$

so that

$$\frac{\dot{\gamma}^p}{\dot{\varepsilon}_n^p} = -\frac{1}{\tan \varphi}$$

Thus, if the yield stress components are plotted, and are superimposed on a similar plot of the corresponding strain rate components (as shown in Fig. 8.6(a)), it will be seen that *the strain rate vector is normal to the yield locus*.

Similarly, the yield condition may be expressed (see Section 6.5(a) above) in the form

$$\sigma_1(1 - \sin \varphi) = \sigma_3(1 + \sin \varphi) + 2c \cos \varphi$$

The associated flow rule is defined by

$$\frac{\dot{\varepsilon}_3^p}{\dot{\varepsilon}_1^p} = \frac{\partial f/\partial \sigma_3}{\partial f/\partial \sigma_1}$$

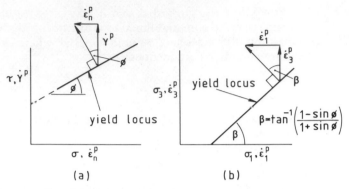

Fig. 8.6　*Normality of the strain rate vector.*

where

$$f = \sigma_3(1 + \sin \varphi) - \sigma_1(1 - \sin \varphi) + 2c \cos \varphi$$

Then

$$\frac{\dot{\varepsilon}_3^p}{\dot{\varepsilon}_1^p} = -\frac{1 + \sin \varphi}{1 - \sin \varphi}$$

But the slope of the yield locus is given by

$$\frac{d\sigma_3}{d\sigma_1} = \frac{1 - \sin \varphi}{1 + \sin \varphi}$$

showing again (Fig. 8.5(b)) that the strain rate vector is normal to the yield locus. This *normality condition* can be shown to be a general consequence of adopting an associated flow rule [8.2].

8.11　*Volumetric strain in frictional materials:*　Consider the rectangular element of Mohr–Coulomb material of thickness h shown in Fig. 8.7(a), yielding in simple shear. Let the volume increase be restricted to the vertical direction, and the relevant boundary stress components be τ and σ_n. If the material has an associated flow rule, the horizontal velocity $\dot{\gamma}^p h$ at point A must be accompanied by a vertical velocity $-\dot{\varepsilon}_n^p h = \dot{\gamma}^p h \tan \varphi$. This results in a volumetric strain, continuously increasing at a rate $\dot{\gamma}^p \tan \varphi$, and the velocity of the upper face relative to the lower is inclined at an angle β, where $\beta = \tan^{-1}(-\dot{\varepsilon}_n^p / \dot{\gamma}^p) = \varphi$.

This continuous volumetric strain rate is the direct consequence of the assumption of an associated flow rule for the model soil. Whether or not any real soil would behave in this way will be considered in Section 8.16 below.

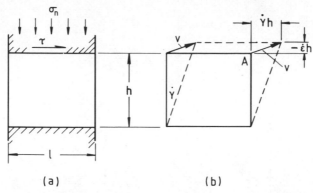

Fig. 8.7 (a) An element in simple shear. (b) Corresponding strain rates and velocities.

ENERGY DISSIPATION IN PLASTIC YIELD

8.12 *Energy dissipation in simple shear* [8.3]: Consider the element shown in Fig. 8.7(a), deforming in simple shear. The rate of dissipation of energy (D) per unit volume is obtained by multiplying the stresses by the corresponding strain rates, and hence

$$D = \tau \dot{\gamma}^p + \sigma_n \dot{\varepsilon}_n^p$$

The flow rule requires that $\dot{\varepsilon}_n^p = -\dot{\gamma}^p \tan \varphi$. Also, since the whole element is yielding,

$$\tau = c + \sigma_n \tan \varphi$$

so that

$$D = (c + \sigma_n \tan \varphi)\dot{\gamma}^p - \sigma_n \dot{\gamma}^p \tan \varphi$$
$$= c\dot{\gamma}^p$$

The total energy dissipation in the element is therefore

$$Dlh = c\dot{\gamma}^p lh = clv \cos \varphi$$

where $v \cos \varphi$ is the component of the velocity v of point A parallel to the element boundary.

8.13 *Rigid-body sliding on a thin deforming layer:* Collapse of a body of the soil may take the form of rigid-body sliding on a thin layer of material deforming in simple shear. The expression for total energy dissipation, derived in the last section, is independent of the thickness h of the element and can therefore be used to analyse sliding on a very thin layer. Since the material is assumed to have an associated flow rule, the velocity vector v must be inclined at an angle $\beta = \varphi$ to the boundary of the rigid body. It follows that, for translation

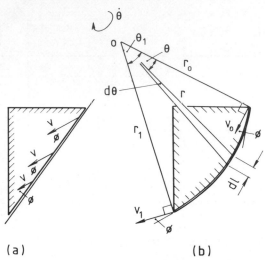

Fig. 8.8   Rigid-body sliding on a thin deforming layer. (a) Translation. (b) Rotation about a fixed centre.

(without rotation), the only admissible surface is a plane (Fig. 8.8(a)). For rotation about a fixed centre O, the only admissible surface is a *logarithmic spiral* (Fig. 8.8(b)). A logarithmic spiral is a curve drawn so that the radius vector from a fixed centre is always inclined at a constant angle (in this case $\pi/2 + \varphi$) to the tangent to the curve. It may be shown that, if the radius vector r_0 is rotated through an angle θ_1, the length increases from r_0 to $r_1 = r_0 \exp(\theta_1 \tan \varphi)$. Where φ is zero, the radius vector is of constant length, and the spiral degenerates into a circle.

For translation (Fig. 8.8(a)) the total rate of energy dissipation is $clv \cos \varphi$, as before. For a rigid-body rotation about O (Fig. 8.8(b)) at constant angular velocity $\dot{\theta}$, the velocity v is proportional to the radius r, so that

$$v_0/r_0 = v_1/r_1 = v/r = \dot{\theta}$$

The velocity v must therefore increase from v_0 at one end of the spiral to $v_1 = v_0 \exp(\theta_1 \tan \varphi)$ at the other. The length of the element dl is $r\, d\theta/\cos \varphi$, so that the total rate of energy dissipation along the whole deforming layer is

$$\int_0^{\theta_1} cv \cos \varphi \, \frac{r\, d\theta}{\cos \varphi} = \int_0^{\theta_1} cr_0 \exp(\theta \tan \varphi) v_0 \exp(\theta \tan \varphi)\, d\theta$$

$$= cr_0 v_0 \int_0^{\theta_1} \exp(2\theta \tan \varphi)\, d\theta$$

$$= \tfrac{1}{2} cr_0 \cot \varphi \left[\exp(2\theta_1 \tan \varphi) - 1 \right]$$

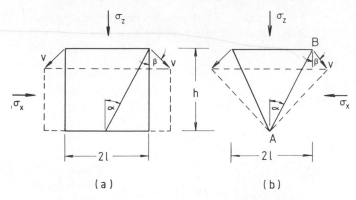

Fig. 8.9   Energy dissipation in direct compression. (a) A rectangular element. (b) A triangular element.

8.14   *Deformation of an element in direct compression:* Consider a rectangular element (Fig. 8.9(a)), yielding to the principal stresses σ_z, σ_x where $\sigma_z > \sigma_x$. As shown in Section 8.10 above,

$$\frac{\dot{\varepsilon}_3^p}{\dot{\varepsilon}_1^p} = \frac{\dot{\varepsilon}_x^p}{\dot{\varepsilon}_z^p} = -\frac{1 + \sin \varphi}{1 - \sin \varphi}$$

Then the rate of dissipation of energy per unit volume (D) is given by

$$D = \sigma_z \dot{\varepsilon}_z^p + \sigma_x \dot{\varepsilon}_x^p$$

$$= \dot{\varepsilon}_z^p \left(\sigma_z - \sigma_x \frac{1 + \sin \varphi}{1 - \sin \varphi} \right)$$

The yield condition must be satisfied within the element, and may be written in the form

$$\sigma_z(1 - \sin \varphi) - \sigma_x(1 + \sin \varphi) - 2c \cos \varphi = 0$$

or

$$\sigma_z - \sigma_x \left(\frac{1 + \sin \varphi}{1 - \sin \varphi} \right) = \frac{2c \cos \varphi}{(1 - \sin \varphi)}$$

Then

$$D = \frac{2c \cos \varphi}{(1 - \sin \varphi)} \dot{\varepsilon}_z^p$$

Simple trigonometry shows that this may be written

$$D = 2c \dot{\varepsilon}_z^p \tan (\pi/4 + \varphi/2)$$

or

$$D = -2c \dot{\varepsilon}_x^p \tan (\pi/4 - \varphi/2)$$

The direction of the velocity vector may be simply derived from the strain rate vectors, as follows.
At point B,

$$v_x = -l\dot{\varepsilon}_x^p \quad \text{and} \quad v_z = -h\dot{\varepsilon}_z^p$$

Then

$$\tan \beta = -\frac{v_x}{v_z} = -\frac{l}{h} \times \frac{\dot{\varepsilon}_x^p}{\dot{\varepsilon}_z^p} = \tan \alpha \frac{1 + \sin \varphi}{1 - \sin \varphi}$$

where

$$\alpha = \tan^{-1}\left(\frac{l}{h}\right)$$

Since the rate of dissipation per unit volume is constant within the element, the expression may be applied to elements of any shape—for example, the triangular element shown in Fig. 8.9(b).

8.15 *Energy dissipation in a deforming wedge bounded by a logarithmic spiral:* Consider the wedge of soil bounded by the logarithmic spiral AB (Fig. 8.10). Let the boundary AO move with velocity v_A normal to AO. The wedge may be conveniently considered to consist of a series of thin deforming layers bounded by logarithmic spirals (PQ, RS, etc.) each rotating about 0 at an angular velocity v_A/r. Then

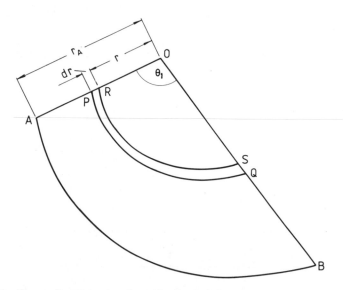

Fig. 8.10   Energy dissipation in a logarithmic spiral shear zone.

the velocity of R relative to P as the layer $PQSR$ deforms, is $v_A \, dr/r$, and the energy dissipated in this layer is (see Section 8.13)

$$\tfrac{1}{2}c(v_A \, dr/r)r \cot \varphi \, [\exp(2\theta_1 \tan \varphi) - 1]$$

Then the total energy dissipation in the wedge is

$$\tfrac{1}{2}cv_A \cot \varphi \, [\exp(2\theta_1 \tan \varphi) - 1] \int_0^{r_A} dr$$

$$= \tfrac{1}{2}cr_A v_A \cot \varphi \, [\exp(2\theta_1 \tan \varphi) - 1]$$

In addition, energy is dissipated in the layer on AB, across which the relative velocity is v_A at A. Then the energy dissipated here is also

$$\tfrac{1}{2}cr_A v_A \cot \varphi \, [\exp(2\theta_1 \tan \varphi) - 1]$$

and the total dissipation rate within the wedge and on the boundary AB is

$$cr_A v_A \cot \varphi \, [\exp(2\theta_1 \tan \varphi) - 1]$$

For the particular case where $\varphi = 0$, the length of the arc RS is $r\theta_1$. Then the energy dissipated in unit time in the element $PQSR$ is

$$cr\theta_1 \frac{dr}{r} v_A$$

Then the total energy dissipated in the wedge is

$$c\theta_1 v_A \int_0^{r_A} dr = c\theta_1 v_A r_A$$

In addition, the energy dissipated in the layer on AB, across which the relative velocity is v_A, is also

$$c\theta_1 v_A r_A$$

so that the total dissipation rate within the wedge and on the boundary AB is

$$2c\theta_1 v_A r_A$$

8.16   *The flow rule in real frictional soils:* Consider an element deforming in simple shear (Fig. 8.7(a)). The total rate at which work is done by the shear stress, per unit volume of the element, is

$$\tau \dot{\gamma}^p = c\dot{\gamma}^p + \sigma_n \tan \varphi \, \dot{\gamma}^p$$

The associated flow rule requires that there should be a continuous increase in volumetric strain at a rate of $\dot{\gamma} \tan \varphi$, and this requires energy to be absorbed at a rate of $\sigma_n \dot{\varepsilon}_n^p = \sigma_n \tan \varphi \dot{\gamma}^p$. Then rate of energy dissipation per unit volume is

$$D = \tau \dot{\gamma}^p - \sigma_n \tan \varphi \, \dot{\gamma}^p = c \dot{\gamma}^p$$

This implies that the whole of the frictional component of the yield strength is absorbed in expanding the element, and there is no true internal friction. If there is no volume change, φ is zero.

A comparison with the analysis given in Section 7.5 above shows that real soils cannot be expected to behave in this way. Although some increase in volume may occur at first, this cannot account for the whole of the friction, and it cannot continue indefinitely. Even at constant volume, some frictional resistance will generally remain. Thus, we are hardly ever justified in assuming an associated flow rule for a real soil. Almost the only exception to this is the case of a saturated undrained soil, where we would expect both φ_u and the volumetric strain rate to be insignificant.

Often however, there are advantages in applying an associated flow rule to our model soils. In the first place, many of the theorems of plasticity (including the limit theorems discussed in Sections 8.17 to 8.19 below) require the normality condition for their proof. Although limit theorems applicable to non-associated flow rules can be derived, they are generally less helpful. Secondly, if the flow rule is not associated, there may be no unique solution, since the magnitude of the collapse load may depend on the order in which its various components are applied. While the extent of the uncertainty is often small, it can be very inconvenient, particularly in computer solutions where the steps taken by the programme in solving a particular problem may not be easily traced. Finally, in many soil problems there are few if any restrictions on the velocity field at the boundaries. Under these conditions, the collapse load is generally rather insensitive to the small changes in the form of the velocity field which would be implied by changes in the flow rule. For example, analysis of rigid-body sliding on a thin deforming circular arc (implying that β is zero) may give nearly the same collapse load as on a logarithmic spiral (implying that $\beta = \varphi$).

A word of warning is needed here. The collapse load is often insensitive to the form of the sliding surface, but the position of the critical surface is not. Thus, the circular arc giving the least value of the collapse load may be far away from the critical logarithmic spiral, and neither may appear very similar to the surface of sliding observed in real soil. It is important to be consistent in one's assumptions. If the model soil is presumed to have an associated flow rule, and to collapse by sliding on a thin deforming layer, the correct collapse load is the least value given by any logarithmic spiral. This may not be the logarithmic spiral which appears most nearly to match a sliding surface observed in the field, since real soils are unlikely to have associated flow rules.

LIMIT ANALYSIS

8.17 Upper and lower bound solutions: In theory, the equilibrium and yield conditions and the flow rule, are sufficient to determine the velocity field, stress distribution, and collapse load. In practice, however, it is seldom possible to obtain a closed solution, except in very simple cases. In other cases, it is possible to obtain useful approximate solutions by means of the lower and upper bound limit theorems described in the next two sections.

8.18 The lower bound theorem: Consider a body carrying a known load (i.e. a known system of surface and body forces). Now suppose that we choose an arbitrary stress field—that is, an arbitrary pattern of stresses, defining the stress state *everywhere* within the body. If the stress nowhere exceeds the yield limit, the stress field is said to be *stable.* If the stress is everywhere in equilibrium with the surface and body forces, the stress field is said to be *statically admissible.* For a perfectly plastic material with an associated flow rule, it can be shown [8.2] that, if any stable statically admissible stress field can be found, collapse will not occur under the given load.

This does not mean that the given load is the most that the body can support. It *may* be possible to find another stable stress field, in equilibrium with a larger load. The collapse load, therefore, cannot be less than the given load but may be greater. Thus, *a load which is in equilibrium with any stable statically admissible stress field is a lower bound to the collapse load.*

Example 8.1
Consider a foundation of width B exerting a uniform pressure q on the surface of a clay soil having a density γ, and an undrained shear strength c_u. We will select the arbitrary stress field shown in Fig. 8.11. Since there are no shear stresses on the soil surface, vertical and horizontal stresses

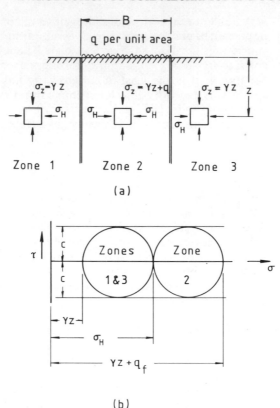

Fig. 8.11   Example 8.1. The lower bound solution. (a) The chosen stress field. (b) Mohr stress circles for the three zones at failure.

are principal stresses. Notice that there are discontinuities (i.e. sudden changes) in the vertical component of stress, along vertical lines through the soil at the edges of the foundation. This is admissible provided that the horizontal stress component is in equilibrium across the discontinuity. It is true that the chosen stress field is most unlikely to occur in practice, but this does not affect our argument. All that concerns us here is that the stress field is in equilibrium with the surface and body forces, and is therefore statically admissible.

If the yield stress is not to be exceeded, the Mohr stress circles show (Fig. 8.11(b)) that

$$q + \gamma z \leq \sigma_H + 2c_u \quad \text{and} \quad \sigma_H \leq 2c_u + \gamma z$$

so that

$$q_{\max} = 4c_u$$

Thus the largest load that can be in equilibrium with the chosen stress field is $4c_u B$, and this is a lower bound to the collapse load.

8.19 *The upper bound theorem:* Now suppose that we choose an arbitrary velocity field—that is, an arbitrary pattern of velocities defining the motion everywhere within the body. If the motion is everywhere compatible with the continuity of the body, and with any restrictions on the motion at the boundary, the velocity field is said to be *kinematically* admissible. If the work done by the displacement of the load exceeds the energy dissipated in deforming the body, the velocity field is said to be *unstable* [8.4]. For a perfectly plastic material with an associated flow rule, it can be shown [8.2] that, if any unstable kinematically admissible velocity field can be found, collapse must occur under the given load or under some smaller load. Thus, *for any kinematically admissible velocity field, the load computed by equating the internal energy dissipation to the external work done is an upper bound to the true collapse load.*

Example 8.2

For the problem of Example 8.1, consider collapse by sliding on a thin circular deforming layer, as shown in Fig. 8.12. Since φ is zero, this motion is compatible with the flow rule. Also, there are no restrictions on displacements at the boundary. The velocity field is therefore kinematically admissible. The total rate of energy dissipation (see Section 8.12) is $clv \cos \varphi$.

But,

$$l = 2\theta \frac{z}{\cos \varphi}; \quad v = r\dot{\theta} = \frac{z\dot{\theta}}{\cos \theta}; \quad \varphi = 0$$

Then the total rate of dissipation is $2c_u \theta \dot{\theta} z^2 / \cos^2 \theta$. The rate of external work done is

$$\tfrac{1}{2} q \dot{\theta} z^2 (\tan^2 \theta - \tan^2 \alpha)$$

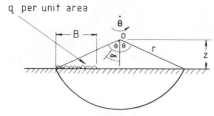

q per unit area

Fig. 8.12 *Example 8.2. An upper bound solution.*

Equating external work and internal energy dissipation, and dividing both sides by $\dot{\theta}z^2$, gives

$$2c_u\theta/\cos^2\theta = \tfrac{1}{2}q(\tan^2\theta - \tan^2\alpha)$$

$$q = \frac{4c_u\theta}{\cos^2\theta(\tan^2\theta - \tan^2\alpha)}$$

Clearly, for any positive value of θ, q is a minimum if α is zero, so that we need only consider the case where

$$q = 4c_u\theta\operatorname{cosec}^2\theta$$

This has a minimum value if

$$0 = \frac{\partial}{\partial\theta}(\theta\operatorname{cosec}^2\theta)$$

$$= \operatorname{cosec}^2\theta - 2\theta\operatorname{cosec}^2\theta\cot\theta$$

Then

$$2\theta = \tan\theta$$
$$\theta = 1\cdot1657 \text{ radians}$$

and

$$q = 4c_u\theta\operatorname{cosec}^2\theta$$
$$= 5\cdot52c_u$$

Then $5\cdot52c_uB$ is an upper bound to the collapse load, which must lie in the range $5\cdot52c_uB \geqslant q_fB \geqslant 4c_uB$.

This range might well be sufficiently small for practical purposes, but it could be reduced by using other stress and velocity fields.

OTHER METHODS OF ANALYSIS

8.20 *Slip-line field solutions:* The stress components in the co-ordinate directions x and z within the plastic zone, may be conveniently defined in terms of two variables $\bar{\sigma}$ and ψ (Fig. 8.13). The value of $\bar{\sigma}$ defines the centre of the Mohr stress circle, and, in conjunction with the yield condition, determines all the stress components. The angle ψ defines the direction of the greatest principal stress σ_1 with reference to the x axis. It has been shown (Section 6.4) that the slip lines are inclined at $\pm(\pi/4 - \varphi/2)$ to the direction of σ_1. The value of ψ therefore defines the directions of the slip lines, which are inclined at $\psi \pm (\pi/4 - \varphi/2)$ to the x axis. The stress components σ_x, σ_z, and τ_{xz} may be expressed in terms of $\bar{\sigma}$ and ψ as follows (see Fig. 8.13):

$$CX = CT = \bar{\sigma}\sin\varphi$$
$$OC = \bar{\sigma} - c\cot\varphi$$

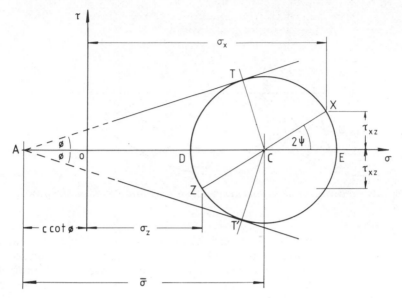

Fig. 8.13   Slip-line field solutions. The stress components at failure in terms of $\bar{\sigma}$ and ψ.

so that

$$\sigma_x = OC + CX \cos 2\psi$$
$$= \bar{\sigma}(1 + \sin \varphi \cos 2\psi) - c \cot \varphi$$
$$\sigma_z = \bar{\sigma}(1 - \sin \varphi \cos 2\psi) - c \cot \varphi$$
$$\tau_{xz} = \bar{\sigma} \sin \varphi \sin 2\psi$$

The condition of equilibrium within the plastic zone may be written (see Section 8.6)

$$\frac{\partial \sigma_x}{\partial_x} + \frac{\partial \tau_{xz}}{\partial_z} + X = 0$$

$$\frac{\partial \sigma_z}{\partial_z} + \frac{\partial \tau_{xz}}{\partial_x} + Z = 0$$

where X and Z are the body forces in the negative directions of x and z respectively.

Substituting the values of the stress components given above, we can obtain two simultaneous differential equations in $\bar{\sigma}$ and ψ. Given sufficient boundary conditions, the solution of these equations gives the magnitude of the stress components and the direction of the slip lines everywhere within the plastic zone. Kötter [8.5] first obtained equations of this type (although not quite in this form) for cohesionless soils. Jáky [8.6] showed that Kötter's equations are equally valid for cohesive materials.

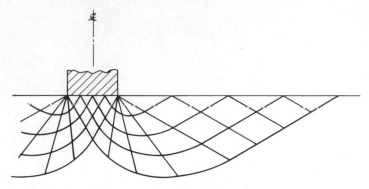

Fig. 8.14   Prandtl's slip-line field for a uniformly distributed load on the surface of a weightless soil.

Prandtl [8.7] used Kötter's equations to obtain a closed solution to the problem of a footing on the surface of *weightless* soil. The slip-line field he obtained for this problem is shown in Fig. 8.14. Prandtl's solution for the collapse load of the footing is

$$q_f = \cot \varphi \, [\exp(\pi \tan \varphi) \tan^2(\pi/4 + \varphi/2) - 1]c$$

where

$$\varphi > 0$$

and

$$q_f = (2 + \pi)c = 5 \cdot 14c$$

where

$$\varphi = 0$$

Notice that the latter is within the range of results obtained in Sections 8.18 and 8.19 using limit analysis. This is to be expected, since these results are independent of the weight of the soil, and are therefore equally valid for a weightless material.

There are very few closed solutions to Kötter's equations, but many approximate answers have been obtained, mostly using a finite difference approach developed by Sokolovskii [8.8]. Some of these solutions will be used in later chapters.

8.21   Validity of slip-line field solutions: A slip-line field solution determines the stress only within the plastic zone, and it has therefore not been proved that the soil is everywhere stable. There might be some place, outside the plastic zone defined by the solution, where the yield condition was exceeded. In practice, however, this objection is more apparent than real. In most soil structures it is intuitively obvious that a stable stress field could be found outside the plastic zone, although it might be tedious to do so. It can usually be accepted therefore that the solution is a lower bound.

The principal limitation of the method stems from the fact that it takes no account of the flow rule. It can therefore only deal with statically determinate problems, in which the boundary conditions are stated in terms of stress. In all such cases, the solutions are kinematically admissible, and can be accepted as being exact. Where, however, there are restrictions on the boundary displacements (as in the case, for example, of a rough rigid foundation) assumptions may have to be made about the distribution of stress on the boundary before a solution is possible. The solution may prove to be very sensitive to these assumptions.

8.22   *Failure in a semi-infinite soil mass with a horizontal surface:*   We will now consider two simple cases for which the slip-line fields may be obtained directly from the failure condition. These solutions were first obtained by Rankine [8.9] for cohesionless soil. The extension to cohesionless soil is generally attributed to Bell [8.10] although apparently first published by Résal [8.11].

Consider an element (Fig. 8.15(a)) in a semi-infinite mass of uniform soil, having a horizontal surface. Let x and z be measured horizontally and vertically respectively. From considerations of symmetry, it is clear that there is no shear stress on any horizontal or vertical surface, so that σ_x and σ_z are principal stress components and τ_{xz} is zero. When the soil is everywhere in a state of failure,

$$\sigma_3 = \sigma_1 \left(\frac{1 - \sin\varphi}{1 + \sin\varphi}\right) - 2c \left(\frac{1 - \sin\varphi}{1 + \sin\varphi}\right)^{\frac{1}{2}}$$

Thus there are two possible failure states, corresponding to the two Mohr stress circles shown in Fig. 8.15(b).

(a)   If $\sigma_x = \sigma_3 < \sigma_1 = \sigma_z$, failure occurs when

$$(\sigma_x)_{min} = \sigma_z \left(\frac{1 - \sin\varphi}{1 + \sin\varphi}\right) - 2c \left(\frac{1 - \sin\varphi}{1 + \sin\varphi}\right)^{\frac{1}{2}}$$

This is called an *active pressure failure*.

(b)   If $\sigma_x = \sigma_1 > \sigma_3 = \sigma_z$, failure occurs when

$$(\sigma_x)_{max} = \sigma_z \left(\frac{1 + \sin\varphi}{1 - \sin\varphi}\right) + 2c \left(\frac{1 + \sin\varphi}{1 - \sin\varphi}\right)^{\frac{1}{2}}$$

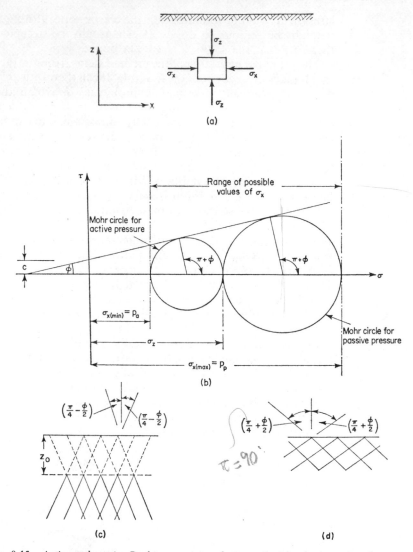

Fig. 8.15   *Active and passive Rankine states in cohesive soil with a horizontal surface.*

This is called a *passive pressure failure*. If the soil is not in a state of failure, σ_x may have any value between these limits.

8.23   *Active earth pressure (p_a):*   If the soil mass is allowed to expand horizontally, failure will occur when the lateral pressure is reduced to the minimum value

$$p_a = (\sigma_x)_{min} = \sigma_z K_a - 2c(K_a)^{\frac{1}{2}}$$

where

$$K_a = \left(\frac{1 - \sin\varphi}{1 + \sin\varphi}\right) = \tan^2(\pi/4 - \varphi/2)$$

(see Section 6.5(b)). K_a is called the *coefficient of active earth pressure*.

It was shown in Section 6.4 that the slip lines are inclined at $\pm(\pi/4 - \varphi/2)$ to the direction of σ_1. In this case, $\sigma_1 = \sigma_z$ and is vertical. Failure occurs on two families of slip lines inclined at $\pm(\pi/4 - \varphi/2)$ to the vertical, as shown in Fig. 8.15(c).

However, $(\sigma_x)_{min}$ is negative (that is, the soil is in horizontal tension) if $\sigma_z < 2c/(K_a)^{\frac{1}{2}}$. If the bulk density (γ) is constant with depth, and the soil surface is unloaded, $(\sigma_x)_{min}$ is zero at a depth z_0 such that

$$\sigma_z = \gamma z_0 = 2c/(K_a)^{\frac{1}{2}}$$

so that

$$z_0 = \frac{2c}{\gamma(K_a)^{\frac{1}{2}}}$$

If the soil surface carries a surcharge load q per unit area,

$$\sigma_z = \gamma z_0 + q = 2c/(K_a)^{\frac{1}{2}}$$

so that

$$z_0 = \frac{2c}{\gamma(K_a)^{\frac{1}{2}}} - \frac{q}{\gamma}$$

If the soil is incapable of sustaining tension, vertical cracks will develop to a depth z_0.

8.24 *Passive earth pressure* (p_p):   If the horizontal pressure in the soil mass is increased, failure will occur when

$$p_p = (\sigma_x)_{max} = \sigma_z K_p + 2c(K_p)^{\frac{1}{2}}$$

where

$$K_p = 1/K_a = \frac{1 + \sin\varphi}{1 - \sin\varphi} = \tan^2(\pi/4 + \phi/2)$$

(see Section 6.5(b)).

K_p is called the *coefficient of passive earth pressure*. In this case, $\sigma_1 = \sigma_x$ and is horizontal, so that the slip lines are inclined at $\pm(\pi/4 + \varphi/2)$ to the vertical (Fig. 8.15(d)). The passive earth pressure (p_p) is always positive, and no cracks develop.

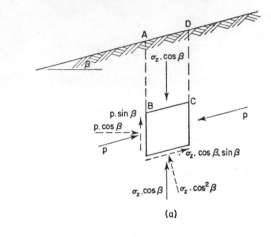

(a)

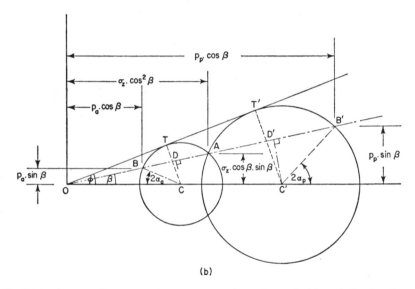

(b)

Fig. 8.16 Active and passive Rankine states in cohesionless soil with an inclined surface.

8.25 Failure in a cohesionless soil mass with an inclined surface:
Consider an element (Fig. 8.16(a)) in a semi-infinite mass of cohesionless soil with a surface inclined at β to the horizontal. Let the vertical and inclined surfaces of the element be of unit area.

Since, at any constant depth below the surface, all such elements are similar, the forces on the vertical surfaces AB and CD are equal and parallel to the ground surface. Then the force on the inclined surface of the element is vertical and is equal to $\sigma_z \cos \beta$. This force may be resolved into two components

$$\sigma_z \cos^2 \beta \text{ normal to } BC$$

and

$$\sigma_z \cos \beta \sin \beta \text{ parallel to } BC.$$

Similarly, the stresses (p) on the vertical surfaces of the element may be resolved into,

$$p \cos \beta \text{ horizontally,}$$
$$p \sin \beta \text{ vertically.}$$

The relationship between the lateral stress (p) and the vertical stress (σ_z) is most easily derived from the Mohr circle diagram shown in Fig. 8.16(b). In this diagram, the points representing the stresses on the vertical and inclined surfaces of the element lie on a line inclined at β to the axis of normal stress.

For the active pressure case:

$$\frac{p_a \cos \beta}{\sigma_z \cos^2 \beta} = \frac{OB}{OA} = \frac{OD - DB}{OD + DA}$$

But

$$OD = OC \cos \beta$$

Also

$$
\begin{aligned}
DB = DA &= [BC^2 - DC^2]^{\frac{1}{2}} \\
&= [TC^2 - DC^2]^{\frac{1}{2}} \\
&= OC (\sin^2 \varphi - \sin^2 \beta)^{\frac{1}{2}} \\
&= OC (\cos^2 \beta - \cos^2 \varphi)^{\frac{1}{2}}
\end{aligned}
$$

Therefore

$$p_a = \sigma_z \cos \beta \frac{\cos \beta - (\cos^2 \beta - \cos^2 \varphi)^{\frac{1}{2}}}{\cos \beta + (\cos^2 \beta - \cos^2 \varphi)^{\frac{1}{2}}}$$

$$= \sigma_z K_a$$

The rupture surfaces are planes inclined at $(\pi/4 - \varphi/2)$ to the direction of the major principal stress (σ_1), which is inclined at an angle α_a to the vertical.

Similarly, for the passive pressure case it may be shown that

$$p_p = \sigma_z \cos \beta \frac{\cos \beta + (\cos^2 \beta - \cos^2 \varphi)^{\frac{1}{2}}}{\cos \beta - (\cos^2 \beta - \cos^2 \varphi)^{\frac{1}{2}}}$$

The major principal stress is inclined at α_p to the horizontal.

If β is equal to φ, the active and passive states coincide, and the soil is always in limiting equilibrium. The major principal stress is inclined at $(\pi/4 - \varphi/2)$ to the ground surface, so that one family of rupture surfaces is parallel to the slope. The slope angle (β) cannot exceed φ. The *angle of repose* of the soil, which is the slope angle (β) at which the soil will stand when tipped, is equal to the angle of shearing resistance of the soil in its loosest state.

A similar analysis might be made for a *cohesive* soil mass with an inclined surface. In this case, however, the inclination of the principal stress component σ_1 varies with σ_z and therefore with depth. The slip lines are curved, and no simple and convenient expressions for lateral pressure are obtainable.

8.26 *Validity of the Rankine-Bell solutions:*   Each of the solutions given above provides a statically admissible stress field *everywhere* within the soil mass. It is therefore a valid lower bound solution. Since no restriction has been placed on displacements at the ground surface (the only boundary), the solutions are also kinematically admissible, and are therefore upper bounds. Since they are both upper and lower bounds, the solutions must be exact.

8.27 *Limiting equilibrium methods:*   By far the largest number of solutions have been obtained using the methods of *limiting equilibrium*. The procedure is as follows:

(a)   Collapse is assumed to occur as a result of sliding on a rupture surface within the soil mass. The failure condition is assumed to be satisfied on this surface.

(b)   The form of the rupture surface is assumed to be known, and a number of surfaces of the chosen form are examined.

(c)   For each such surface, we compute the collapse load which will just cause the soil mass to slide.

(d)   A systematic search is made to find the particular surface for which the computed collapse load is least.

8.28 *Validity of limiting equilibrium solutions:*   Used with care, the method can give reliable predictions of collapse load. The following points should, however, be constantly remembered:

(a)   Nothing is said about the stresses anywhere except on the rupture surface. Since the stress field is not defined, it is not known that the solution is a lower bound. Also, the boundary stresses are generally unknown. Although the external forces (forming part of the collapse load system) may be computed, the distribution of these forces over the boundary is generally indeterminate.

(b) The method has some similarities with the upper bound method (Section 8.18). Unless, however, it can be positively proved that the chosen rupture surface implies a kinematically admissible velocity field, it is not known that the solution is an upper bound.

(c) Since the method does not invoke the flow rule, it can only be used to solve statically determinate problems.

(d) The computed collapse load is generally not very sensitive to small changes in the chosen *form* of the rupture surface. Provided, therefore, that the form of the rupture surface is chosen with reasonable care, the error in the computed collapse load should be small. If, however, the assumed surface differs greatly from the true form, the error may be considerable. Although it is not definitely known that the computed value is an upper bound, the error will usually be found to be on the unsafe side —that is, the computed collapse load will be too large.

8.29   An illustrative example: As an illustration of the methods described above, we will consider the problem of determining the critical height of an unsupported vertical bank. Consider a bank of height H_c on the point of collapse, with a crack running down to a depth nH_c within the bank, as shown in Fig. 8.17.

(a)   *A limiting equilibrium solution.* Consider first the

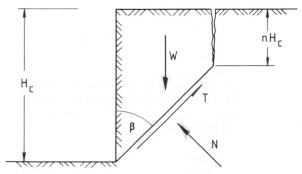

Fig. 8.17   The critical height of a vertical bank. A limiting equilibrium solution.

condition of limiting equilibrium on the plane rupture surface of length l shown in Fig. 8.17

$$W = \tfrac{1}{2}\gamma H_c^2 \tan \beta (1 - n^2)$$
$$W \sin \beta = N$$
$$W \cos \beta = T = cl + N \tan \varphi$$
$$= c\frac{H_c}{\cos \beta}(1 - n) + W \sin \beta \tan \varphi$$

Then

$$W(\cos \beta - \sin \beta \tan \varphi) = c\frac{H_c}{\cos \beta}(1 - n)$$

$$\tfrac{1}{2}\gamma H_c^2 \tan \beta (1 - n^2)(\cos \beta - \sin \beta \tan \varphi)$$
$$= c\frac{H_c}{\cos \beta}(1 - n)$$

$$H_c = \frac{2c}{\gamma(1 + n)} \times \frac{1}{\sin \beta (\cos \beta - \sin \beta \tan \varphi)}$$

$$= \frac{2c}{\gamma(1 + n)} \times \frac{\cos \varphi}{\sin \beta (\cos \beta \cos \varphi - \sin \beta \sin \varphi)}$$

$$= \frac{2c}{\gamma(1 + n)} \times \frac{\cos \varphi}{\sin \beta \cos (\beta + \varphi)}$$

It may be shown that H_c has a minimum value when

$$\beta = \pi/4 - \varphi/2$$

and

$$H_c = \frac{4c}{\gamma(1 + n)} \tan (\pi/4 + \varphi/2)$$

Since the plane rupture surface chosen in this case is kinematically admissible, and since there are no restrictions on movement on any other boundary, the solution is an upper bound.

(b) *A solution by the upper bound method.* The result obtained above may be confirmed by equating the external work done to the internal energy dissipation. Let the wedge move with velocity v. Since the motion of the sliding wedge is inclined at an angle φ to the sliding surface, the vertical velocity is $v \cos (\beta + \varphi)$. Then, the rate of work done by the gravitational force is

$$\tfrac{1}{2}\gamma H_c^2 \tan \beta (1 - n^2)v \cos (\beta + \varphi)$$

and the rate of dissipation of energy is

$$clv\cos\varphi = c\frac{H_c}{\cos\beta}(1-n)v\cos\varphi$$

Hence

$$\tfrac{1}{2}\gamma H_c\tan\beta(1-n^2)v\cos(\beta+\varphi)$$

$$= c\frac{H_c}{\cos\beta}(1-n)v\cos\varphi$$

$$H_c = \frac{2c}{\gamma(1+n)} \times \frac{\cos\varphi}{\sin\beta\cos(\beta+\varphi)}$$

for which the minimum value was shown above to be

$$H_c = \frac{4c}{\gamma(1+n)}\tan(\pi/4+\varphi/2)$$

(c) *A lower bound solution.* Now consider the stress field shown in Fig. 8.18. If this is not to violate the failure condition, the Mohr diagram (Fig. 8.18(b)) shows that, at the base of Zone 1,

$$\frac{\sigma_z}{2} = \frac{\gamma H_c}{2} = CE = TC = AC\sin\varphi$$

$$= \left(c\cot\varphi + \frac{\gamma H_c}{2}\right)\sin\varphi$$

Then

$$H_c(1-\sin\varphi) = \frac{2c}{\gamma}\cos\varphi$$

and

$$H_c = \frac{2c}{\gamma}\tan(\pi/4+\varphi/2)$$

(see Section 6.5 above).

We will now consider two particular cases.

(i) *Immediately after excavation,* while the bank is nearly undrained, we will assume that φ_u is zero and that the soil is temporarily capable of sustaining tension, as a result of negative pore pressures, so that n is zero.

The limiting equilibrium method and the upper

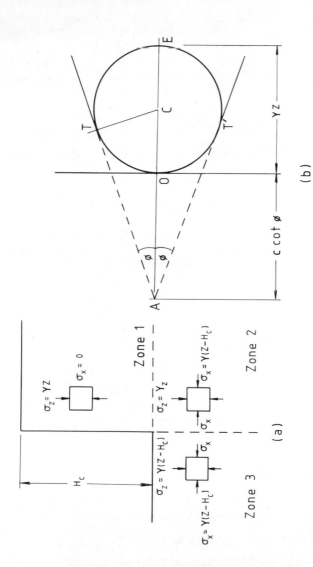

Fig. 8.18 The critical height of a vertical bank—a lower bound solution. (a) The chosen stress field. (b) The Mohr stress circle at failure at the base of Zone 1.

bound method both give $H_c = 4c_u/\gamma$, and this is an upper bound. Fellenius [8.12], however, using a circular arc rupture surface and the limiting equilibrium method, obtained the solution $H_c = 3\cdot83c_u/\gamma$. Since the circular arc is kinematically admissible if φ_u is zero, this is also an upper bound and is an improvement on the previous value. The same result may also be obtained by the upper bound method [Chen, 8.3]. The lower bound solution gives $H_c = 2c_u/\gamma$ so that we have

$$3\cdot83c_u/\gamma \geqslant H_c \geqslant 2c_u/\gamma$$

(ii) *In the long term*, after full drainage, we will assume that the pore pressures are everywhere zero, and that the soil is incapable of taking tension, so that cracks will develop within the bank. In this case calculations must be made in terms of effective stress.

In the stress field shown in Fig. 8.18(a), σ_x is zero within the bank, so that the field is still admissible after cracking, and $(2c'/\gamma) \tan(\pi/4 + \varphi'/2)$ is a lower bound for H_c.

The limiting equilibrium method (Section (a) above) yields

$$H_c = \frac{4c'}{\gamma(1 + n)} \tan(\pi/4 + \varphi'/2)$$

and this has a minimum value (as n tends to $1\cdot0$) of

$$H_c = \frac{2c'}{\gamma} \tan(\pi/4 + \varphi'/2)$$

This is an upper bound and, being identical with the lower bound, must be the exact solution. Sokolovskii [8.8], using the slip-line field method, also obtained this value as a closed solution.

REFERENCES

8.1   VON MISES, R. 1938. Mechanik der plastischen Formänderung von Kristallen. *Z. Angew. Math. Mech.*, **8**.

8.2   DRUCKER, D. C. and PRAGER, W. 1952. Soil mechanics and plastic analysis or limit design. *Q. Jour. Appl. Math.*, **10**.

8.3   CHEN, W-F. 1975. *Limit analysis and soil plasticity.* Elsevier (Amsterdam).

8.4   DRUCKER, D. C. 1959. A definition of stable inelastic material. *Trans. ASCE*, **81**.

8.5   KÖTTER, F. 1888. Über das Problem der erddruckbestimmung. *Verh. Phys. Ges. Berlin*, **7**.

8.6   JÁKY, J. 1936. The stability of earth slopes. *Proc. 1st Int. Conf. Soil Mech.*, **2**.

8.7   PRANDTL, L. 1920. Über die Härte plastischer Körper. *Nachr. Kgl. Ges. Wiss. Göttingen, Math.-Phys. Kl.*

8.8   SOKOLOVSKII, V. V. 1965. *Statics of granular media.* Pergamon (London).

8.9   RANKINE, W. M. J. 1857. On the stability of loose earth. *Phil. Trans. Roy. Soc. London*, **147**.

8.10   BELL, A. L. 1915. Lateral pressure and resistance of clay and the supporting power of clay foundations. *Proc. Inst. Civil Eng.*, **199**.

8.11   RÉSAL, J. 1910. *La poussée des terres.* Béranger (Paris).

8.12   FELLENIUS, W. 1927. *Erdstatische berechnungen.* W. Ernst und Sohn (Berlin).

CHAPTER 9

Lateral earth pressure

LATERAL PRESSURE IN A SOIL MASS

9.1 Active and passive earth pressure, and earth pressure at rest: In Section 8.22, we showed that, for material stability, the lateral pressure in the soil may have any value between the maximum or passive pressure p_p and the minimum or active pressure p_a, the actual value depending on the lateral strains. The condition of zero lateral strain is of particular interest, since it closely approximates to the state of strata of large lateral extent, consolidating under a uniform increase of overburden. The *coefficient of earth pressure at rest* (K_0) is defined as the ratio σ'_x/σ'_z for this condition of no lateral strain.

The variation of K_0 with changes of the vertical effective stress may be illustrated by plotting σ'_z against σ'_x, as shown in Fig. 9.1. If a point A in the figure represents the state of stress at some point in the ground, then K_0 is 1/(the slope of OA).

For a normally consolidated soil, K_0 is often found to agree closely with the empirical value $1 - \sin \varphi'$. During deposition of the soil, therefore, the stresses will follow the path OA in the figure. If the vertical effective stress is reduced by erosion, the horizontal effective stress is not reduced in proportion, and K_0 increases until it approaches the value of K_p (AB in Fig. 9.1). If the vertical effective stress is then increased, there may again be little change in σ'_x, and K_0 may be considerably reduced. It is clear therefore that the value of K_0 is critically dependent on the stress history of the soil, and is unlikely to have a constant value throughout any large soil mass. An estimate of the value of K_0 in fine-grained soils may be made by measuring the swelling pressures or by other means. The measured values are, however, very sensitive to disturbance of the soil during sampling and specimen preparation, and great care is needed if reliable results are to be obtained.

9.2 Analysis of lateral earth pressure on a retaining wall: Consider the section through a simple retaining wall shown in Fig. 9.2. Unless the soil masses both in front of, and behind, this wall are on the point of collapse, the wall is stable and the thrusts

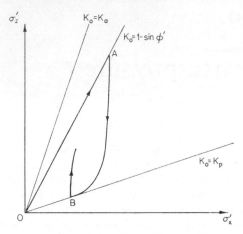

Fig. 9.1   Earth pressure at rest—variation of K_0 with stress history.

exerted by the soil on the vertical faces are indeterminate. However, forward displacement of the wall (towards the left), or anti-clockwise rotation about a point near the base, allows horizontal strains to develop in the soil, reducing the pressure behind the wall and increasing that in front. A relatively small (but finite) movement or rotation of the wall is sufficient to reduce the lateral pressure of the soil behind it to the minimum or active value. A similar (but perhaps rather larger) movement will increase the resistance of the soil in front of the wall to the maximum or passive value. If under these conditions, there is a resultant forward force or overturning moment on the wall, large uncontrolled movements will occur. In examining the stability of the wall, therefore, these two limiting states of minimum or active thrust and maximum or passive resistance must be examined.

The classical methods for determining these limiting values of lateral thrusts on retaining walls are those due to Rankine [9.1] and Coulomb [9.2]. The former uses an

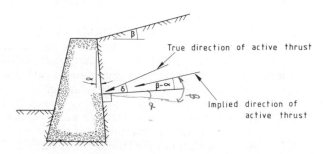

Fig. 9.2   Lateral earth pressure on a retaining wall—implications of Rankine's analysis.

adaptation of the analysis given in Sections 8.22 to 8.26. The latter is a limiting equilibrium method (Section 8.27) using plane rupture surfaces.

In this chapter, the methods will be developed first in terms of effective stress. Some cases will then be examined in which a total stress analysis may be used. It will be convenient to consider Rankine's method first.

EARTH PRESSURE ON RETAINING WALLS BY RANKINE'S METHOD

9.3   *Lateral earth pressure of cohesionless soil:*   Rankine [9.1] suggested that the replacement of part of a semi-infinite soil mass by a rigid retaining wall would not significantly alter the stress field in the remaining soil. Then, as the wall moves or tilts forward, the lateral effective pressures are $p'_a = K_a \sigma'_z$ on the back of the wall and $p'_p = K_p \sigma'_z$ on the front, where K_a and K_p are the coefficients of active and passive earth pressure defined in Sections 8.23 and 8.24 above. There is an approximation implicit in this assumption which we must now examine. Consider the wall supporting a cohesionless soil shown in Fig. 9.2. The derivation of the expression for K_a (Section 8.25) requires that the direction of the active pressure p'_a be parallel to the ground surface. As the wall moves from right to left, the soil mass behind it expands and moves downwards. Unless the wall can also move downwards, slip must occur between the soil and the wall. The true direction of the thrust is therefore determined by the angle of friction (δ) between the soil and the wall. If $|\beta - \alpha| < \delta$, Rankine's method implies a shear force at the back of the wall which is less than the available frictional resistance. Rankine's stress field is therefore still statically admissible. However, if the limiting friction at the back of the wall is not developed, there can be no slip, and this is incompatible with the required relative movement of the soil and the wall. The solution is therefore kinematically inadmissible, and provides a lower bound only. A lower bound solution gives a lower limit to the collapse load, and therefore underestimates the safety of the structure under any given system of loads. It follows, therefore, that the computed value of the active pressure is either correct or too large. Similarly, the computed value of the passive pressure is correct or too small.

Where $|\beta - \alpha| > \delta$, the method implies a shear force on the back of the wall which cannot be developed, and the method provides an upper bound solution* (that is, the computed active pressure may be too small and the passive pressure too large).

For the particular case of a vertical wall and a horizontal ground surface ($\alpha = \beta = 0$), the method can only err on the safe side. It provides an exact solution if the shearing resistance between the soil and the wall is to be ignored.

In practice, this latter assumption is not always unjustified. In the case of a sheet-pile wall, the resistance to penetration of the bottom of the wall into the ground may be insufficient to resist the downward shearing force on the back, so that the wall moves down with the sliding soil. Where the wall is used as the abutment of a bridge, traffic vibration may make it doubtful whether the friction between the soil and the wall can be sustained. In many cases, also, all that is needed is a safe estimate of the lateral pressure. In all such cases, Rankine's method may be justifiably used.

9.4 *Applications of Rankine's method:* The applications of Rankine's method can best be seen in a series of examples. First, consider the simplest possible case—a wall supporting a uniform cohesionless soil with horizontal surface (Fig. 9.3(a)). For the moment, the pore pressure (u) will be assumed to be zero everywhere.

Example 9.1

If $\varphi' = 30°$, $K_a = (1 - \sin 30°)/(1 + \sin 30°) = 0.333$. At any depth z below the surface,

$$p'_a = K_a\sigma'_z = K_a\gamma z$$

Thus, the effective active pressure (p'_a) increases linearly with depth, and may be represented by the triangular pressure diagram ABC. Then, per metre run of wall, the total thrust on the back of the wall is

$$P_a = \int_0^H p'_a dz = \tfrac{1}{2}K_a\gamma H^2$$
$$= 0.5 \times 0.333 \times 18 \times 10^2 = 300 \, kN/m \, run$$

This is equal to the area of the pressure diagram ABC. The line of action of the force P_a is at the centroid of the pressure diagram, which is 3.33 m above the base. Now consider a uniform surcharge load, q per unit area, applied over the whole of the soil surface. The vertical effective stress (σ'_z) is

---

* Although this may appear intuitively obvious, we are not entitled, in this case, to invoke the limit theorems for proof that this is an upper bound. The reason for this is explained in Section 9.18.

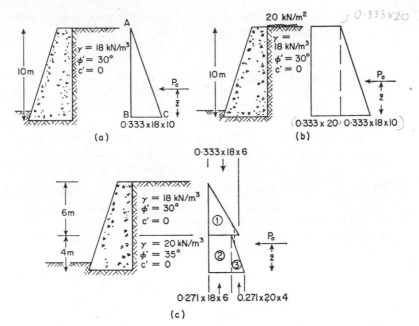

Fig. 9.3   Rankine's solution applied to the calculation of earth pressures on a retaining wall.

everywhere increased by q, and the effective active pressure (p'_a) is therefore increased by $K_a q$.

Example 9.2
The pressure diagram is shown in Fig. 9.3(b). Then

$$P_a = 0.333 \times 20 \times 10 + 0.5 \times 0.333 \times 18 \times 10^2$$
$$= 66.7 + 300$$
$$= 366.7 \text{ kN/m run}$$

By moments about the base of the wall,

$$\bar{z} = (66.7 \times 5 + 300 \times 3.33)/366.7 = 3.6 \text{ m}$$

Where the soil is stratified (Fig. 9.3(c)), and where the value of φ' is different in the two materials, a sharp break occurs in the pressure diagram at the interface between the two soils. Immediately above the interface $p'_a = K_{a1}\sigma'_z$, while immediately below this point $p'_a = K_{a2}\sigma'_z$. It may reasonably be argued that such a pressure distribution could never really develop, and that the shear forces at the interface (which have been implicitly ignored) would result in a modified distribution similar to that indicated by the broken line in Fig. 9.3(c). While this objection is perfectly valid, the effect on the computed values of P_a and $\bar{z}$ is negligible and is generally ignored.

Example 9.3

In computing the values of P_a and $\bar{z}$, it is convenient to divide the pressure diagram into a number of triangular and rectangular areas, as shown in Fig. 9.3(c).

$$K_{a1} = 0\cdot333; \quad K_{a2} = (1 - \sin 35°)/(1 + \sin 35°) = 0\cdot271$$

Area	P_a (kN/m run)	Moment arm about the base (m)	Moment about the base (kN-m/m run)
1	$0\cdot5 \times 0\cdot333 \times 18 \times 6^2 = 108$	6	648
2	$0\cdot271 \times 18 \times 6 \times 4 = 118$	2	236
3	$0\cdot5 \times 0\cdot271 \times 20 \times 4^2 = 43$	1·33	58
	269		942

$P_a = 269$ kN/m; $\bar{z} = 942/269 = 3\cdot5$ m.

9.5 *Application of Rankine's method where the pore pressure (u) is not zero:* Where the pore pressure is not zero, the effective stress and pore pressure resultants must be computed separately. The derivation of Rankine's method requires that the stresses are constant at any given depth below the ground surface, which implies that the water table is parallel to the ground surface. Where the soil surface is horizontal the method can therefore be used only for hydrostatic conditions. The pore pressure is then defined by the position of the water table, at which $u = 0$. Above the water table, water may be held in the pores by capillary forces, and the pore pressure in this capillary zone is negative. This negative pore pressure increases the effective stress, and consequently increases the shear strength also. However, this increase in strength is unreliable, for two reasons. Firstly, the maximum height of the capillary zone depends on the capillary suction. This is a function of the pore size, and is generally unknown. Secondly, downward seepage of surface water towards the water table may increase the pore pressure in the capillary zone above the value computed for the strictly hydrostatic condition. It is therefore usual to take a conservative view, and to assume that the pore pressure is zero at all points above the water table. Below the water table, the body force must be taken to be $(\gamma_{sat} - \gamma_w)$ (see Section 8.6) when calculating the effective stress.

Example 9.4 (Fig. 9.4)

$$K_a = 0.33$$

Area	P_a (kN/m run)	Moment arm about the base (m)	Moment about the base (kN-m/m run)
1	$0.5 \times 0.333 \times 18 \times 6^2 = 108$	6	648
2	$0.333 \times 18 \times 6 \times 4 \quad = 144$	2	288
3	$0.5 \times 0.333 \times 9 \times 4^2 \quad = 24$	1.33	32
W	$0.5 \times 9.18 \times 4^2 \qquad = 79$	1.33	105
	355		1 073

$P_a = 355$ kN/m; $\bar{z} = 1\,073/355 = 3.0$ m.

9.6 Passive pressure of cohesionless soils: This is calculated in exactly the same way as the active pressure.

Example 9.5 (Fig. 9.5)

$$K_p = (1 + \sin 30°)/(1 - \sin 30°) = 3.0$$

Area	P_p (kN/m run)	Moment arm about the base (m)	Moment about the base (kN-m/m run)
1	$3.0 \times 20 \times 2 \qquad = 120$	1	120
2	$0.5 \times 3.0 \times 18 \times 1^2 = 27$	1.33	36
3	$3.0 \times 18 \times 1 \times 1 \quad = 54$	0.5	27
4	$0.5 \times 3.0 \times 9 \times 1^2 = 13$	0.33	4
W	$0.5 \times 9.81 \times 1^2 \qquad = 5$	0.33	2
	219		189

$P_a = 219$ kN/m; $\bar{z} = 189/219 = 0.86$ m.

9.7 Rankine's method extended to cohesive soils: In Section 8.23 it was shown that, for a soil mass with a horizontal surface and apparent cohesion c',

$$p'_a = K_a\sigma'_z - 2c'(K_a)^{\frac{1}{2}}$$
$$p'_p = K_p\sigma'_z + 2c'(K_p)^{\frac{1}{2}}$$

The effective active pressure (p'_a) is negative if

$$\sigma'_z < 2c'/(K_a)^{\frac{1}{2}}$$

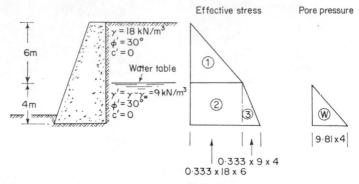

Fig. 9.4   Example 9.4.

For a soil having a density of γ, supporting a uniform surcharge load q per unit area of the surface, p'_a is zero at a depth z_0 where

$$z_0 = \frac{2c'}{\gamma(K_a)^{\frac{1}{2}}} - \frac{q}{\gamma}$$

Since the soil cannot sustain tension indefinitely, it is usual to assume that cracks develop to a depth z_0, and to ignore the negative part of the pressure diagram.

Cohesive soils commonly have low permeabilities. If surface water can drain into a tension crack, it may be some time before it drains away, and the crack may be completely filled. This water therefore exerts an additional thrust on the wall, although the pore pressures within the soil are not immediately affected.

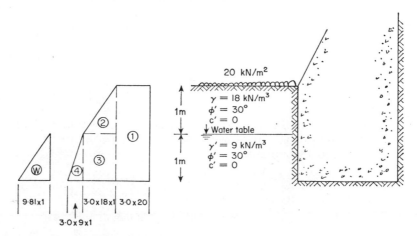

Fig. 9.5   Example 9.5.

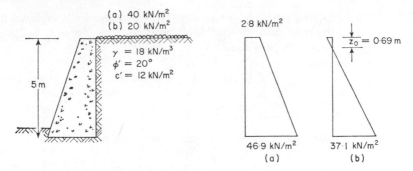

Fig. 9.6 Example 9.6.

Example 9.6 (Fig. 9.6)

$$K_a = (1 - \sin 20°)/(1 + \sin 20°) = 0.49; (K_a)^{\frac{1}{2}} = 0.70$$

(a) If $q = 40\ kN/m^2$
At the top of the wall

$$p'_a = 40 \times 0.49 - 2 \times 12 \times 0.70 = 2.8\ \text{kN/m}^2$$

At the bottom of the wall

$$p'_a = 2.8 + 5 \times 18 \times 0.49 = 46.9\ \text{kN/m}^2$$

Thus there is no negative pressure, and no cracking.

$$P_a = 5 \times 0.5(2.8 + 46.9) = 124\ \text{kN/m run}$$

(b) If $q = 20\ kN/m^2$,
At the top of the wall

$$p'_a = 20 \times 0.49 - 2 \times 12 \times 0.7 = -7.0\ \text{kN/m}^2$$

Since this is negative, cracks must be allowed for to a depth z_0 where

$$z_0 = \frac{2 \times 12}{18 \times 0.7} - \frac{20}{18} = 0.79\ \text{m}$$

At the base of the wall

$$p'_{a.} = (5 - 0.79) \times 18 \times 0.49 = 37.1\ \text{kN/m}^2$$

Allowing for water in the tension cracks, the total thrust is

$$P_a + P_w = 37.1 \times 0.5 \times (5 - 0.79) + 0.5 \times 9.81 \times 0.79^2$$
$$= 78.0 + 3.06$$
$$= 81.1\ \text{kN/m run}$$

9.8 *Passive resistance of cohesive soils:* This may also be calculated by the extension of Rankine's method. For uniform soil and zero pore pressure,

$$P_p = \int_0^H p'_p \, dz = \int_0^H [K_p(\gamma z + q) + 2c'(K_p)^{\frac{1}{2}}].dz$$

$$= \tfrac{1}{2}K_p\gamma H^2 + [qK_p + 2c'(K_p)^{\frac{1}{2}}]H$$

The passive pressure is positive at all depths.

LATERAL PRESSURE ON RETAINING WALLS—COULOMB'S ANALYSIS

9.9 *Coulomb's method:* Coulomb determined the thrust on the wall by considering the stability of the wedge of soil bounded by the wall, the ground surface, and some rupture surface. (Fig. 9.7). If the form and position of this rupture surface is known, the thrust on the wall may be determined by considering the equilibrium of the sliding soil wedge. This is an example of a limiting equilibrium method, the validity of which was discussed in Section 8.28 above.

Although Coulomb realised that the back of the sliding soil wedge may be curved, he applied his method to plane rupture surfaces only. This greatly simplifies the analysis, but affects the accuracy of the solution. The magnitude of the errors which results from this are discussed in Section 9.18 below.

9.10 *Analytical and graphical methods:* Coulomb developed an analytical solution for a special case. An extension of this solution

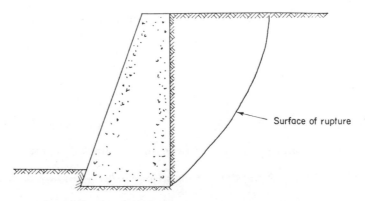

Fig. 9.7 Coulomb's solution for lateral earth pressure on a retaining wall.

will be considered later (Section 9.15), but in many practical problems (particularly those with complicated geometry) a graphical method is simpler.

The position of the critical rupture surface is not known. In the graphical method, force diagrams are drawn for a number of possible rupture surfaces, and a systematic search is made for that surface which gives the greatest value of P_a or the least value of P_p.

9.11   *Active thrust of cohesionless soil:*   Consider a simple case of a wall with a flat back supporting a bank of cohesionless soil. The back of the wall need not be vertical, and the ground surface may have any profile. Fig. 9.8 shows one potential rupture surface, and the force diagram for the corresponding wedge of soil. The reaction (R) which is the resultant of the normal and frictional stresses on the rupture surface, is inclined at φ' to the normal to that surface. The active thrust (P_a) is inclined at an angle δ to the normal to the back

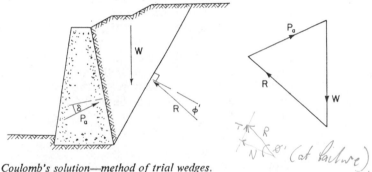

Fig. 9.8   Coulomb's solution—method of trial wedges.

of the wall, where δ is the angle of friction between the soil and the wall. Unlike Rankine's approach, Coulomb's method implies nothing about the angle δ, which may be freely chosen to suit the nature of the materials. The angle δ is commonly found to lie between $\frac{1}{3}\varphi'$ and $\frac{2}{3}\varphi'$.

The critical wedge is most easily located by superimposing the force diagrams for a number of wedges, as shown in Fig. 9.9. The maximum value of P_a may then be determined by drawing the locus of all points A_1, A_2, A_3, etc.

Surcharge loads on the soil surface are easily dealt with. Any such load falling within the limits of a wedge must be added to the weight of that wedge, while any loads beyond these limits are ignored. The loads need not be uniformly

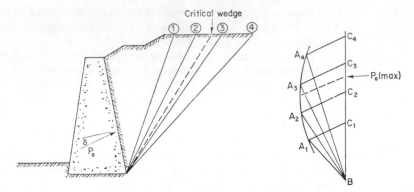

Fig. 9.9   Method of trial wedges—determination of the critical wedge.

distributed. Dockside cranes or foundation beams may impose linear loads parallel to the wall. Where such a load occurs, there are two almost identical wedges, one just including and one just excluding the load. If the force diagrams for these two wedges are drawn, it may be seen that there is a sudden break in the locus of A at this point. The effect of the line load on the active thrust depends on the proximity of the load to the wall. If the load is close in (*e.g.* Q_1 in Fig. 9.10), the critical wedge is likely to be that which just includes the load. If the load is further from the wall, the critical wedge may reach the ground surface at a point between the load and the wall (*e.g.* Q_2 in Fig. 9.10). This load does not affect the thrust on the wall.

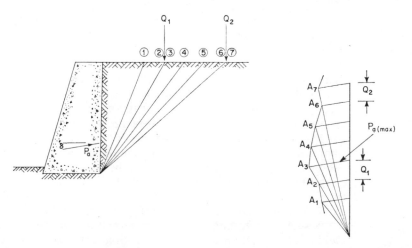

Fig. 9.10   Method of trial wedges—effect of a linear surcharge load parallel to the wall.

9.12   Effect of pore pressure: If the pore pressure is not everywhere zero, the effective stress and pore pressure resultants must be considered separately, as shown in Fig. 9.11. Where there is no flow, the vector sum of the two pore pressure resultants P_w and R_w is equal to the weight (W_w) of the volume of water which would just fill that part of the wedge below the water table. The length BC on the force diagram represents

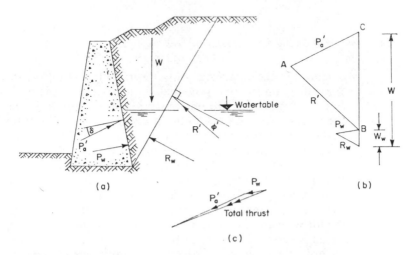

Fig. 9.11   *Method of trial wedges—effect of pore pressures on the lateral thrust.*

the weight of the wedge less the weight of this volume of water, and may be calculated by assuming the soil below the water table has an effective density of $(\gamma_{sat} - \gamma_w)$. The effective stress resultant (P'_a) may therefore be determined by drawing the triangle ABC for each of the wedges in the manner described in the preceding section. The pore pressure resultant (P_w) may be easily calculated, and the lower triangle of the force diagrams need not be drawn. The total thrust is the vector sum of P'_a and P_w. Since P_w is the same for each wedge, the maximum thrust results from the same wedge as the maximum value of P'_a.

The stability of a retaining wall is adversely affected by large pore pressures in the soil behind it. For this reason, a drain of coarse permeable material is, wherever possible, placed in the soil behind the wall. The purpose of this drain is to cause flow downwards and towards the wall, and so to reduce the pore pressures on the rupture surface. This increases the effective stress, and therefore the shearing resistance, on this surface (see Fig. 9.12).

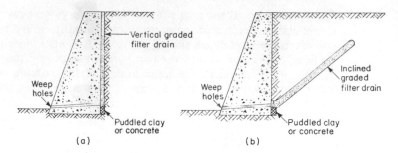

Fig. 9.12 Drainage of the soil behind a retaining wall.

Consider the retaining wall shown in Fig. 9.13(a). The force diagram for one potential sliding wedge is shown in Fig. 9.13(b). In order to demonstrate the effect of drainage, the forces have been taken in an order which differs from that in the previous example. If, due to a fault in the drainage system, the water table rises to the level shown, the force diagram is A_1ECBD_1, and the total thrust on the wall is CA_1. If the water is suddenly removed from the drain without changing the pore pressure on the potential rupture surface, P_w is reduced to zero but R_w is unchanged. Then the force diagram is A_2CBD_1, and the total thrust is CA_2, which is very little less than before. In time, however, flow towards the wall will reduce the pore pressure on the rupture surface. If R_w is reduced by half, the force diagram is A_3CBD_3, and the total thrust is reduced to CA_3, which is considerably less than before. It is clear, therefore, that it is the reduction in the pore pressure on the rupture surface which has the greatest effect in reducing the thrust on the wall. The drainage system should be designed with this end in view.

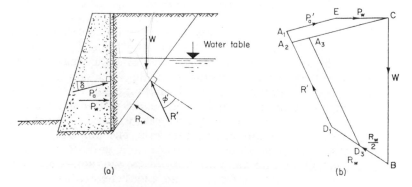

Fig. 9.13 Effect of drainage on the active thrust.

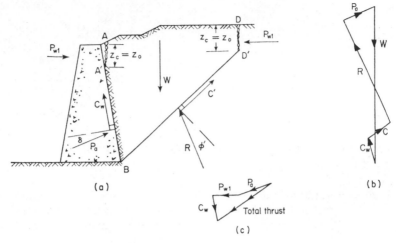

Fig. 9.14   Method of trial wedges—cohesive soils.

9.13   Coulomb's method applied to cohesive soils: If the apparent cohesion is not zero, there are additional forces (Fig. 9.14) acting on the soil wedge. These forces result from adhesion of the soil to the back of the wall, and cohesion on the rupture surface. In addition, tension cracks may occur in the upper zone of the soil, where the lateral pressure is negative. It is generally assumed that cracks develop to a depth z_0 at which p_a is zero. Then

$$z_c = z_0 = \frac{2c'}{\gamma(K_a)^{\frac{1}{2}}} - \frac{q}{\gamma}$$

It is assumed that there is no cohesion in the soil and no adhesion between the soil and the wall, above this level. In soils of low permeability, the cracks may fill with surface water. If this is possible, allowance must be made for the resulting force (P_{w1}) on the wall.

The cohesive force C is equal to the product of the apparent cohesion (c') and the length of the rupture surface below the tension zone (BD' in Fig. 9.14). The adhesion (c_w per unit area) between the soil and the wall results in a force C_w equal to $c_w \times BA'$. The adhesion (c_w) depends on the soil and wall materials, and, in some cases, on the method of construction of the wall. It can never be greater than c'. The force diagram for the wedge is shown in Fig. 9.14. The total thrust on the wall is the vector sum of C_w, P_{w1} and the maximum value of P_a.

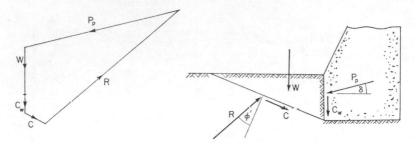

Fig. 9.15   Method of trial wedges—passive resistance of cohesive soil.

9.14   Passive resistance determined by Coulomb's method: Coulomb's method may be used in a similar manner, to determine the passive resistance (P_p). The critical wedge in this case is that for which P_p is a minimum. The force diagram for a typical wedge is shown in Fig. 9.15. The accuracy of Coulomb's method in this case is discussed in Section 9.18 below.

9.15   An analytical solution: Consider a vertical wall of height H, supporting a mass of cohesive soil with a horizontal surface (Fig. 9.16). The soil surface carries a surcharge load q per unit area. For the moment we will assume that no tension cracks develop. Consider a potential plane rupture surface BD defined by the angle α. Equating forces parallel to and normal to BD, acting on the wedge ABD,

$$T = W\cos\alpha - P_a\sin(\alpha + \delta) - c_wH\cos\alpha$$
$$N = W\sin\alpha + P_a\cos(\alpha + \delta) - c_wH\sin\alpha$$

Since the failure condition must be satisfied on BD,

$$T = c'H\sec\alpha + N\tan\varphi'$$

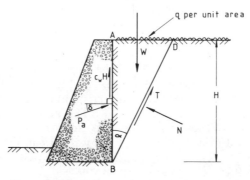

Fig. 9.16   Coulomb's method for active earth pressure—an analytical solution.

so that

$$c'H \sec \alpha + [W \sin \alpha + P_a \cos(\alpha + \delta) - c_w H \sin \delta] \tan \varphi'$$
$$= W \cos \alpha - P_a \sin(\alpha + \delta) - c_w H \cos \alpha$$

$$P_a[\sin(\alpha + \delta) + \cos(\alpha + \delta)\tan \varphi']$$
$$= W(\cos \alpha - \sin \alpha \tan \varphi')$$
$$- c'H \sec \alpha - c_w H(\cos \alpha - \sin \alpha \tan \varphi')$$

Writing

$$W = \tfrac{1}{2}\gamma H^2 \tan \alpha + qH \tan \alpha$$

and

$$a_n = c_w/c'$$

$$P_a[\sin(\alpha + \delta) + \cos(\alpha + \delta)\tan \varphi']$$
$$= (\tfrac{1}{2}\gamma H^2 \tan \alpha + qH \tan \alpha)(\cos \alpha - \sin \alpha \tan \varphi')$$
$$- c'H[\sec \alpha + a_n(\cos \alpha - \sin \alpha \tan \varphi')]$$

Then

$$P_a = \tfrac{1}{2}\gamma H^2 K_{a\gamma} + qH K_{aq} - c'H K_{ac}$$

where $K_{a\gamma}$, K_{aq} and K_{ac} are coefficients having the following values:

$$K_{a\gamma} = \frac{(\cos \alpha - \sin \alpha \tan \varphi')\tan \alpha}{\sin(\alpha + \delta) + \cos(\alpha + \delta)\tan \varphi'} = K_{aq}$$

$$K_{ac} = \frac{\sec \alpha + a_n(\cos \alpha - \sin \alpha \tan \varphi')}{\sin(\alpha + \delta) + \cos(\alpha + \delta)\tan \varphi'}$$

Values of $K_{a\gamma}$, K_{aq} and K_{ac} may also be derived for inclined walls, sloping ground surfaces, and curved rupture surfaces. Notice that in such cases, K_{aq} is not generally equal to $K_{a\gamma}$, although the difference is often small.

No allowance has been made for tension cracks in this analysis. If $q < 2c/(K_a)^{\frac{1}{2}}$, and the soil cannot sustain tension, cracks will develop to a depth $z_0 = 2c/\gamma(K_a)^{\frac{1}{2}} - q/\gamma$. The effective height (H) of the wall should be measured to the bottom of the crack, and the weight of soil above this level should be included in the surcharge load q.

9.16 *Superposition:* The expression for P_a has been stated in terms of the angle α which defines the position of BD. The required value of α is that which gives the greatest computed value of P_a.

$$P_a = (\tfrac{1}{2}\gamma H^2 K_a + qH K_{aq} - c'H K_{ac})_{\max}$$

It would be inconvenient to determine the critical value of α in every individual case. However, *approximately*,

$$P_a = (\tfrac{1}{2}\gamma H^2 K_{a\gamma})_{max} + (qHK_{aq})_{max} - (c'HK_{ac})_{min}$$

The critical values of $K_{a\gamma}$, K_{aq} and K_{ac} depend only on φ, a_n and δ and may be readily tabulated. The values of the three terms may then be computed separately and may be superposed to give P_a.

Superposition in this manner is not valid in an analysis of plastic behaviour. However, the error is always on the safe side (that is, the computed value of P_a is too large). Trial calculations show that it is usually insignificant.

9.17 *Passive resistance:*   We may similarly derive an expression for the passive force

$$P_p = \tfrac{1}{2}\gamma H^2 K_{p\gamma} + qHK_{pq} + c'HK_{pc}$$

For a vertical wall, horizontal ground surface and plane rupture surface,

$$K_{p\gamma} = K_{pq} = \frac{(\cos\alpha + \sin\alpha\tan\varphi')\tan\alpha}{\sin(\alpha - \delta) - \cos(\alpha - \delta)\tan\varphi'}$$

$$K_{pc} = \frac{\sec\alpha + a_n(\cos\alpha + \sin\alpha\tan\varphi')}{\sin(\alpha - \delta) - \cos(\alpha - \delta)\tan\varphi'}$$

9.18 *Comparison with other solutions:*   The solution given above is an extension of Coulomb's analysis [9.2]. It is an example of the limiting equilibrium method, the validity of which was discussed in Section 8.28. The computed values of P_a and P_p probably (but not certainly) represent upper bounds to the factor of safety (that is, P_a may be too small and P_p too large). Before using the method, therefore, we need to be certain that the error is not unacceptably large.

Table 9.1 shows a comparison of the values of $K_{a\gamma}$ and $K_{p\gamma}$ obtained by four methods. The values in Column 1 were computed from the expressions given in Sections 9.15 and 9.17. Sokolovskii [9.13] used a slip-line field method, and his results are generally accepted as being the most reliable. Caquot and Kerisel [9.4] used a method which directly integrates the equilibrium equations along curved rupture surfaces. Their results are nearly identical with those of Sokolovskii. Chen [9.14] used the upper bound method (Section 8.19) to analyse a number of rather complex collapse mechanisms. A special problem arises in this case. Collapse of the soil mass behind the wall implies sliding between the back of the wall and the soil wedge. Although the

TABLE 9.1 *Earth pressure coefficients $K_{a\gamma}$ and $K_{p\gamma}$ for a vertical wall and horizontal soil surface*

φ' δ (Degrees)		(1) Coulomb (Sections 9.15 and 9.17)	(2) Sokolovskii [9.13]	(3) Caquot and Kerisel [9.4]	(4) Chen [9.14]
$K_{a\gamma}$ 0	0	1·0	1·0	1·0	1·0
10	0	0·704	0·70	0·704	0·704
	5	0·662	0·66	—	0·664
	10	0·635	0·65	0·649	0·642
20	0	0·490	0·49	0·490	0·490
	10	0·447	0·45	—	0·448
	20	0·427	0·44	0·440	0·434
30	0	0·333	0·33	0·333	0·333
	15	0·301	0·30	—	0·302
	30	0·297	0·31	0·308	0·303
40	0	0·217	0·22	0·217	0·217
	20	0·199	0·20	—	0·200
	40	0·210	0·22	0·219	0·214
$K_{p\gamma}$ 0	0	1·0	1·0	1·0	1·0
10	0	1·42	1·42	1·42	1·42
	5	1·57	1·56	1·55	1·56
	10	1·73	1·66	1·64	1·68
20	0	2·04	2·04	2·04	2·04
	10	2·63	2·55	2·59	2·58
	20	3·52	3·04	3·01	3·17
30	0	3·00	3·00	3·00	3·00
	15	4·98	4·62	4·78	4·71
	30	10·10	6·55	6·42	7·10
40	0	4·60	4·60	4·59	4·60
	20	11·77	9·69	10·36	10·10
	40	—	18·20	17·50	20·90

shearing force on this surface can be determined, the rate of doing work cannot, since it depends on the relative velocity of the wall and the soil wedge. This in turn depends on the direction of motion of the wall, which is not known. This illustrates the general principal (Drucker [9.15], Heyman [9.16]) that *the limit theorems are not valid where there is friction on a boundary.* However, the theorems are valid where δ is zero, because the frictional work done is zero. The theorems are also valid if $\delta > \varphi$, since in this case slip occurs in a thin deforming layer just within the soil mass, and not at the boundary. We may therefore hope that the upper bound method will give useful answers for intermediate values of δ, even though we cannot prove that such answers are upper bounds.

The comparison of the values given in Table 9.1 shows the following:

(a) All methods give identical values of $K_{a\gamma}$ and $K_{p\gamma}$ where δ is zero. This is to be expected, since in this case all the methods reproduce the Rankine–Bell solution which we know to be exact (see Section 9.3).

(b) The curvature of the critical rupture surface increases with increasing values of δ. As a result, the Coulomb method under-estimates $K_{a\gamma}$ and over-estimates $K_{p\gamma}$ for values of δ greater than zero. For an active pressure failure, the curvature is always small, and the error in $K_{a\gamma}$ does not exceed 5%. For a passive pressure failure, however, the curvature increases sharply for large values of δ, and the error in $K_{p\gamma}$ may be 50% or more. The error in each case is on the unsafe side.

(c) Similar discrepancies occur in Chen's values, but the errors in this case are very much smaller.

From this and other comparisons, we find that the Coulomb method may generally be used to compute P_a with sufficient accuracy. It may also be used to compute P_p if δ does not exceed $\varphi'/3$. In practice, there are often good reasons for using small values of δ in predicting the passive resistance:

(a) Large horizontal movements may be required to develop the passive pressure forces implied by large values of δ. Such movements may be intolerable.

TABLE 9.2 *Earth pressure coefficients for the horizontal components of active and passive pressure in cohesionless soil (vertical wall and horizontal ground surface)*

	δ	25°	30°	φ' 35°	40°	45°
$K_{a\gamma(H)} = K_{aq(H)}$	0°	0·41	0·33	0·27	0·22	0·17
	10°	0·37	0·31	0·25	0·20	0·16
	20°	0·34	0·28	0·23	0·19	0·15
	30°	—	0·26	0·21	0·17	0·14
$K_{p\gamma(H)} = K_{pq(H)}$	0°	2·5	3·0	3·7	4·6	
	10°	3·1	4·0	4·8	6·5	
	20°	3·7	4·9	6·0	8·8	
	30°	—	5·8	7·3	11·4	

After CECP No. 2 [9.3].

TABLE 9.3 *Earth pressure coefficients for the horizontal components of active and passive earth pressures in cohesive soil (vertical wall and horizontal ground surface)*

	δ	$\dfrac{c_w}{c}$	φ'					
			0°	5°	10°	15°	20°	25°
$K_{a\gamma(H)} = K_{aq(H)}$	0	All	1·00	0·85	0·70	0·59	0·48	0·40
	φ	values	1·00	0·78	0·64	0·50	0·40	0·32
$K_{ac(H)}$	0	0	2·00	1·83	1·68	1·54	1·40	1·29
	0	1·0	2·83	2·60	2·38	2·16	1·96	1·76
	φ	0·5	2·45	2·10	1·82	1·55	1·32	1·15
	φ	1·0	2·83	2·47	2·13	1·85	1·59	1·41
$K_{p\gamma(H)} = K_{pq(H)}$	0	All	1·0	1·2	1·4	1·7	2·1	2·5
	φ	values	1·0	1·3	1·6	2·2	2·9	3·9
$K_{pc(H)}$	0	0	2·0	2·2	2·4	2·6	2·8	3·1
	0	0·5	2·4	2·6	2·9	3·2	3·5	3·8
	0	1·0	2·6	2·9	3·2	3·6	4·0	4·4
	φ	0·5	2·4	2·8	3·3	3·8	4·5	5·5
	φ	1·0	2·6	2·9	3·4	3·9	4·7	5·7

After CECP No. 2 [9.3].

(b) A large value of δ implies a large upward force on the face of the wall. If this upward force exceeds the available downward force (including the weight of the wall) it cannot be developed.

In many cases, therefore, the Coulomb method is sufficient. Where a value of δ greater than $\varphi'/3$ must be used, any of the other methods should provide an acceptable solution. For the particular case of a vertical wall and horizontal soil surface, the values of the *horizontal* components of the earth pressure coefficient given in Tables 9.2 and 9.3 may be used. The Coulomb method has been used to compute $K_{a\gamma(H)}$ and $K_{ac(H)}$. $K_{p\gamma(H)}$ and $K_{pc(H)}$ have been determined by the friction circle method (Section 10.9). This is a limiting equilibrium method, which uses a combination of plane and circular rupture surfaces.

9.19 *Distribution of pressure on the wall:* Limiting equilibrium methods only define the stress state on a single rupture surface, and the stress state at the back of the wall is therefore unknown. Although the forces on the back of the wall may be computed using the Coulomb method, their distribution over the wall surface is indeterminate. However, it seems reasonable to assume that the distribution implied by the Rankine–Bell solution (which we know to be correct in one case) can be generally applied without serious error. The components $qHK_{aq} - c'HK_{ac}$ (or $qHK_{pq} + c'HK_{pc}$) are assumed to be uniformly distributed over the surface of the wall. The pressure resulting from the component $\frac{1}{2}\gamma H^2 K_{a\gamma}$ (or $\frac{1}{2}\gamma H^2 K_{p\gamma}$) is assumed to increase linearly with depth.

Then the pressures at any depth z below the ground surface are

$$p'_a = \gamma z K_{a\gamma} + q K_{aq} - c' K_{ac}$$
$$p'_p = \gamma z K_{p\gamma} + q K_{pq} + c' K_{pc}$$

Comparison with Section 9.7 shows that, for the case considered there,

$$K_{a\gamma} = K_{aq} = K_a \quad \text{and} \quad K_{ac} = 2(K_a)^{\frac{1}{2}}$$

and

$$K_{p\gamma} = K_{pq} = K_p \quad \text{and} \quad K_{pc} = 2(K_p)^{\frac{1}{2}}$$

9.20 *Total and effective stress analysis:* So far, the analysis has been developed in terms of effective stress, but the expressions have been derived from the Mohr–Coulomb failure condition, and are therefore equally applicable to a total stress analysis (see Section 8.7). A total stress analysis may be used where the soil is saturated, and where the most critical condition occurs with the soil undrained.

Where the soil permeability is low, so that little drainage takes place during construction, the condition at the end of construction closely approximates to the undrained state. A total stress analysis, with φ equal to zero, may be used to examine the stability at this stage.

Where saturated, or nearly saturated, fill material is placed behind a wall, high pore pressures develop during construction. These will be reduced by subsequent drainage, and the most dangerous situation is likely to occur immediately construction is complete. Provided that the wall is safe at this stage, it may not be necessary to investigate the stability in the long term.

Where the wall supports the side of an excavation, the reduction in total stress during the removal of the soil results in an immediate drop in pore pressure. The latter may subsequently increase, and the most dangerous state may then occur when the soil is fully drained. The stability in the final condition must be analysed in terms of effective stress, the pore pressures being determined from a flow net. If the wall supports the side of a temporary excavation, and if negligible drainage is to be expected during the life of the structure, a total stress analysis may be used. However, it is necessary to remember that the stability of the wall is deteriorating with time. The greatest difficulty in such a case often lies in deciding how long the undrained state may be assumed to continue, rather than in demonstrating the initial stability of the structure.

STABILITY OF GRAVITY RETAINING WALLS

9.21 Forms of failure of gravity retaining walls: Gravity retaining walls depend for their stability principally on their own weight. Failure of such a wall may take the form of

    (a) forward movement
    (b) overturning about some point near the base, or
    (c) a foundation failure, due to excessive pressure on the soil beneath the toe.

Example 9.7
Consider the simple mass concrete gravity wall shown in Fig. 9.17. If the concrete is placed *in situ,* the angle of friction (δ) may be taken to be equal to φ' on the base of the wall. In this example, δ is assumed to be zero on vertical faces. Rankine's coefficients may be used to calculate the

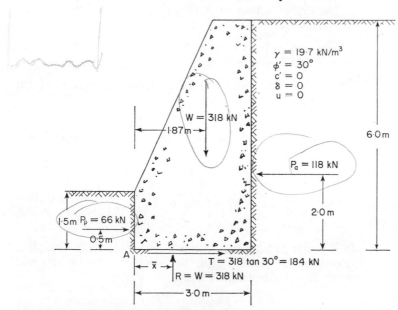

Fig. 9.17   Stability of a gravity retaining wall.

earth pressures on the wall. The forces acting on the wall are shown in Fig. 9.17.

Forward movement. The factor of safety (F_1) against forward movement will be defined as the ratio of the sum of the horizontal stabilising forces to the sum of the horizontal disturbing forces. Then

$$F_1 = (T + P_p)/P_a = (184 + 66)/118 = 2\cdot12$$

The Code of Practice for Earth Retaining Structures [9.3] suggests that this factor of safety should not be less than 2·0.

Overturning. If the wall overturns about the toe, the reaction R may be assumed to act approximately through A. The factor of safety against overturning will be defined as the ratio of the stabilising moments about A to the disturbing moments. Then

$$F_2 = (318 \times 1\cdot87 + 66 \times 0\cdot5)/(118 \times 2\cdot0) = 2\cdot65$$

The code suggests that this factor of safety should also be not less than 2·0.

Pressure beneath the toe. In calculating the actual pressure under the toe of the wall, it will be assumed that

(a) the actual value of the passive resistance is $P_p/F_1 = 66/2\cdot12 = 31$ kN/m, and
(b) the normal pressure varies linearly across the base.

The line of action of R may be determined by equating moments about A.

$$R\bar{x} = 318 \times 1\cdot87 + 31 \times 0\cdot5 - 118 \times 2\cdot0 = 375 \text{ kN-m}$$
$$\bar{x} = 375/318 = 1\cdot18 \text{ m}$$

Then, the eccentricity (e) of R from the centre of the foundation is

$$3\cdot0/2 - 1\cdot18 = 0\cdot32 \text{ m}$$

The maximum pressure under the toe of the foundation is

$$p_{max} = \frac{R}{B}\left(1 + \frac{6e}{B}\right)$$

$$= \frac{318}{3\cdot0}\left(1 + \frac{6 \times 0\cdot32}{3\cdot0}\right) = 175 \text{ kN/m}^2$$

If R passes outside the middle third of the base, since it is reasonable to assume that the interface between soil and concrete cannot sustain tension,

$$p_{max} = \frac{2}{3}\frac{R}{\bar{x}}$$

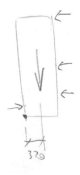

320

(01

9.22 Progressive development of passive failure: The forward movement required to bring the soil behind the wall to the state of full active failure is very small, and there is evidence that the peak strength is generally developed at one time through almost the whole of this body of soil. The passive failure in the soil in front of the wall, however, develops progressively from the toe of the wall. Large movements are required to cause general shear failure, and, by the time that this has developed, the strains in much of the soil mass are so large that the strength has been reduced to the residual value. The peak strength is only developed locally. To allow for this loss in strength, and to limit displacements of the structure in service, it is usual to assume a very conservative value for the passive resistance.

STABILITY OF SHEET PILE WALLS

9.23 Forms of failure: Unlike a rigid gravity retaining wall, a sheet pile wall is to some extent flexible. It depends for its stability on the passive resistance of the soil in front of, and behind, the lower part of the wall. It may also be supported by struts or anchors. The commonest forms of failure are

(a) forward movement of the base, as a result of inadequate passive resistance of the soil in front of the wall

(b) failure of the piles in bending, and

(c) failure of the anchors.

When failure takes one of these forms, the distribution of earth pressure on the wall at failure may differ from that derived in Sections 9.3 to 9.19 above. The consequences of this will be examined in Section 9.28. For the moment, it will be assumed that these earth pressure calculations are valid.

A sheet pile wall is relatively light, and there is little resistance to penetration of the toe into the ground. The wall can therefore sustain only limited vertical forces. In computing the earth pressures on the wall, it is unwise to rely on the vertical shearing forces between the soil and the wall, unless the latter is firmly restrained against vertical movement.

9.24 Cantilever sheet pile walls: Walls of this type may be used to support low banks of coarse-grained soils. They may also be used for temporary works in silts and clays. Over a long period, however, they are liable to develop large displacements in these soils, and they are therefore generally unsuitable for permanent works.

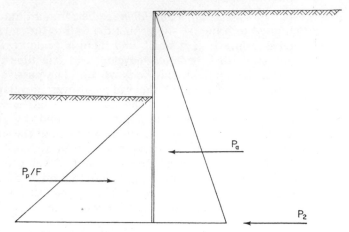

Fig. 9.18   Cantilever sheet pile wall—assumed pressure distribution.

A cantilever wall generally fails by rotation about some point above the base. The wall depends for its stability on the passive resistance of the soil in front of the wall above the point of rotation, and behind the wall below this point. The exact distribution of pressure near the base of the wall is not easily determined, and the approximate distribution shown in Fig. 9.18 is used in calculating the length of pile required. This distribution is not correct, since P_2, which represents the passive resistance below the point of rotation, must act some distance above the bottom of the wall. In practice, therefore, the calculated driving depth (d) is increased by an arbitrary amount (usually about 20%) to allow for this discrepancy. The calculated passive resistance of the soil in front of the wall is divided by a suitable factor of safety—usually about 2·0.

Example 9.8 (Fig. 9.19)
A cantilever wall is to support the sides of an excavation 1·8 m deep in a dry cohesionless sandy gravel, having a bulk density of 19 kN/m³ and an angle of shearing resistance of 30°. Since the wall is not restrained against vertical movement, shearing forces between the soil and the wall will be ignored. The passive resistance of the soil in front of the wall will be divided by a factor of safety (F) of 2·0.
    Then

$$K_a = 0.333; \qquad K_p/F = 3.0/2.0 = 1.5$$

Considering a section of the wall 1·0 m long, and equating moments about the base,

$$\tfrac{1}{6} \times 1.5 \times 19 \times d^3 = \tfrac{1}{6} \times 0.333 \times 19(d + 1.8)^3$$

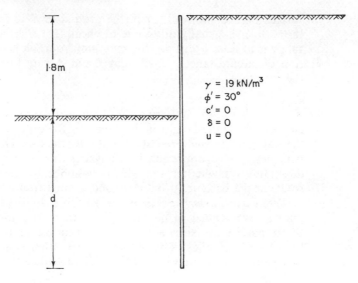

Fig. 9.19   Example 9.8.

so that

$$\left(\frac{d + 1{\cdot}8}{d}\right)^3 = 4{\cdot}5$$

and

$$d = 2{\cdot}76\,\text{m}$$

Adding 20% to the calculated embedded length, the total length of pile required is

$$1{\cdot}8 + 1{\cdot}2 \times 2{\cdot}76 = 5{\cdot}12\,\text{m}$$

The maximum bending moment occurs at a depth (d_0) at which the shear force in the wall is zero. Then

$$\tfrac{1}{2} \times 1{\cdot}5 \times 19 \times d_0^{\,2} = \tfrac{1}{2} \times 0{\cdot}333 \times 19(d_0 + 1{\cdot}8)^2$$

so that

$$\left(\frac{d_0 + 1{\cdot}8}{d_0}\right)^2 = 4{\cdot}5$$

and

$$d_0 = 1{\cdot}49\,\text{m}$$

At this depth, the bending moment in the wall is

$$\tfrac{1}{6} \times 19\,[0{\cdot}333(1{\cdot}49 + 1{\cdot}8)^3 - 1{\cdot}5 \times 1{\cdot}49^3]$$
$$= 21{\cdot}6\ \text{kN-m per metre run of wall.}$$

The factor of safety in this type of structure is very sensitive to small changes in the surcharge load on the ground surface behind the wall, or in the ground level in front of the wall. It is usually advisable to check that the factor of safety is at least 1·0 under the most unfavourable combination of circumstances which may occur during the life of the structure.

9.25   *Anchored sheet pile walls:*   Where a cantilever wall is not suitable, either because the height is too great, or for other reasons, the wall may be anchored at a point near the top. There are two methods of approach in analysing this type of wall, depending on whether the wall is assumed to be free to rotate at the base or to be restrained against rotation.

   'Free earth support' assumptions. In this case, the base of the wall is assumed to be entirely free to rotate, and there is no passive resistance to backward movement of the bottoms of the piles. The pressure distribution is assumed to be that shown in Fig. 9.20(a). The calculated value of

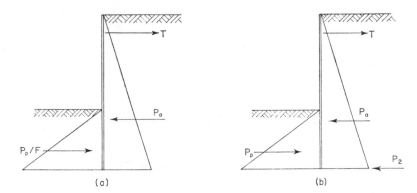

Fig. 9.20   *Anchored sheet pile wall: (a) 'free earth support' assumptions. (b) 'fixed earth support' assumptions.*

the passive resistance of the soil in front of the wall is divided by a suitable factor of safety—usually not less than 2·0.

   'Fixed earth support' assumptions. In this case, the base of the wall is assumed to be entirely prevented from rotating by the passive resistance of the soil behind it. The assumed pressure distribution is shown in Fig. 9.20(b). The passive resistance of the soil behind the wall is represented by the force P_2, which is assumed to act at the base of the wall. As in the case of the cantilever wall, the calculated length of embedment is arbitrarily increased to allow for the effect of this approximation. Since failure by forward movement of the toe is unlikely in a wall designed in this

way, no factor of safety is applied to the passive resistance of the soil in front of the wall.

9.26 'Free earth support' method: This is best illustrated by an example.

Example 9.9
   Consider a wall supporting the side of an excavation, 8 m deep, in the same sandy gravel described in the previous

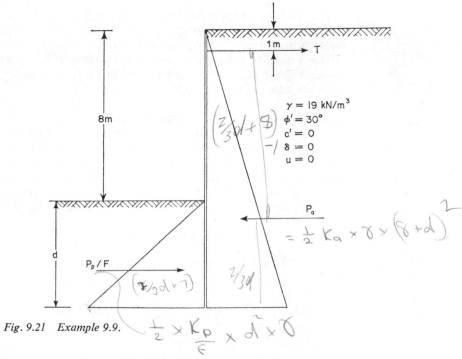

Fig. 9.21   Example 9.9.

example. The wall is to be anchored at a point 1 m below the top and shearing forces between the soil and the wall are to be ignored. The assumed pressure distribution is shown in Fig. 9.21.
   Then

$$K_a = 0.333; \qquad K_p/F = 3.0/2.0 = 1.5$$
$$P_a = \tfrac{1}{2} \times 19 \times 0.333(d + 8)^2$$
$$\frac{P_p}{F} = \tfrac{1}{2} \times 19 \times 1.5 \times d^2$$

Equating moments about the anchor,

$$[\tfrac{2}{3}(d + 8) - 1]P_a = [\tfrac{2}{3}d + 7]\frac{P_p}{F}$$

Then
$$[\tfrac{2}{3}(d + 8) - 1](d + 8)^2 = 4 \cdot 5[\tfrac{2}{3}d + 7]d^2$$
Hence
$$d = 5 \cdot 5 \text{ m}$$
Equating horizontal forces,

$$
\begin{aligned}
T &= P_a - P_p/F \\
&= \tfrac{1}{2} \times 19[0 \cdot 333(5 \cdot 5 + 8)^2 - 1 \cdot 5 \times 5 \cdot 5^2] \\
&= 147 \text{ kN per metre run of wall.}
\end{aligned}
$$

The anchors are usually spaced 2 m to 3 m apart, the load being distributed along the wall by walings running either behind or in front of the piles, and bolted to them.

The point of zero shear and the maximum bending moment in the piles may be determined in the same manner as in the case of the cantilever wall.

9.27 *'Fixed earth support' analysis—equivalent beam method:* The forces acting on the wall in this case are not statically determinate, and the required length of embedment can only be determined by a rather laborious process of successive approximation. However, the deflected form of the wall is shown approximately in Fig. 9.22, and from this it may be seen that there is a contraflexure at some point (C)

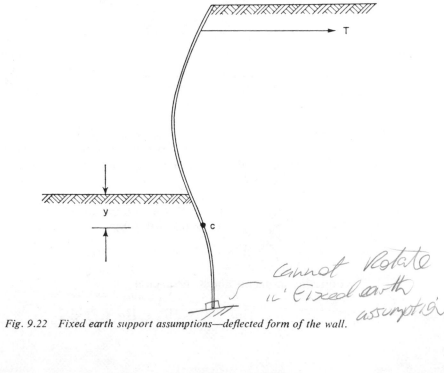

Fig. 9.22   Fixed earth support assumptions—deflected form of the wall.

close to the ground level in front of the wall. If the position of the point C is known, the forces acting on the wall may be determined.

Terzaghi has shown that, in uniform cohesionless soils, and provided that there is no surcharge load nor a high water table behind the wall, the depth (y) of the point of contra-flexure below the ground level in front of the wall is almost

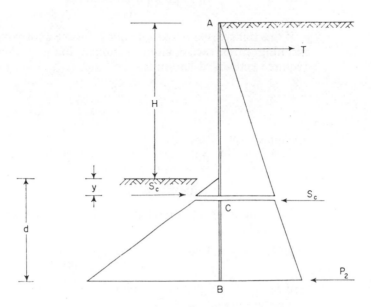

Fig. 9.23   Fixed earth support assumptions—equivalent beam method.

entirely dependent on the value of φ', and has the approximate values given below [9.5]:

φ'	20°	30°	40°
y	0·25H	0·08H	−0·007H

For the purpose of this analysis, the wall may now be assumed to be formed of two beams AC and CB, hinged together at C, the beam AC being equivalent to the section above the point of contraflexure of a beam encastré at the base (Fig. 9.23). There is a shear force S_c at C, but no bending moment. The values of T and S_c may be determined by considering the equilibrium of the beam AC.

Now consider the beam CB, assumed to be simply supported at B and C. Let the net pressure $(p_p - p_a)$ at C be p_n. Equating moments about B, we have

$$S_c(d - y) = \tfrac{1}{6}\gamma(K_p - K_a)(d - y)^3 + \tfrac{1}{2}p_n(d - y)^2$$

and

$$(d - y) = \frac{-3p_n + [9p_n{}^2 + 24(K_p - K_a)\gamma S_c]^{\frac{1}{2}}}{2\gamma(K_p - K_a)}$$

The value of p_n is often small. If it may be taken to be nearly zero, then

$$(d - y) \simeq \left(\frac{6S_c}{(K_p - K_a)\gamma}\right)^{\frac{1}{7}}$$

If this is increased by 20 % to allow for the approximation resulting from the position assumed for P_2, then the required embedded length is

$$y + 1\cdot2\left(\frac{6S_c}{(K_p - K_a)\gamma}\right)^{\frac{1}{2}}$$

Example 9.10

Consider the wall described in the previous example. The depth y to the point of contraflexure is

$$0\cdot08 \times 8 = 0\cdot64 \text{ m}$$

Equating moments about the anchor, for the upper part of the wall,

$$7\cdot64S_c = \tfrac{1}{2} \times 19[8\cdot64^2 \times 0\cdot333(\tfrac{2}{3} \times 8\cdot64 - 1)$$
$$- 0\cdot64^2 \times 3\cdot0(\tfrac{2}{3} \times 0\cdot64 + 7)]$$

and $S_c = 136$ kN per metre run of wall.

The net pressure at C is

$$p_n = (0\cdot64 \times 3\cdot0 - 8\cdot64 \times 0\cdot333)19\cdot0 = -18\cdot3 \text{ kN/m}^2$$

Hence

$$(d - y)$$
$$= \frac{3 \times 18\cdot3 + [9 \times 18\cdot3 + 24(3\cdot0 - 0\cdot333)19\cdot0 \times 136]^{\frac{1}{2}}}{2 \times 19\cdot0(3\cdot0 - 0\cdot333)}$$
$$= 4\cdot56 \text{ m}$$

and the required penetration is $0\cdot64 + 1\cdot2 \times 4\cdot56 = 6\cdot1$ m. If the net pressure p_n had been assumed to be zero, the required penetration would be

$$0\cdot64 + 1\cdot2\left(\frac{6 \times 136}{19\cdot0(3\cdot0 - 0\cdot333)}\right)^{\frac{1}{2}} = 5\cdot5 \text{ m}$$

9.28 *Effect of restraint at the anchorage:* Before the soil adjacent to the wall reaches the state of limiting equilibrium, some forward movement of the wall is necessary. If the anchorage is immovable, the soil immediately behind the point of support is not on the point of failure, and the pressure distribution on the wall is not determinate. However, if the wall is flexible, some small forward movement occurs below the point of support, due to bending of the piles. This tends to reduce the load on the centre of the wall, and to increase the load at the top and bottom. This phenomenon is commonly known as arching.

It has long been known that the conventional methods of analysis—and in particular the 'free earth support' method—tend to over-estimate the real bending moments in the piles. It is often assumed that this is due to arching, and a number of design methods (*e.g.* Stroyer's method [9.6] and the Danish rules [9.7]) allow a reduction in the computed bending moment on this assumption.

In a series of model tests, Rowe [9.8] showed that the forward movement of the anchorage which is necessary to promote complete active pressure failure in the soil behind the wall is very small. It seems unlikely, therefore, that arching can occur unless the top of the wall is rigidly secured to some massive structure. Rowe demonstrated that the lower bending moments may be explained by the incorrect assumptions about the distribution of passive pressure in front of the wall.

Figure 9.24 shows the assumed pressure distribution compared with the type of distribution found by Rowe. It will

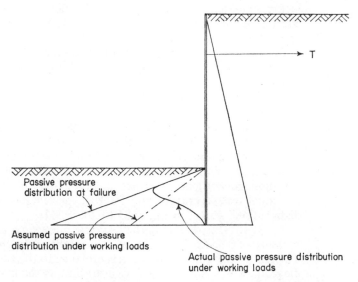

Fig. 9.24   Passive pressure distribution in front of anchored sheet pile walls.

be seen that the centroid of the passive pressure diagram is appreciably higher than has been assumed, and that this reduces the bending moment considerably. The true shape of the passive pressure diagram depends on the flexibility of the pile and on the rigidity of the soil. Rowe suggested a method of estimating the reduced bending moment, based on his experimental results.

Rowe has shown that the assumptions of the free earth support method generally represent the true distribution of pressure on the wall as it fails by rotation about the anchor. It is doubtful if the fixed earth support method represents the true state of affairs at any stage of loading. It seems logical therefore to use the former method to determine the required depth of penetration. All other things being equal, the fixed earth support method will generally require a greater depth. The actual loads in the tie-rods are generally rather less than those computed using the free earth support method, but as there is often some variation in the loads carried by adjacent anchors, it is reasonable to design them for the full computed load. Both methods tend to over-estimate the bending moments in the piles, and it is best to use a semi-empirical method such as that suggested by Rowe.

Some handbooks contain simple formulae for computing the required depth of penetration, such as that derived in Section 9.27 above. Since these formulae usually derive ultimately from the fixed earth support method, they could be expected to yield conservative values. However, they often incorporate simplifying assumptions which may have the effect of reducing the required driving depth. Although the computed values do not usually differ greatly from those obtained using the free earth support method, they should be treated with caution if the assumptions are not clearly stated.

9.29   Strutted trench sheeting: In a strutted excavation (Fig. 9.25(a)), it is clear that little or none of the soil against the sheeting is on the point of failure. The pressure distribution can only be estimated empirically, and is to some extent dependent on the manner in which the struts are installed. Based on measurements of load on trench sheeting, and on the results of model tests, Terzaghi and Peck suggested the pressure distributions shown in Fig. 9.25 ([9.9], [9.12]).

For sands (Fig. 9.25(b)), they suggest allowing a uniform pressure of $0.65 K_a \gamma H$ over the whole height of the sheeting. The real distribution bears no resemblance to the suggested diagram, which is merely a device for finding the maximum load on any strut. The real bending moment is also likely to

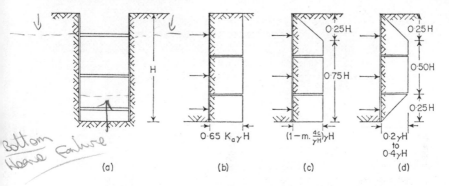

Bottom
Heave Failure

Fig. 9.25   Earth pressures on trench sheeting. (After Terzaghi and Peck [9.9, 9.12].)

be substantially less than the value calculated from the diagram.

For clays where $\gamma H/c_u$ is more than 6 (Fig. 9.25(c)), the soil is approaching a condition of plastic failure, and the maximum value of the lateral pressure is taken to be $(\gamma H - 4mc_u)$, where m is generally 1·0. However, where there is a considerable depth of soft material below the base of the excavation, so that the soil is approaching a failure of the type described in Section 11.18 below, a much smaller value of m is appropriate, and they suggest 0·4.

Figure 9.25(d) shows a tentative suggestion for the pressure to be allowed where $\gamma H/c_u$ is less than 4, so that the soil is not approaching plastic failure. Cases where $\gamma H/c_u$ is between 4 and 6 should be treated as being transitional between these two conditions.

TEMPORARY SUPPORT OF TRENCHES USING A BENTONITE SLURRY

9.30   *Trenches in clay soils:* Trenches for diaphragm walls may be temporarily supported by keeping them filled with a bentonite slurry. Mechanical grabs are used to excavate the trenches, which are kept filled with slurry while this work is carried out. Subsequently, concrete is placed in the trenches through a tremie, and the slurry is displaced.

Where the excavation is in a clay soil, the reduction of the normal stress during excavation tends to lower the pore pressure in the soil adjacent to the sides of the trench, and water is drawn into the clay from the slurry. However, the penetration of this water is very limited, since the

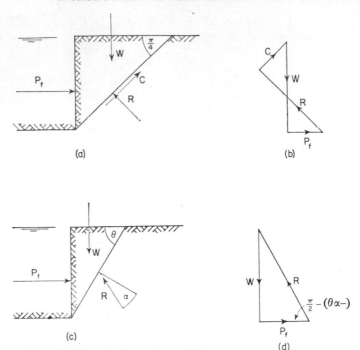

After Nash and Jones [9·10]

Fig. 9.26 Support of trenches using bentonite grout.

extraction of water from the slurry quickly deposits a thin but almost impervious filter cake of bentonite on the excavated face. This effectively seals the surface of the clay, and practically prevents the penetration of any more water. Nash and Jones made tests on the diaphragm walls constructed for the Hyde Park underpass [9.10]. These tests showed that the clay was softened to a depth of less than 25 mm, while laboratory tests on the bentonite filter cake indicated a permeability of only $2 \cdot 3 \times 10^{-11}$ m/s.

Since the support is only required temporarily, clay soils may be assumed to be effectively undrained, and calculations may be made in terms of total stress.

$$\tau_f = c_u; \qquad \varphi_u = 0$$

and the critical rupture surface is inclined at $\pi/4$ to the horizontal (Fig. 9.26(a)). Then, considering the force diagram for the critical wedge (Fig. 9.26(b)),

$$C = (W - P_f)/(2)^{\frac{1}{2}}$$
$$(2)^{\frac{1}{2}}H\tau = (\tfrac{1}{2}\gamma H^2 - \tfrac{1}{2}\gamma_f H^2)/(2)^{\frac{1}{2}}$$

where γ_f is the bulk density of the slurry. Then, if the factor of safety is defined as

$$F = \frac{\tau_f}{\tau} = \frac{c_u}{\tau}$$

where τ is the actual stress on the potential rupture surface,

$$F = \frac{4c_u}{(\gamma - \gamma_f)H}$$

Example 9.11

A trench 10 m deep is to be excavated in a saturated clay having a bulk density of 19·0 kN/m³, and a shear strength of 60 kN/m². The trench is to be supported by filling it with a slurry of density 11·5 kN/m³.

Then

$$F = \frac{4 \times 60}{(19\cdot0 - 11\cdot5) \times 10} = 3\cdot2$$

9.31   *Trenches in sand:*   In coarse dry sand, the slurry can penetrate a considerable distance into the pores. In a short time, however, the slurry forms a thixotropic gel within the pores, and effectively seals them. Assuming that the pore pressure in the soil is everywhere zero, the factor of safety may be calculated as follows:

Consider the force diagram shown in Fig. 9.26(d). Let α be the angle between the normal to the potential rupture surface and the direction of the reaction R. Then

$$\frac{W}{P_f} = \frac{\frac{1}{2}\gamma H^2 \tan (\pi/2 - \theta)}{\frac{1}{2}\gamma_f H^2} = \tan [\pi/2 - (\theta - \alpha)]$$

In Section 8.22, it was shown that the critical plane is inclined at $\pi/4 + \alpha/2$ to the horizontal.

Then

$$\frac{\gamma}{\gamma_f} = \tan^2 (\pi/4 + \alpha/2)$$

Hence

$$\tan \alpha = \frac{\gamma - \gamma_f}{2(\gamma . \gamma_f)^{\frac{1}{2}}}$$

Then, if the factor of safety is defined as $\tan \varphi'/\tan \alpha$,

$$F = \frac{2(\gamma . \gamma_f)^{\frac{1}{2}}}{\gamma - \gamma_f} . \tan \varphi'$$

This is independent of the depth of the trench.

Example 9.12

A trench is to be excavated in a dense sand having an angle of shearing resistance of 35°, and a bulk density of 20 kN/m^3. The trench is to be supported by filling it with a slurry whose bulk density is 12 kN/m^3.

Then

$$F = \frac{2(20 \times 12)^{\frac{1}{2}}}{20 - 12} \cdot \tan 35° = 2 \cdot 7$$

This factor of safety is very much reduced if the water level rises in the soil, since only the submerged density of the slurry and of the soil are effective below the water table. In the extreme case, where the water table is at the ground surface,

$$F = \frac{2(\gamma' \cdot \gamma'_f)^{\frac{1}{2}}}{\gamma' - \gamma'_f} \cdot \tan \varphi'$$

This calculation ignores the effect of the shear strength of the slurry. There is evidence that this effect may increase the stability of the trench quite considerably, but it cannot be relied upon for design purposes. Owing to the highly sensitive nature of the bentonite slurry, the shear strength is greatly reduced by disturbance (for example, by the operation of the excavating tools).

Morgenstern has shown that, owing to its thixotropic properties, the bentonite will retain considerable quantities of soil in suspension. This considerably increases the density of the slurry, and this increase may be allowed for in analysing the stability of the trench [9.11].

REFERENCES

9.1   RANKINE, W. J. M. 1857. On the stability of loose earth. *Phil. Trans. Royal Soc.*, **147**.

9.2   COULOMB, C. A. 1776. Essai sur une application des règles des maximis et minimis à quelques problèmes de statique relatif à l'architecture. *Mem. Acad. Royal Prés. Divers Sav.* (Paris).

9.3   Civil Engineering Codes of Practice Joint Committee. 1951. *Earth retaining structures.* (CECP No. 2) Inst. Struct. Eng.

9.4   CAQUOT, A. and KERISEL, J. 1948. *Tables for calculation of passive pressure, active pressure and bearing capacity of foundations.* Gautier-Villars (Paris).

9.5   TERZAGHI, K. 1943. *Theoretical soil mechanics.* Wiley (New York).

9.6   STROYER, J. P. R. N. 1935. Earth pressure on flexible walls. *Jour. Inst. Civil Eng.*, **1**.

9.7 Dansk Ingeniorforening. 1952. *Normer for bygnings-kontructioner.* Teknisk Forlag (Copenhagen).

9.8 ROWE, P. W. 1952. Anchored sheet pile walls. *Proc. Inst. Civil Eng.*, **1.**

9.9 TERZAGHI, K. and PECK, R. B. 1967. *Soil mechanics in engineering practice.* Wiley (New York) 2nd Ed.

9.10 NASH, J. K. T. L. and JONES, G. K. 1963. The support of trenches using fluid mud. *Proc. Symp. on Grouts and drilling muds in engineering practice.* Butterworths. (London).

9.11 MORGENSTERN, N. and AMIR-TAHMASSAB, I. 1965. The stability of a slurry trench in cohesionless soils. *Géotechnique*, **15.**

9.12 PECK, R. B. 1969. Deep excavations and tunnelling in soft ground. *Proc. 7th Int. Conf. on Soil Mech. and Found. Eng.*

9.13 SOKOLOVSKII, V. V. 1965. *Statics of granular media.* Pergamon (London).

9.14 CHEN, W-F. 1975. *Limit analysis and soil plasticity.* Elsevier (Amsterdam).

9.15 DRUCKER, D. C. 1954. Coulomb friction, plasticity, and limit loads. *Trans. ASME*, **76.**

9.16 HEYMAN, J. 1972. *Coulomb's memoir on statics.* Cambridge Univ. Press (London).

CHAPTER 10

Stability of slopes

FAILURE MECHANISMS AND THE FACTOR OF SAFETY

10.1   Failure mechanisms and methods of analysis: Failure of a slope frequently takes the form of quite large translational or rotational movements of a body of soil having a lower boundary deep within the soil mass. The stability of the slope with respect to this type of failure may be analysed by any of the methods described in Chapter 8, but most of the methods in general use depend on considerations of limiting equilibrium (Section 8.27). Failure is assumed to occur as a result of sliding along a rupture surface at the lower boundary of the moving soil mass, and a systematic search is made for the rupture surface which yields the least factor of safety against this form of failure.

In many types of earth structures, such as embankments and cuttings, the length parallel to the base of the slope is much greater than the height or width. The extent of the sliding soil mass parallel to the base is uncertain. While there is some resistance to movement at the ends of the slide, the effect of this on the stability of the whole mass is not easily determined. For these reasons, it is usual to consider only the shearing resistance on the lower boundary of the slip, and to ignore the end effects. This reduces the analysis to one of plane strain.

In a slope of infinite length normal to the base, and in a cohesionless soil, it was shown in Section 8.25 that failure occurs when $\beta = \varphi$, that the slip-lines are straight, and that one family of these is parallel to the ground surface. In cohesive soils, the critical rupture surface is more complex. However, *small* discrepancies in the assumed form of the rupture surface make little difference to the computed value of the factor of safety. We may therefore choose to analyse any convenient form of surface, provided that it is generally similar to the true rupture surface. Except where there are marked variations in the soil properties or pore pressures in different parts of the soil mass, it is often sufficiently accurate to assume that the rupture surface is a circular arc.

It is, however, necessary to be consistent. The factor of safety must be computed for the critical circular surface— that is, the circle giving the least factor of safety. This is not necessarily the circle closest in position to the surface of sliding observed in the ground. In particular, the critical circle and the observed slip surface may reach the ground surface at points a considerable distance apart.

10.2 *Shear strength and the factor of safety:* The shear strength of the soil may be expressed in the form

$$\tau_f = c + \sigma_n \tan \varphi$$

in terms of total stress, or

$$\tau_f = c' + (\sigma_n - u) \tan \varphi'$$

in terms of effective stress.

The factor of safety may be defined in any convenient way, so long as the definition is used consistently. In practice, the most useful definition is

$$\tau = \tau_f / F$$

where τ is the shear stress on the potential rupture surface required to maintain equilibrium, and F is the factor of safety. Put another way, F measures the factor by which the shear strength would have to be reduced to bring the structure to a state of imminent collapse.

It is as well to emphasise at this point that a factor of safety is simply a standard of comparison, and has no physical meaning beyond that given by the definition. The method of computation must also be defined, and no useful comparison can be made between values of F computed in different ways.

SLOPES IN COHESIONLESS SOILS

10.3 *Failure mechanisms in cohesionless soils:* A cohesionless soil is one in which the cohesion intercept c is zero, and may be a clean sand or gravel or a normally consolidated clay. In either case, however, we are only concerned with the drained state. Calculations are made in terms of effective stress, and the pore pressures are determined from a flow net.

In terms of effective stress,

$$\tau = \frac{\tau_f}{F} = (\sigma_n - u) \frac{\tan \varphi'}{F} ; \qquad c' = 0$$

Consider a semi-infinite mass of cohesionless soil, with a surface inclined at β to the horizontal. In Section 8.25 it was shown that the limiting condition for such a slope is that $\beta = \varphi'$, and that one family of plane rupture surfaces lies parallel to the slope.

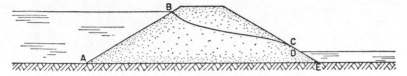

Fig. 10.1 Boundary conditions for a seepage zone.

For convenience, the dimension x will be measured parallel to the slope (and therefore parallel to the rupture surfaces), and z will be measured normal to it.

10.4 Stability of the slope above the free surface (see Fig. 10.1): Here the pore pressure (u) is zero, and the total and effective stresses are the same, so that

$$\tau_f = \sigma_z \tan \varphi'$$

Then

$$\frac{\partial \sigma_x}{\partial x} + \frac{\partial \tau_{xz}}{\partial z} + \gamma \sin \beta = 0$$

$$\frac{\partial \sigma_z}{\partial z} + \frac{\partial \tau_{xz}}{\partial x} + \gamma \cos \beta = 0$$

If the slope is long (so that conditions at the top and bottom may be assumed to have insignificant effects on the value of F) all elements of soil between the ground surface and the potential sliding surface parallel to it must be identical. The normal and shear stresses on such a surface are therefore constant, and

$$\frac{\partial \sigma_x}{\partial x} = \frac{\partial \tau_{xz}}{\partial x} = 0$$

Then

$$\frac{d\tau_{xz}}{dz} + \gamma \sin \beta = 0$$

$$\frac{d\sigma_z}{dz} + \gamma \cos \beta = 0$$

Therefore

$$\frac{d\tau_{xz}}{d\sigma_z} = \frac{-\gamma \sin \beta}{-\gamma \cos \beta} = \tan \beta$$

Since both σ_z and τ_{xz} are zero at the surface,

$$\frac{\tau_{xz}}{\sigma_z} = \frac{d\tau_{xz}}{d\sigma_z} = \tan \beta$$

Then

$$F = \frac{\tau_f}{\tau_{xz}} = \frac{\sigma_z \tan \varphi'}{\tau_{xz}} = \frac{\tan \varphi'}{\tan \beta}$$

10.5 *Stability of the slope below the free surface* $(u > 0)$: Here,

$$\frac{\partial \sigma'_x}{\partial x} + \frac{\partial \tau_{xz}}{\partial z} + (\gamma - \gamma_w) \sin \beta - i_x \gamma_w = 0$$

$$\frac{\partial \sigma'_z}{\partial z} + \frac{\partial \tau_{xz}}{\partial x} + (\gamma - \gamma_w) \cos \beta - i_z \gamma_w = 0$$

For a slope of infinite length,

$$\frac{\partial \sigma'_x}{\partial x} = \frac{\partial \tau_{xz}}{\partial x} = 0$$

Therefore

$$\frac{d\tau_{xz}}{dz} + (\gamma - \gamma_w) \sin \beta - i_x \gamma_w = 0$$

$$\frac{d\sigma'_z}{dz} + (\gamma - \gamma_w) \cos \beta - i_z \gamma_w = 0$$

Since both σ'_z and τ_{xz} are zero at the surface,

$$\frac{\tau_{xz}}{\sigma'_z} = \frac{d\tau_{xz}}{d\sigma'_z} = \frac{(\gamma - \gamma_w) \sin \beta - i_x \gamma_w}{(\gamma - \gamma_w) \cos \beta - i_z \gamma_w}$$

Therefore

$$F = \frac{\tau_f}{\tau_{xz}} = \frac{\sigma'_z \tan \varphi'}{\tau_{xz}}$$

$$= \frac{(\gamma - \gamma_w) \cos \beta - i_z \gamma_w}{(\gamma - \gamma_w) \sin \beta - i_x \gamma_w} \cdot \tan \varphi'$$

Writing $(\gamma - \gamma_w)/\gamma_w = i_c$, where i_c is the critical hydraulic gradient for upward vertical flow, derived in Section 4.18 above,

$$F = \frac{i_c \cos \beta - i_z}{i_c \sin \beta - i_x} \cdot \tan \varphi'$$

A number of special cases will now be examined.

    (a)   *Hydrostatic conditions* $(i_x = i_z = 0)$

$$F = \frac{i_c \cos \beta}{i_c \sin \beta} \cdot \tan \varphi' = \frac{\tan \varphi'}{\tan \beta}$$

Thus, in the absence of seepage forces, submerging the slope does not alter the factor of safety.

(b) *On a submerged permeable boundary.* This boundary is an equipotential, so that flow is normal to the surface and $i_x = 0$. Then

$$F = \frac{i_c \cos \beta - i_z}{i_c \sin \beta} \cdot \tan \varphi'$$

Note that, as a result of our sign convention, a positive value of i_z implies flow out of the bank.

(c) *On a seepage surface.* At the upper end of this surface, the flow is practically parallel to the slope, and the pore pressure on the surface is zero. Then

$$i_z = 0 \quad \text{and} \quad i_x = -\sin \beta$$

and

$$F = \frac{i_c \cos \beta}{(i_c + 1) \sin \beta} \cdot \tan \varphi'$$

Example 10.1 (Fig. 10.1)

Suppose that $\varphi' = 30°$, $\beta = 20°$, $i_c = 0.9$, and $i_z = 0.1$ on the submerged permeable boundary *DE*. Then

(a) *above the free surface* (or for hydrostatic conditions)

$$F = \frac{\tan 30°}{\tan 20°} = 1.57$$

(b) *on the submerged boundary DE*

$$F = \frac{0.9 \cos 20° - 0.1}{0.9 \sin 20°} \cdot \tan 30° = 1.39$$

(c) *at the upper end of the seepage surface C*

$$F = \frac{0.9 \cos 20°}{(0.9 + 1) \sin 20°} \cdot \tan 30° = 0.75$$

which would not be stable. Thus, the existence of a seepage surface can reduce the factor of safety by half. A toe drain or internal filter drain is normally provided, where possible, to prevent discharge of water through the face of the slope.

10.6 *Validity of solutions for cohesionless soils:* We may readily show that, for a slope of infinite length, the solution given above is exact. The solution shows that the failure condition is satisfied on *all* planes parallel to the ground surface, irrespective of depth. A stable statically admissible stress field therefore exists everywhere (Fig. 10.2), and the solution

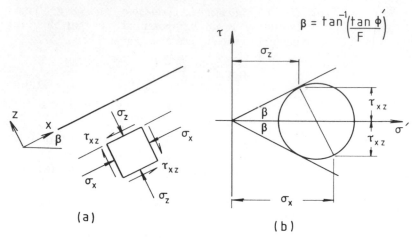

Fig. 10.2   *Failure of an infinite slope in cohesionless soil.* (a) *The stress field.* (b) *The Mohr stress circle at failure.*

is a lower bound. The solution also shows that collapse would occur by translational movements along one or more planes parallel to the ground surface. Since there is no restriction on displacements on this surface, which is the only boundary, the corresponding velocity field is kinematically admissible. The solution is therefore an upper bound. Since it is both an upper and a lower bound solution, it must be exact.

SLOPES IN COHESIVE SOILS—TOTAL STRESS ANALYSIS

10.7   Total stress analysis where $\varphi_u = 0$:   This method is applicable to undrained conditions in saturated clays. In this case

$$\tau = \frac{\tau_f}{F} = \frac{c_u}{F}; \quad \varphi_u = 0$$

In cohesive soils, cracks may develop at the top of the slope, and such cracks are usually the first signs of incipient failure. The depth of such cracks is uncertain, but allowance is commonly made for cracking to a depth $z_c = 1\cdot33c_u/\gamma$. Not only will this reduce the available shearing resistance by reducing the length of the rupture surface: if surface water can drain into this crack, the resulting hydrostatic pressure will increase the disturbing moment.

Provided that the rupture surface may be assumed to be a circular arc, the factor of safety may be determined by equating the moments of all the forces about the centre of the arc. Where the variations in soil strength are such that this

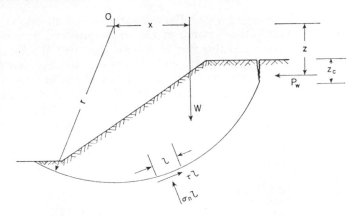

Fig. 10.3 Total stress analysis—$\varphi_u = 0$.

assumption cannot reasonably be made, the problem must be solved by one of the methods described in Sections 10.22 and 10.23 below.

Consider the potential circular rupture surface in Fig. 10.3. Equating moments about the centre 0,

$$Wx + P_w z = r \sum \tau l = r \sum \frac{c_u l}{F}$$

$$F = \frac{r \sum c_u l}{Wx + P_w z}$$

The nearest approach to the true factor of safety is given by the lowest value of F obtained with any circle. Analyses of actual failure (where $F = 1\cdot0$) commonly give computed values of F_{min} on the critical trial circle which are within $\pm 10\%$ of $1\cdot0$, although this circle may lie well outside the observed failure surface (Fig. 10.4). Calculations by this method, but for a circle chosen to be close to the observed failure surface, may give values of F as high as $1\cdot2$ or $1\cdot3$.

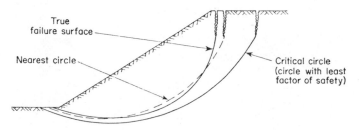

Fig. 10.4 Position of the critical rupture surface.

10.8 Partially submerged slopes: If the lower part of the slope is submerged, the pressure of the water provides an additional moment resisting movement. In Fig. 10.5 it may be seen that this moment exactly balances the moment about O of a mass of water filling the space below the external water level and above the rupture surface. We may therefore allow for the external water pressure by using the submerged density $(\gamma - \gamma_w)$, instead of the bulk density (γ), when calculating

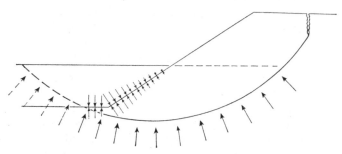

Fig. 10.5   Effect of partially submerging the bank.

the weight of that part of the sliding soil mass which lies below the external water level.

Equating moments about O (Fig. 10.6),

$$(Wx + W'x') + P_w z = r \sum \tau l = r \sum \frac{c_u l}{F}$$

$$F = \frac{r \sum c_u l}{(Wx + W'x') + P_w z}$$

It must be noticed at this point, that the submerged density is used in calculating the weight below the external water level in order to allow for the external water pressure

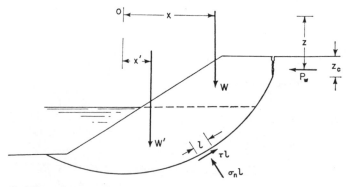

Fig. 10.6   Stability of a partially submerged bank.

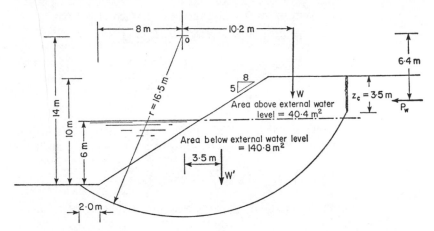

Fig. 10.7   Example 10.2.

on the exposed face of the bank. This implies nothing about the pore pressures within the bank. The latter are unknown and are in any case irrelevant to a total stress analysis.

Example 10.2

A cutting 10 m deep, with sides sloping at 8:5, is to be made in a clay soil having a mean undrained strength of 50 kN/m^2 and a mean bulk density of $19 \cdot 0 \text{ kN/m}^3$. Determine the factor of safety under immediate (undrained) conditions, against failure on the rupture surface shown in Fig. 10.7,

(a)   if the lower 6 m of the bank is submerged, and

(b)   if there is no external water pressure on the face of the bank.

(a)   Allowance will be made for cracks to a depth of

$$z_c = 1 \cdot 33 \times 50/19 \cdot 0 = 3 \cdot 5 \text{ m}$$

The geometry of the sliding soil mass is shown in Fig. 10.7. Then

$$W = 40 \cdot 4 \times 19 \cdot 0 = 769 \text{ kN}$$
$$W' = 140 \cdot 8(19 \cdot 0 - 9 \cdot 81) = 1\ 290 \text{ kN}$$
$$P_w = \tfrac{1}{2} \times 9 \cdot 81 \times 3 \cdot 5^2 = 55 \text{ kN}$$
$$\text{arc length} = 16 \cdot 5 \times \frac{99}{180} \times \pi = 28 \cdot 6 \text{ m}$$

Then

$$F = \frac{16 \cdot 5 \times 28 \cdot 6 \times 50}{769 \times 10 \cdot 2 + 1\ 290 \times 3 \cdot 5 + 55 \times 6 \cdot 4}$$

$$= 1 \cdot 85$$

(b) With no water pressure on the face of the bank,

$$W' = 140{\cdot}8 \times 19{\cdot}0 = 2\ 680\ \text{kN}$$

$$F = \frac{16{\cdot}5 \times 28{\cdot}6 \times 50}{769 \times 10{\cdot}2 + 2\ 680 \times 3{\cdot}5 + 55 \times 6{\cdot}4}$$

$$= 1{\cdot}34$$

10.9 Friction circle method: In developing this method, it will be convenient to define two quantities, F_c and F_φ, such that

$$\tau = \frac{\tau_f}{F} = \frac{c}{F_c} + \sigma_n \frac{\tan \varphi}{F_\varphi}$$

For the moment, it will be assumed that $F_\varphi = 1$, so that

$$\tau = \frac{\tau_f}{F} = \frac{c}{F_c} + \sigma_n \tan \varphi$$

Consider an element of length l on the trial slip circle *ABD* in Fig. 10.8(a). The forces on this element are

(a)   the shearing force $(c/F_c)l$
(b)   the shearing force $\sigma_n \tan \varphi l$, and
(c)   the normal force $\sigma_n l$

Consider the sum of all the forces cl/F_c round the arc *AD*.

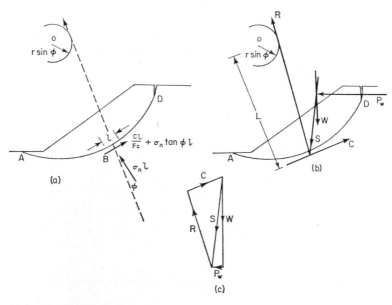

Fig. 10.8   The friction circle method.

The sum of the components parallel to AD

$$= \frac{c}{F_c} (\text{chord } AD)$$

and the sum of the components normal to AD is zero. The sum of the moments of all the forces cl/F_c about O

$$= r \frac{c}{F_c} (\text{arc } AD)$$

Then the resultant of all the forces cl/F_c is equal to

$$\frac{c}{F_c} (\text{chord } AD)$$

and acts parallel to AD at a distance (L) from the centre O, such that

$$L = r \frac{(\text{arc } AD)}{(\text{chord } AD)}$$

The vector sum of $\sigma_n l$ and $\sigma_n \tan \varphi l$ on the element is inclined at an angle φ to the normal to the rupture surface. It is therefore tangential to the circle drawn with centre O and radius $r \sin \varphi$. This is known as the *friction circle*. It will be assumed that the resultant (R) of all the forces $\sigma_n l$ and $\sigma_n \tan \varphi l$ is also tangential to this circle. This is not strictly true, and the error in the computed distance of R from O may be 20% for deep seated circles. It is possible to make allowance for this error, but the effect on the computed value of the factor of safety is generally negligible.

The reaction (R) must also pass through the intersection of C with the resultant (S) of W and P_w. The direction of R is therefore fixed, and the force diagram may be completed (Fig. 10.8(c)). Then

$$F_c = \frac{c(\text{chord } AD)}{C}$$

The factor of safety (F) is only equal to F_c if $\varphi = 0$, or if $F_c = F_\varphi$. If $\varphi \neq 0$, F may be determined by a process of successive approximations. The coefficient F_φ is given an arbitrary value. The radius of the friction circle is then $r \sin \psi$, where $\tan \psi = \tan \varphi / F_\varphi$. F_c is determined as described above. F_φ is then successively adjusted until $F = F_c = F_\varphi$.

10.10 *Taylor's stability numbers [10.1]:* Taylor's stability coefficients for total stress analysis were determined using the friction circle method. For all geometrically similar failure surfaces (Fig. 10.9) the force diagrams obtained by the friction circle method are similar, whatever the height of the bank.

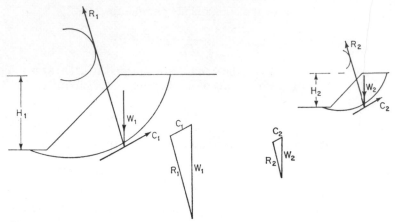

Fig. 10.9   Derivation of Taylor's stability numbers.

(It is necessary to ignore the possibility of tension cracks—
otherwise geometrically similar failure surfaces do not occur
on banks having different heights: z_c depends on c and γ,
and is not proportional to the height (H).) Then

$$\frac{C}{W} \text{ is constant}$$

But $W \propto H^2\gamma$ and $C \propto (c/F_c)H$. Therefore $c/F\gamma H$ is
constant. This coefficient, which depends only on φ and on
the geometry of the bank, is Taylor's stability number (N).
Taylor computed values of N for the critical circles
corresponding to various values of β and φ.

Although the friction circle method can be adapted to
deal with effective stress problems, Taylor's stability numbers were determined from an analysis of total stress only.
Attempts are sometimes made to use the stability numbers
for effective stress analysis, but this may give rise to serious
errors. The use of Taylor's method is therefore practically
restricted to problems involving undrained saturated clays
(for which $\varphi_u = 0$), or to the much less common cases where
the pore pressure is everywhere zero. Stability numbers for
effective stress analysis are discussed in Section 10.22,
below. Taylor's values of N for $\varphi = 0$ may be obtained
from Fig. 10.10.

The position of the critical circle may be limited by either
of two factors:

(a)   The depth of the stratum in which sliding can
occur.
(b)   The possible distance from the toe of the rupture
surface to the toe of the bank.

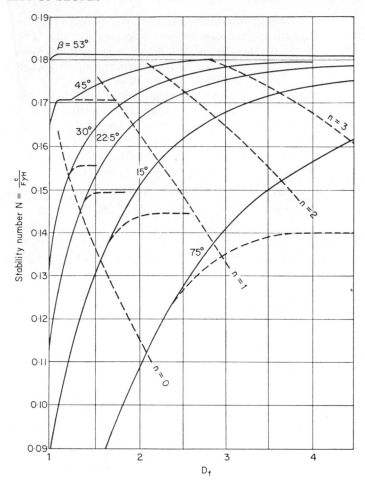

Fig. 10.10 Taylor's stability numbers for $\varphi = 0$ [10.1].

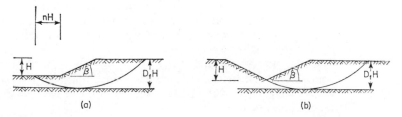

Fig. 10.11 Effect of a hard stratum below the bank. (a) Use the upper (full) curves in Fig. 10.10. The diagonal broken lines give values of n. (b) Use the lower (broken) curves in Fig. 10.10.

Where a hard stratum exists below the bank (Fig. 10.11) the rupture surface cannot pass into this stratum. The factor of safety may be increased in consequence. Taylor computed values of N for various depths to the hard stratum. These are expressed in Fig. 10.10 in terms of the coefficient D_f, where the depth of the hard stratum is $D_f H$ below the top of the bank. The distance from the toe of the slip to the toe of the bank is expressed in terms of the coefficient n, as shown in Fig. 10.11. If the slope angle (β) is greater than 54°, all critical circles pass through the toe of the bank, and N has the values given below:

β	60°	65°	70°	75°	80°	85°	90°
N	0·191	0·199	0·208	0·219	0·232	0·246	0·261

SLOPES IN COHESIVE SOILS—EFFECTIVE STRESS ANALYSIS

10.11 The method of slices: Consider a body of soil on the point of sliding on the surface $ABCD$ shown in Fig. 10.12. For the purpose of analysis, we will divide the whole body of soil

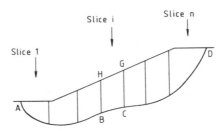

Fig. 10.12 *The method of slices.*

above the surface of sliding into n elementary slices, separated by $n-1$ vertical boundaries. The choice of *vertical* interslice boundaries is merely a matter of convenience, and it is not implied that these are surfaces of sliding: in general, the failure condition is not satisfied on these surfaces, and the relationship between normal and shear forces is not known.

If the body is stable, force and moment equilibrium conditions must be satisfied for each slice, and also for the whole body. Also, if the body is on the point of sliding, the failure condition

$$\tau_f = c' + (\sigma_n - u)\tan\varphi'$$

must be satisfied everywhere on the surface $ABCD$.

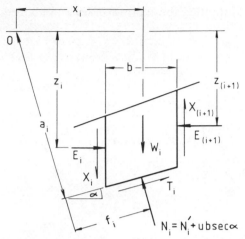

Fig. 10.13   Forces on a single slice.

10.12   The failure condition and the factor of safety: Where the soil body is not on the point of sliding, we may define a factor of safety in the form

$$\tau = \frac{\tau_f}{F} = \frac{c'}{F} + (\sigma_n - u)\frac{\tan \varphi'}{F}$$

where σ_n and τ are the normal and shear stresses at any point on *ABCD*.

For the slice i shown in Fig. 10.13,

$$T_i = [\tau b \sec \alpha]_i; \quad N_i = [\sigma_n b \sec \alpha]_i$$

Then

$$T_i = [\tau b \sec \alpha]_i$$

$$= \frac{1}{F}[c'b \sec \alpha + (N - ub \sec \alpha)\tan \varphi']_i$$

It will be convenient to specify the pore pressure u_i on *BC* in terms of a coefficient r_u equal to $[ub/W]_i$ where W_i is the weight of slice i. Then

$$T_i = \frac{1}{F}[c'b \sec \alpha + (N - Wr_u \sec \alpha)\tan \varphi']_i$$

10.13   Equilibrium of forces: For equilibrium of the vertical forces on slice i,

$$[T \sin \alpha + N \cos \alpha]_i = [W - \Delta X]_i$$

where

$$\Delta X_i = X_{(i+1)} - X_i$$

Substituting the expression for T_i obtained in the previous section,

$$\frac{1}{F}[\{c'b\sec\alpha + (N - Wr_u\sec\alpha)\tan\varphi'\}\sin\alpha + N\cos\alpha]_i$$
$$= [W - \Delta X]_i$$

Rearranging the terms, this yields

$$N_i = \left[W\left(1 + r_u\tan\alpha\frac{\tan\varphi'}{F}\right) - \frac{c'}{F}b\tan\alpha - \Delta X\right]_i \bigg/ m_{\alpha i}$$

where

$$m_{\alpha i} = \left[\cos\alpha + \sin\alpha\frac{\tan\varphi'}{F}\right]_i$$

Substituting this expression for N_i in the expression for T_i obtained in the previous section yields

$$T_i = \frac{1}{F}[c'b + (W(1 - r_u) - \Delta X)\tan\varphi']_i/m_{\alpha i}$$

For equilibrium of the horizontal forces on slice i,

$$[T\cos\alpha - N\sin\alpha]_i = \Delta E_i$$

where

$$\Delta E_i = E_{(i+1)} - E_i$$

Substituting for N_i and T_i in this expression,

$$\frac{1}{F}[c'b + (W(1 - r_u) - \Delta X)\tan\varphi']_i\cos\alpha_i/m_{\alpha i}$$

$$- \left[W\left(1 + r_u\tan\alpha\frac{\tan\varphi'}{F}\right) - \frac{c'}{F}b\tan\alpha - \Delta X\right]_i$$

$$\times \sin\alpha_i/m_{\alpha i} = \Delta E_i$$

After some manipulation, this may be written

$$\frac{1}{F}\left[\{c'b + (W(1 - r_u) - \Delta X)\tan\varphi'\}\frac{\sec\alpha}{m_\alpha}\right]_i$$

$$- [(W - \Delta X)\tan\alpha]_i = \Delta E_i$$

In considering the equilibrium of the whole soil mass, the internal interslice forces (E_2 to E_n and X_2 to X_n) must vanish. Also if there are no external forces on the end slices,

$$E_1 = E_{(n+1)} = X_1 = X_{(n+1)} = 0$$

so that

$$\sum_{i=1}^{i=n} \Delta E_i = 0$$

and

$$\frac{1}{F} \sum_{i=1}^{i=n} \left[\{c'b + (W(1 + r_u) - \Delta X)\tan \varphi'\} \frac{\sec \alpha}{m_\alpha} \right]_i$$

$$- \sum_{i=1}^{i=n} [(W - \Delta X)\tan \alpha]_i = 0$$

10.14 *The moment equilibrium condition:* This is satisfied for the slice i if it can be shown that there is no resultant moment about any one point such as O in Fig. 10.13. Then

$$Wx_i + X_i(x_i - b/2) - X_{(i+1)}(x_i + b/2)$$
$$- E_i z_i + E_{(i+1)} z_{(i+1)} = T_i a_i + N_i f_i$$

where x_i, z_i, $z_{(i+1)}$, a_i, f_i are the moment arms about O, as shown in Fig. 10.13. Since the moments of all the internal forces (E_i, X_i) must also vanish when considering the whole body,

$$\sum_{i=1}^{i=n} T_i a_i = \sum_{i=1}^{i=n} (W_i x_i - N_i f_i)$$

Substituting the expressions for T_i and N_i obtained in the previous section, this becomes

$$F = \frac{\displaystyle\sum_{i=1}^{i=n} [\{c'b + (W(1 - r_u) - \Delta X)\tan \varphi'\}a/m_\alpha]_i}{\displaystyle\sum_{i=1}^{i=n} [Wx]_i - \sum_{i=1}^{i=n} \left[\left\{ W\left(1 + r_u \tan \alpha \frac{\tan \varphi'}{F}\right) - \frac{c'b}{F}\tan \alpha - \Delta X \right\} f/m_\alpha \right]_i}$$

10.15 *Determinacy of the problem:* The two equations for force and moment equilibrium derived in the last two sections contain $2n$ unknown variables as follows:

(a) The factor of safety F.
(b) The $n - 1$ values of X_i on the interslice boundaries.
(c) The n values of f_i defining the position of the line of action of the force N_i on the base of each slice.

A further $2n - 2$ assumptions must be made if the solution is to be statically determinate. Most of the methods described below assume that N_i acts at the mid-point of BC, so that all values of f_i are known. This leaves n unknown quantities, and a further $n - 2$ assumptions must be made if the problem is to be statically determinate. If fewer than this number of assumptions are made, no solution can be obtained from the equilibrium and failure conditions alone: if more assumptions are made, it is not generally possible to find any solution which satisfies all the equilibrium conditions.

PLANAR FAILURE SURFACES

10.16 Slopes of infinite length: It is not uncommon to find a plane of weakness (perhaps caused by an old slip) lying parallel to, and at a comparatively shallow depth below, a long uniform slope. If the slope is so long that end effects may be ignored, it may be treated as if it were of infinite length. Then all slices of the potential sliding soil mass are identical, and the forces on the ends of each slice are equal and opposite. Since there is effectively only one slice, no further assumptions need be made to obtain a solution.

Consider one such potential slip plane (Fig. 10.14) at a depth z below the ground surface. Let the ground water level be at a constant height nz above the slip plane, so that there is flow parallel to the slope. Then for equilibrium of forces

$$W \cos \beta = N = N' + ub/\cos \beta$$
$$W \sin \beta = T = \{c'b/\cos \beta + N' \tan \varphi'\}/F$$

and

$$F = \{c'b/\cos \beta + N' \tan \varphi'\}/W \sin \beta$$

For flow parallel to the ground surface (Fig. 10.14), AB is an equipotential line. Then the pore pressure u at A is

$$AC\gamma_w = AB \cos \beta \gamma_w = nz \cos^2 \beta \gamma_w$$

Also

$$W = bz\gamma$$

Then

$$N' = W \cos \beta - ub/\cos \beta = bz \cos \beta (\gamma - n\gamma_w)$$

$$F = \frac{\{c'b/\cos \beta + bz \cos \beta (\gamma - n\gamma_w) \tan \varphi'\}}{W \sin \beta}$$

$$= \frac{c'}{z\gamma \sin \beta \cos \beta} + \left(\frac{\gamma - n\gamma_w}{\gamma}\right) \frac{\tan \varphi'}{\tan \beta}$$

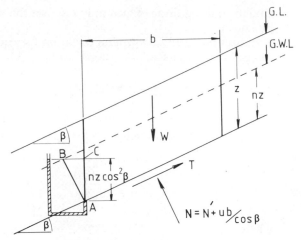

Fig. 10.14 *Failure on a plane parallel to the slope.*

10.17 *Sliding wedge analysis;* In some geological conditions, and in some soil structures, such as embankment dams with sloping cores, planes of weakness within the bank 'may suggest the probability of failure on a rupture surface consisting of two or three nearly straight sections. Figure 10.15 shows an example of such a potential slip surface. This may be analysed by the method of slices, using two slices separated by a vertical boundary.

For two slices, two independent assumptions are required for a complete solution. We may conveniently make these the height and the inclination (δ) of the effective stress resultant (P') on the interslice boundary. The value of the factor of safety (F) may be obtained directly from the force

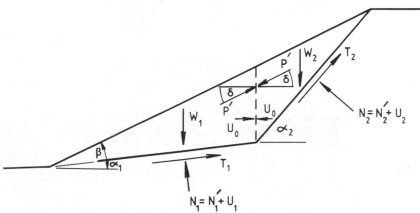

Fig. 10.15 *A sliding wedge analysis.*

equilibrium condition, which only requires the value of δ to be known.

In Section 10.13, the general equation for equilibrium of forces on a single slice was shown to be

$$\frac{1}{F}\left[\{c'b + (W(1 - r_u) - \Delta X)\tan \varphi'\}\frac{\sec \alpha}{m_\alpha} \right]_i$$
$$= [(W - \Delta X)\tan \alpha + \Delta E]_i$$

In this case, we may write

$$l_i = [b \sec \alpha]_i$$
$$U_i = [W r_u \sec \alpha]_i$$
$$\Delta X_i = [k P' \sin \delta]_i$$
$$\Delta E_i = [k(P' \cos \delta + U_0)]_i$$

where

$$k_1 = -1 \quad \text{for slice 1}$$
$$k_2 = +1 \quad \text{for slice 2}$$

Then

$$\frac{1}{F}[\{c'l + (W \sec \alpha - U - k P' \sin \delta)\tan \varphi'\}/m_\alpha]_i$$
$$= [(W - k P' \sin \delta)\tan \alpha + k(P' \cos \delta + U_0)]_i$$

The problem is now reduced to finding values of F and P which satisfy this equation for both wedges.

CYLINDRICAL FAILURE SURFACES

10.18   *General expressions:*   Where the soil properties and the pore pressure ratio r_u do not vary greatly within the bank, it is often sufficiently accurate to assume that the potential failure surface takes the form of a circular arc. Both the methods described below assume that the line of action of the force N_i (Fig. 10.13) passes through the mid-point of the base of each slice. A further $(n - 2)$ assumptions are therefore required to make the problem determinate.

If the pole O is taken to be at the centre of the arc whose radius is r, then f_i is zero, $a_i = r$, and $x_i = r \sin \alpha_i$. The force equilibrium equation is that derived in Section 10.13 above, but the moment equilibrium equation of Section 10.14 reduces to

$$F = \frac{\sum\limits_{i=1}^{i=n} [\{c'b + \tan \varphi'(W(1 - r_u) - \Delta X)\}/m_\alpha]_i}{\sum\limits_{i=1}^{i=n} [W \sin \alpha]_i}$$

10.19 Fellenius' method [10.2]: In this method, it is assumed that

    (a) the force N_i acts at the centre of the base of each of the n slices, and

    (b) both X_i and E_i are zero on each of the $(n-1)$ interslice boundaries, so that ΔX_i and ΔE_i are zero for each slice.

Thus, $(3n-2)$ assumptions are made where only $(2n-2)$ are required for determinacy. Strictly speaking, therefore, this should not be described as a limiting equilibrium method, as it is not generally possible to find a solution which satisfies both the equilibrium equations.

The failure condition requires that, on the base of each slice,

$$T_i = [\{c'l + (N - ul)\tan\varphi'\}/F]_i$$

so that, for the whole sliding soil mass,

$$F = \frac{\displaystyle\sum_{i=1}^{i=n} [\{c'l + (N - ul)\tan\varphi'\}]_i}{\displaystyle\sum_{i=1}^{i=n} T_i}$$

The values of N_i and T_i may be obtained by drawing the force diagrams for each slice. Example 10.3 shows, in tabular form, the calculation of the factor of safety for the potential failure surface shown in Fig. 10.16.

Alternatively, if $\Delta X_i = \Delta E_i = 0$,

$$N_i = [W\cos\alpha]_i \quad \text{and} \quad T_i = [W\sin\alpha]_i$$

so that

$$F = \frac{\displaystyle\sum_{i=1}^{i=n} [c'l + (W\cos\alpha - ul)\tan\varphi']_i}{\displaystyle\sum_{i=1}^{i=n} [W\sin\alpha]_i}$$

Bishop [10.3] has shown that the computed factor of safety is generally too small and, where the circle is deep seated, the error may be as much as 20%. Further significant errors may also occur as a result of the way in which the pore pressures are usually handled.

Since

$$N_i = [W\cos\alpha]_i \qquad \text{if } \Delta X_i = \Delta E_i = 0$$
$$N'_i = N_i - ul$$
$$= [W\cos\alpha - ub\sec\alpha]_i$$
$$= [W(\cos\alpha - r_u\sec\alpha)]_i$$

Example 10.3

Slice	b (m)	h (m)	W (kN)	c' (kN/m²)	$\tan \varphi'$	l (m)	N (kN)	T (kN)	$c'l$ (kN)	ul (kN)	$(N - ul)$ (kN)	$(N - ul)$ $\times \tan \varphi'$ (kN)
1	2	0·7	27·7	8·0	0·466	2·3	24·5	−13·0	18·3	4·8	19·7	9·2
2	2	2·6	96·5	8·0	0·466	2·1	93·9	−22·5	16·5	14·8	79·1	36·9
3	2	4·0	148·0	8·0	0·466	2·0	148·0	0	16·0	22·2	125·8	58·6
4	2	5·1	188·7	8·0	0·466	2·1	183·0	44·4	16·5	29·0	154·0	72·8
5	2	5·4	199·8	8·0	0·466	2·3	176·1	94·0	18·3	34·2	141·9	61·1
6	2	4·0	148·0	15·0	0·364	2·8	105·5	103·9	42·0	31·4	74·1	27·0
7	1	2·0	37·0	15·0	0·364	2·0	16·5	32·6	30·4	11·4	5·1	1·9
								274·9	158·0			267·5
								−35·5				
								239·4				

$$F = \frac{158 \cdot 0 + 267 \cdot 5}{239 \cdot 4} = 1 \cdot 78$$

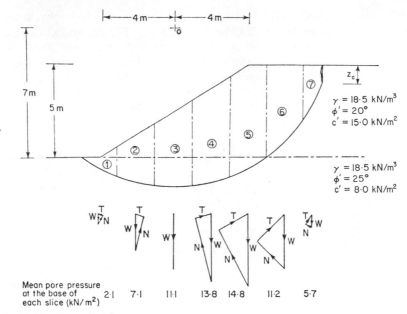

Fig. 10.16   Example 10.3.

If r_u approaches 1.0, and if $\alpha > 45°$ (so that $\sec \alpha_i > \cos \alpha_i$), N'_i may be negative. This implies that there is a negative effective stress on the base of the slice, which is physically impossible. In order to complete the calculation, it is usual to set N'_i equal to zero for all such slices, but this can result in substantial errors. These are particularly serious if

(a)   the circle is deep seated, so that α_i is large for several of the slices, and

(b)   the pore pressures are large so that r_u is nearly 1.0.

In extreme cases, the computed factor of safety may be less than half that computed by more rigorous methods.

10.20   *Bishop's method:*   In this method, the rupture surface is assumed to be an arc of a circle, and the normal force N_i is assumed to be at the centre of the base of each slice. The equilibrium equations are those derived in Sections 10.13 and 10.18 above, and a further $(n - 2)$ assumptions are required to make them determinate.

In Bishop's general method, it is assumed that the relative magnitudes of the X forces on the interslice boundaries are known, so that

$$X = \lambda f(x)$$

where λ is an unknown constant, and $f(x)$ is a function of the horizontal dimension x and is assumed to be known at each of the $(n-1)$ interslice boundaries. The known values of $f(x)$ represent $(n-1)$ assumptions (which is one more than required), but there is now one further unknown quantity (λ) so that the equations are determinate.

The solution of the equilibrium equations may be obtained by the following procedure:

(a) *Stage 1.* The value of λ is set equal to zero, so that all values of ΔX_i vanish, and the moment equilibrium equation (Section 10.18) is reduced to

$$F = \frac{\displaystyle\sum_{i=1}^{i=n} [\{c'b + \tan\varphi'(W(1-r_u))\}/m_\alpha]_i}{\displaystyle\sum_{i=1}^{i=n} [W\sin\alpha]_i}$$

The value of F is determined from this equation as follows:

    (i)   A trial value of F is chosen, and is used to compute values of $[m_\alpha]_i$ for each slice.

    (ii) These values of $[m_\alpha]_i$ are used in the moment equilibrium equation above to compute F.

    (iii) If this value of F differs substantially from the value originally chosen, the process is repeated, using the new value of F to compute $[m_\alpha]_i$. Since $[m_\alpha]_i$ is fairly insensitive to the value of F, acceptable accuracy is usually obtained after two or three iterations.

(b) *Stage 2.* It will usually be found that the value of F, computed as above, does not satisfy the force equilibrium equation (Section 10.13) when X_i is everywhere zero. A suitable distribution $f(x)$ of the X forces is therefore assumed, and various values of λ are tried until both equations are satisfied by the same values of λ and F. The solution obtained above may still not be exact, as the distribution of the X forces may not be correct. However, for reasons which are explained in Section 10.23 below, the value of F is found to be very insensitive to variations in $f(x)$, provided two conditions are satisfied. These are as follows:

    (i) No X force may be so large that it violates the failure condition on the interslice boundary.

(ii)   The line of thrust (that is, the line of action of the resultant of the E and X (forces) on each boundary must lie within the limits of the sliding soil mass. If it did not do so, the soil would be in tension.

Provided these conditions are satisfied, any apparently probable distribution of X forces may be assumed, and the solution may be accepted as being practically exact *for the particular rupture surface chosen.*

The method must now be repeated with other circular rupture surfaces, to find the one with the least value of F.

10.21   *Bishop's simplified method:*   Where the soil properties and the pore pressure ratio r_u are nearly uniform, so that the critical rupture surface is nearly circular, the computed value of F is found to be rather insensitive to the values of X_i. As a result, the value of F, computed at the end of Stage 1 above, is usually only 2–3 % different from the final value. This discrepancy is usually much less than the probable error in the measured values of c' and φ'. For circular arc solutions, therefore, we can generally assume that X_i is everywhere zero, and can determine F from the moment equilibrium equation as in Stage 1 above. The resulting solution does not satisfy the force equilibrium condition, but the error in the computed value of F is insignificant.

This is known as *Bishop's simplified method.* Example 10.4 shows, in tabular form, the calculation of the factor of safety for the same problem as that considered in Example 10.3 above. Notice that, as expected, Fellenius' method gives a significantly smaller value of F in this case.

Several computer programs based on this method are available commercially (see for example reference [10.4]), and the method is the one most commonly used for the routine analysis of the stability of embankments, cuttings and natural slopes. The method has also been used to prepare non-dimensional stability coefficients, such as those described in the next section.

10.22   *Effective stress stability coefficients:*   Bishop and Morgenstern [10.5] have computed stability coefficients, similar in principle to Taylor's coefficients, but applicable to effective stress calculations. The factor of safety (F) depends on

(a)   the slope angle β
(b)   the angle of shearing resistance φ'
(c)   the depth factor (D_f) (see Fig. 10.11)

Example 10.4

Slice	W (kN)	α	$W\sin\alpha$ (kN)	c' (kN/m²)	b (m)	(1) $c'b$ (kN)	$\tan\varphi'$	(2) $W(1-r_u)$ × $\tan\varphi'$ (kN)	(3) (1)+(2) (kN)	$F=1.5$ (4) m_α	(3) × (4)	$F=2.0$ (4) m_α	(3) × (4)
1	27·7	−28·0°	−13·0	8	2	16	0·466	11·0	27·0	1·350	36·5	1·275	34·5
2	96·5	−13·5°	−22·5	8	2	16	0·466	38·2	54·2	1·110	60·2	1·089	59·1
3	148·0	0	0	8	2	16	0·466	58·7	74·7	1·000	74·7	1·000	74·7
4	188·7	13·5°	44·4	8	2	16	0·466	75·0	91·0	0·950	86·5	0·975	88·6
5	199·8	28·0°	94·0	8	2	16	0·466	79·2	95·2	0·976	93·0	1·008	95·9
6	148·0	44·5°	103·9	15	2	30	0·364	45·9	75·9	1·131	81·6	1·180	89·5
7	37·0	61·5°	32·6	15	1	15	0·364	11·4	26·4	1·448	38·2	1·576	41·4
			274·9								470·7		483·7
			−35·5										
			239·4										

1st trial $F = \dfrac{470\cdot7}{239\cdot4} = 1\cdot97$

2nd trial $F = \dfrac{483\cdot7}{239\cdot4} = 2\cdot02$

(d) the pore pressure coefficient r_u, and

(e) the non-dimensional parameter $c'/\gamma H$.

The factor of safety is not very sensitive to changes in the value of the depth factor (D_f), and curves for three values only have been calculated $(D_f = 1\cdot0, 1\cdot25, 1\cdot5)$. Also, there are approximately linear relationships between $c'/\gamma H$ and F, and between r_u and F. Curves (Fig. 10.17) are provided from which coefficients m and n may be determined as functions of β and φ', for $c'/\gamma H = 0$, $0\cdot025$, and $0\cdot05$, and for $D_f = 1\cdot0$, $1\cdot25$, and $1\cdot5$. Then

$$F = m - nr_u$$

A coefficient r_{ue} may be read from the broken lines on the lower sets of curves. This permits a rapid determination of the approximate depth of the critical circle. The coefficient r_{ue}, read from any set of curves, is such that, if $r_u > r_{ue}$, the critical circle is at a greater depth than that to which this set of curves refers. If there are no broken lines on the set of curves, the critical circle never passes below the corresponding level, for any value of r_u.

Example 10.5

A cutting, 21 m deep, is to be made at a slope of $3\frac{1}{2}$:1 in an overconsolidated clay, having shear strength parameters (with respect to effective stress) of $c' = 14\cdot0 \text{ kN/m}^2$ and $\varphi' = 25°$. The bulk density of the clay is $19\cdot0 \text{ kN/m}^3$, and the value of r_u is $0\cdot15$. Then

$$\frac{c'}{\gamma H} = \frac{14\cdot0}{19\cdot0 \times 21} = 0\cdot035$$

For $c'/\gamma H = 0\cdot05$ and $D_f = 1\cdot0$ (Fig. 10.17(a)) $r_{ue} = 0 < r_u$. Then the critical circle passes below this level. For

$$\frac{c'}{\gamma H} = 0\cdot05 \quad \text{and} \quad D_f = 1\cdot25 \quad \text{(Fig. 10.17(b))} \quad r_{ue} = 0\cdot7 > r_u$$

Then the critical circle is at this level. From Fig. 10.17(b),

$$m = 2\cdot46 \quad \text{and} \quad n = 2\cdot09$$

so that

$$F = 2\cdot46 - 0\cdot15 \times 2\cdot09 = 2\cdot15$$

For

$$\frac{c'}{\gamma H} = 0\cdot025 \quad \text{and} \quad D_f = 1\cdot0 \quad \text{(Fig. 10.17(d))} \quad r_{ue} = 0\cdot5 > r_u$$

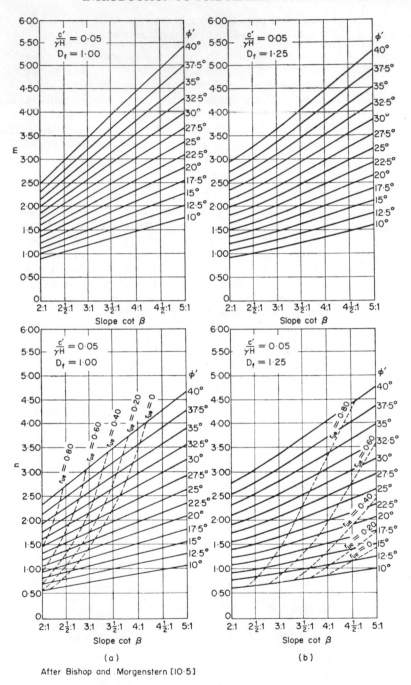

After Bishop and Morgenstern [10·5]

Fig. 10.17   Effective stress stability coefficients.

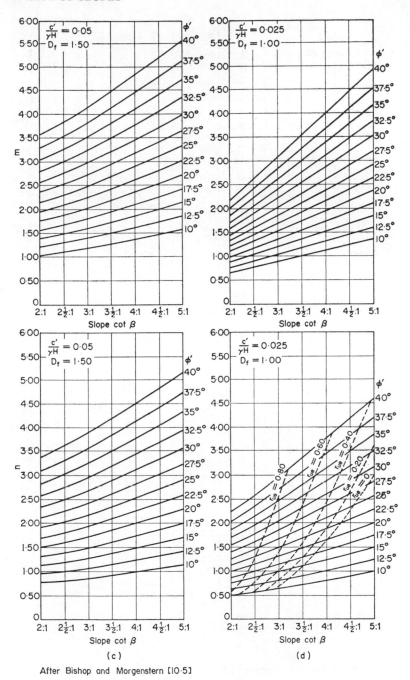

After Bishop and Morgenstern [10·5]

Fig. 10.17 continued

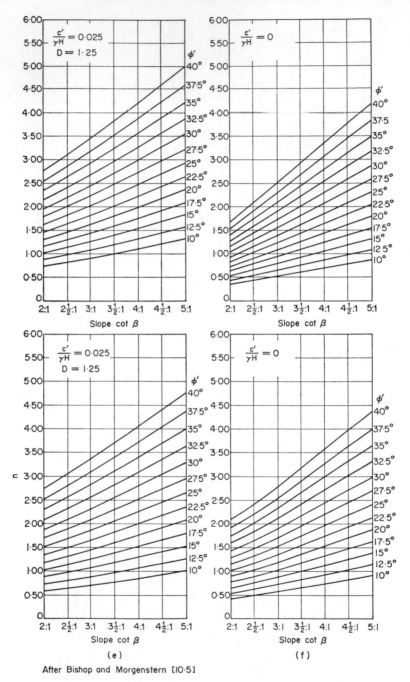

After Bishop and Morgenstern [10·5]

Fig. 10.17 continued

Then the critical circle is at this level. From Fig. 10.17(d),

$$m = 2 \cdot 12 \quad \text{and} \quad n = 1 \cdot 91$$

so that

$$F = 2 \cdot 12 - 0 \cdot 15 \times 1 \cdot 91 = 1 \cdot 83$$

Then, for $c'/\gamma H = 0 \cdot 035$,

$$F = 1 \cdot 83 + 0 \cdot 4(2 \cdot 15 - 1 \cdot 83)$$
$$= 1 \cdot 96$$

NON-CIRCULAR FAILURE SURFACES

10.23   Analysis of surfaces of general shape: The assumption of a circular failure surface is sufficiently accurate for the solution of many problems. However, there are cases where the critical rupture surface departs so far from the circular form that this assumption would lead to substantial errors. This usually occurs

  (a) where there are large variations in the soil properties in the pore pressures in different parts of the soil mass, or
  (b) where the soil is markedly anisotropic, or
  (c) where there are drainage blankets or similar layers of hard material embedded in the bank.

Where this occurs, it is necessary to analyse a non-circular surface which approximates more closely to the form of the true critical surface. In this case, however, an analysis which fails to satisfy the equilibrium conditions may lead to unacceptable errors.

The equations of equilibrium for failure surfaces of general shape were derived in Sections 10.13 and 10.14. Several methods of solution have been suggested. Essentially, these differ only in the assumptions made to render the equations determinate, and in the iterative procedures used to obtain the solution.

Nonveiller's method [10.6] is essentially the same as Bishop's general method, but is applied to surfaces of general shape. Morgenstern and Price [10.7] and Spencer [10.8] assume that the variation of the ratio $X_i : E_i$ is known, as that

$$X_i = \lambda f(x) E_i$$

where λ is an unknown constant and $f(x)$ is a known function. An iterative procedure is used to determine the values of F and λ which satisfy both equilibrium equations.

Janbu [10.9] assumes that the position of the line of thrust is known on all but one of the $(n - 1)$ interslice boundaries.

Sarma [10.10] makes the same general assumptions as Bishop and Nonveiller. His method was developed for the analysis of seismic conditions, where there are horizontal as well as vertical body forces. A *seismic coefficient k* may be defined as the maximum ratio of the horizontal and vertical forces. Sarma's method computes the value of k which will cause collapse if the shear strength is reduced by a known factor of safety F. The value of F where k is zero may be determined by computing k for several values of F, and extrapolating.

10.24   *Logarithmic spiral surfaces:*   Let the potential rupture surface be in the form of a logarithmic spiral, along which the tangents are inclined at a constant angle $\pi/2 + \psi$ to the radius vectors, which all pass through a common pole O (Fig. 10.18). Then on the base of the slice i, the resultant force normal to the radius vector is

$$R = [T\cos\psi - N\sin\psi]_i$$

$$= \left[\left(\frac{c'l}{F} + \frac{N'\tan\varphi'}{F}\right)\cos\psi - (N' + ul)\sin\psi\right]_i$$

If $\tan\psi = \tan\varphi'/F$

$$R = \left[\left(\frac{c'l}{F} - ul\tan\psi\right)\cos\psi\right]_i$$

$$= \frac{1}{F}[(c'l - ul\tan\varphi')\cos\psi]_i$$

Then, if there are no external forces, the equation for moment equilibrium of the whole sliding mass about O is

$$\sum_{i=1}^{i=n} \frac{1}{F}[(c'l - ul\tan\varphi')\cos\psi]_i - \sum_{i=1}^{i=n} [Wx]_i = 0$$

$$F = \frac{\displaystyle\sum_{i=1}^{i=n} [(c'l - ul\tan\varphi')\cos\psi]_i}{\displaystyle\sum_{i=1}^{i=n} [Wx]_i}$$

This solution is independent of the distribution of the normal forces N_i, since, for each slice, the resultant of N_i and the tangential force $N_i\tan\varphi'/F$ passes through the pole O. It is

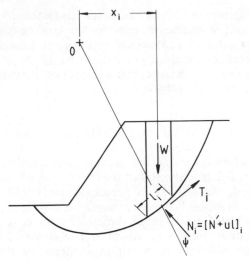

Fig. 10.18   Failure on a logarithmic spiral surface.

also independent of the values of the internal forces X_i and E_i, since these can have no resultant moment about O. The solution can therefore be obtained exactly from the moment equilibrium equation, without making any further assumptions about the internal forces.

If the soil is not homogeneous, so that φ' varies, the rupture surface must be formed of a series of logarithmic spirals, each having $\tan \psi$ equal to the *local* value of $\tan \varphi'/F$, and all having a common pole O.

This solution is of considerable interest. It has been shown (Baker and Garber, 10.11) that, for the factor of safety to be a minimum, the rupture surface must be such that the resultant of N_i and $N_i \tan \varphi'/F$ for each slice either

(a) is parallel to a common direction, or
(b) passes through a common pole.

The first condition implies a rupture surface consisting of a series of straight lines. The second requires a composite logarithmic spiral as described above.

If any other form of rupture surface is chosen, we can expect the solutions to become increasingly insensitive to the distribution of the E and X forces as the chosen surface approaches that on which the factor of safety is least. This explains why solutions obtained by the methods described in Sections 10.20 and 10.23 are very insensitive to the chosen form of the function $f(x)$.

A method using logarithmic spiral surfaces would yield a rigorous solution without any assumptions about the distribution of internal forces. Unfortunately, the angle ψ which determines the required shape of the spiral depends on the value of F. It would therefore be necessary to use an iterative procedure which adjusted the shape of the rupture surface at each stage. While this is possible in principle, it has not yet been achieved.

APPLICATION TO PRACTICAL PROBLEMS

10.25 Methods of analysis: In the case of a small embankment placed rapidly, or of the soil in the foundation of such an embankment, the material is practically undrained at the end of construction if the permeability is low. If the soil is nearly saturated, the angle of shearing resistance (φ_u) with respect to total stress is nearly zero. The factor of safety may be determined in terms of total stress by the methods of Sections 10.7 and 10.10. The pore pressures at the end of construction are high, and they subsequently decrease. As this happens, the effective stress and the shear strength increase. Thus, the undrained state at the end of construction is critical, and there may be no need to examine the stability in the long term.

In the case of a large embankment dam, the rate of construction is slower, and an appreciable drop in the pore pressure occurs during construction. An unduly expensive design would result if this dissipation of pore pressure were not taken into account in calculating the stability. Calculations must therefore be made in terms of effective stress, by the methods of Sections 10.11 to 10.24, the pore pressure being determined as a balance between the pressure rise due to the increasing load and the drop due to dissipation. It is not easy to predict the pore pressures with any great precision, partly because of the complexity of the calculation, but also because the pressures depend on factors which are not accurately known at the design stage. It is now common practice to install piezometers within the embankment to check the pore pressures actually developed during construction. If high pore pressures threaten the stability of the embankment, the placing of fill may have to be slowed down or temporarily stopped until dissipation has reduced the pressures to safe levels.

The stability of a dam during draw-down (that is, when the water level in the reservoir is being rapidly lowered) must also be investigated. The draw-down condition is usually critical for the stability of the upstream face of the dam. The pore pressures in this case are a balance of

(a) the initial pore pressure before draw-down commenced (determined from a flow net)

(b) the pore pressure changes which result from unloading as the external water pressure is removed from the embankment face, and

(c) dissipation during the draw-down period, if the latter is sufficiently long.

Since rapid dissipation of pore pressure is important for stability during both construction and draw-down, and since the dam may be constructed of material with a very low permeability, it is common practice to place horizontal drainage blankets of permeable material within the embankment. These blankets drain freely to the upstream and downstream faces, but do not pass through the central impervious core of the dam. Vertical sand drains may also be used to increase the rate of dissipation of pore pressure in the foundation.

The initial stability of excavated slopes in saturated clay soils may be determined by a total stress analysis (Sections 10.7 and 10.10). Immediately after excavation, the pore pressures are low, and are likely to increase with time. The final condition may therefore be less stable than at the end of construction, and must also be investigated. This must be done by means of an effective stress analysis, and the pore pressures must be determined from a flow net. In overconsolidated fissured clays, there may be a gradual reduction in strength to the critical state value. Where there are old slip surfaces in the soil, the residual strength on these surfaces may be smaller still [10.12]. This was discussed in Section 7.35 above.

10.26 *Flow slides:* Where the natural soil has a void ratio greater than the critical value, a potentially dangerous situation exists. Once the soil has been disturbed, the void ratio tends to be reduced. However, if the water content is more than sufficient to saturate the soil at the critical void ratio, the full reduction in volume is not possible. As a result, there is a sharp rise in the pore pressure and a corresponding fall in the effective stress. The shear strength is temporarily much reduced. If the permeability of the soil is low, so that the pore pressure is not rapidly dissipated, a flow slide may result. The disturbed soil flows out in a nearly liquid state, and may travel a large distance, even on nearly level ground. Such flow slides have been observed in loess soils, in quick clays and in certain types of industrial waste tips.

Loess soils are wind blown deposits which are in an

extremely loose state. When saturated with water, extensive flow slides may occur on quite gentle slopes, 'triggered' by seismic shocks which break down the loose structure of the soil.

The formation of *'quick' clays* was discussed in Section 2.26. When these materials are disturbed, the flocculated structure is broken down, and the particles tend to take up a dispersed arrangement. The natural water content of the clay is commonly greater than the liquid limit of the disturbed soil. A small slip, perhaps caused by a stream under-cutting the toe of the slope, may cause the liquefaction of a small body of soil. This soil flows out, leaving an exposed bank, the upper part of which is almost vertical. This bank is itself unstable, and a further slip occurs. A progressive sequence of such failures can result in a very large volume of soil flowing out, even on slopes of only a few degrees.

In the case of *spoil tips*, the void ratio of the material as tipped is approximately equal to the critical void ratio at zero effective stress. Within the tip, however, the effective stresses are large and the critical void ratio is correspondingly low. Moreover, if the tipped material is of a weak or friable nature (either naturally or as a result of weathering), the critical void ratio after failure is considerably reduced as a result of the crushing of the particles and the consequent degrading of the material. The permeability, also, is greatly reduced by the degrading of the material, and pore pressures which occur within the soil mass are not immediately dissipated. The tip failure at Aberfan [10.13] resulted from this sequence of events. Artesian pressure in the ground immediately below the tip reduced the effective stress and reactivated an old slip near the toe. This in turn released a considerable mass of material above, which flowed out in virtually liquid condition down the mountain side and into the village. This flow slide caused most of the damage and loss of life. Finally, the action of the slip removed a layer of boulder clay from the ground surface beneath the tip and released a considerable quantity of water from the sandstone below. This caused a 'mud run' which brought down more material, and caused further damage and serious flooding.

REFERENCES

10.1   TAYLOR, D. W. 1948. *Fundamentals of soil mechanics.* Wiley (New York).
10.2   FELLENIUS, W. 1927. *Erdstatische berechnungen.* W. Ernst und Sohn (Berlin).

10.3 BISHOP, A. W. 1955. The use of the slip circle in the stability analysis of earth slopes. *Géotechnique*, **5**.

10.4 LITTLE, A. L. and PRICE, V. E. 1958. The use of an electronic computer for slope stability analysis. *Géotechnique*, **8**.

10.5 BISHOP, A. W. and MORGENSTERN, N. 1960. Stability coefficients for earth slopes. *Géotechnique*, **10**.

10.6 NONVEILLER, E. 1965. The stability analysis of slopes with a slip surface of general shape. *Proc. 6th Int. Conf. Soil Mech. and Found. Eng.*

10.7 MORGENSTERN, N. and PRICE, V. E. 1965. The analysis of the stability of general slip surfaces. *Géotechnique*, **15**.

10.8 SPENCER, E. 1973. Thrust line criterion in embankment stability analysis. *Géotechnique*, **23**.

10.9 JANBU, N. 1957. Earth pressure and bearing capacity calculations by generalised procedure of slices. *Proc. 4th Int. Conf. Soil Mech. and Found. Eng.*

10.10 SARMA, S. K. 1973. Stability analysis of embankments and slopes. *Géotechnique*, **23**.

10.11 BAKER, R. and GARBER, M. 1978. Theoretical analysis of the stability of slopes. *Géotechnique*, **28**.

10.12 SKEMPTON, A. W. 1970. First-time slides in overconsolidated clays. *Géotechnique*, **20**.

10.13 Report of the tribunal appointed to inquire into the disaster at Aberfan on October 21st 1966. 1967. HMSO.

CHAPTER 11

Bearing capacity of foundations

FAILURE OF FOUNDATIONS

11.1   Forms of failure: The purpose of a structural foundation is to transfer the structural loads safely to the ground below. Failure of a foundation may take one of two forms [11.9]:

(a)   Catastrophic collapse of the soil beneath the foundation, if the shear strength is inadequate to support the applied load. Failures of this type are not very common, but, where they do occur, they may lead to large movements and distortion of the superimposed structures, and they may take place very rapidly.

(b)   Excessive settlement of the structure, due in part to distortion of the soil mass as a result of the applied shear stresses, and in part to consolidation of the soil as a result of the increased normal stresses.

Both the settlement and the resistance to shear failure depend on the size and shape of the foundation, and on its depth below the surface, in addition to the properties of the soil. In designing a foundation, it is generally necessary to examine the possibility of both forms of failure. The ultimate resistance to catastrophic failure is considered in this chapter. Settlement problems will be examined in Chapter 12.

11.2   Types of foundation: The type of foundation selected depends on the type and magnitude of the loads to be supported, and on the properties of the soil below. Where a stratum having a high shear strength and adequate thickness exists near the surface, the structural loads may be supported on small separate pad or strip footings. The structure offers little resistance to relative movement of such foundations, and considerable distortion of the structure can result if individual footings settle differentially. This form of foundation is therefore not generally satisfactory unless the compressibility of the soil is low or the structural loads are small.

289

Where the total area of the footings would exceed about half the plan area of the structure, the possibility of using a raft foundation should be examined. The principal purpose of such a foundation is to spread the load over the largest possible area, so as to increase the load carrying capacity and to reduce the settlement. A raft, by restricting the relative movements of column bases, also limits the differential movements in the structure. However, a raft is itself generally much more rigid, and therefore much more susceptible to damage by differential settlement, than the structure above it. The need to strengthen the raft to resist such distortion usually makes this form of foundation unduly expensive if the differential settlements are likely to be large.

Where large settlements are likely, or where no stratum of sufficient bearing capacity exists close to the surface, it may be necessary to construct a piled foundation. The purpose of this is to carry the structural loads down to a lower stratum which is capable of supporting them. The design of piled foundations is discussed in Chapter 13.

Both shear stresses and settlements depend on the net increase in the load applied to the ground—that is, the difference between the load imposed by the structure and the weight of any excavated soil. The net load may be reduced by increasing the amount of excavation. In extreme cases, where large basements are incorporated in the structure, the imposed loads may be made equal to the weight of the excavated soil, so that there is no net increase in the load on the ground. There should then be negligible ground movements and no possibility of a catastrophic failure, regardless of the shear strength or compressibility of the soil.

11.3 *Slip-line fields in the soil beneath a foundation:* Fig. 11.1 shows a typical pattern of slip-lines in the soil beneath a foundation on the point of collapse. The regions ACD and $A'C'D'$ are zones of passive Rankine failure, as described in Section 8.24 above. In each of these zones, there are two families of slip-lines inclined to each other at an angle $(\pi/2 + \varphi)$. The

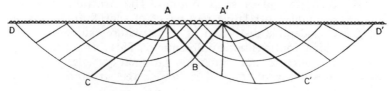

Fig. 11.1 Typical rupture surfaces beneath a foundation at failure.

regions ABC and $A'BC'$ are zones of radial shear, in each of which one family of slip-lines originates at the corner of the foundation (A or A'). The soil in the region ABA' may or may not be in a state of plastic failure, depending on the roughness of the underside of the foundation. If the foundation is absolutely smooth, so that there is no shear stress on AA', the region ABA' is a zone of active Rankine failure. If the foundation is absolutely rough, so that no slip takes place on AA', the zone ABA' moves downwards as a rigid wedge with the foundation. The lines AB, $A'B$, AC, and $A'C'$ are discontinuities between the plastic zones. Any slip-line which crosses one of these lines must be continuous, but the shape of the slip-line differs on either side of the discontinuity.

11.4 *The bearing capacity of a foundation:* The bearing capacity q_f of a foundation is the mean total stress on the surface of the underlying soil when the latter is on the point of collapse. It is a function of the foundation geometry, the soil weight, and the soil strength which will be assumed to be defined by the Mohr–Coulomb failure condition

$$\tau_f = c + \sigma_n \tan \varphi$$

If the soil strength is defined in terms of effective stress, in the form

$$\tau_f = c' + (\sigma_n - u) \tan \varphi'$$

the same analytical methods may be used to compute the *effective* bearing capacity q'_f, which is the mean *effective* contact stress when the underlying soil is on the point of collapse. In sections 11.5 to 11.12 below, the primes denoting effective stress have for convenience been omitted, but the methods used and the coefficients derived are equally applicable to total and effective stress analyses.

The bearing capacity of an infinitely long strip foundation on the surface of an ideal weightless soil may be determined exactly. This will be derived first using limit analysis (Section 8.17), although any of the methods described in Chapter 8 could have been used. The effects of soil weight, and of the depth and shape of the foundation will then be examined.

BEARING CAPACITY OF STRIP FOUNDATIONS AT GROUND LEVEL

11.5 *Lower bound solutions for smooth foundations on weightless soil:*
(a) *A foundation on frictionless soil.* Consider a foundation (Fig. 11.2) of width B and infinite length, resting on

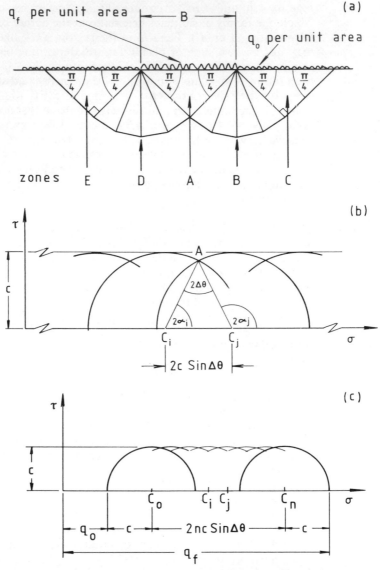

Fig. 11.2   A smooth foundation on weightless frictionless soil. (a) The chosen stress field. (b) Mohr circles for stresses at a discontinuity. (c) The complete Mohr stress circles.

the surface of a uniform frictionless weightless soil. The stress field beneath the foundation at failure will be assumed to consist of five zones separated by stress discontinuities as follows:

(i) A zone of active pressure failure (Zone A)

immediately below the foundation. In this zone, the greatest principal stress component σ_1 is vertical, so that there is no shear stress on the interface between soil and foundation, which may be assumed to be perfectly smooth.

(ii) Two zones of passive pressure failure (Zones C and E), in which σ_1 is horizontal. The upper surface of each of these zones carries a uniform surcharge pressure q_0.

(iii) Two zones of radial shear failure (Zones B and D), each consisting of several triangular wedges of uniform stress, separated by stress discontinuities.

The stress fields on either side of a discontinuity are different, but the normal and shear stress components on the discontinuity must be equal in order to satisfy the condition of equilibrium. The greatest principal stress component σ_1 is vertical in Zone A and horizontal in Zone C. The principal stress direction is therefore rotated through a total angle $\pi/2$ in passing across all the discontinuities between Zones A and C.

Let C_i and C_j (Fig. 11.2(b)) be the centres of Mohr circles defining the stresses to the left and right respectively of one such discontinuity. The intersection of the circles (Point A) defines the normal and shear stress components on the discontinuity (since these must be the same on both sides). The rotation of the principal stress directions in passing across the discontinuity is therefore

$$(\alpha_j - \alpha_i) = \Delta\theta$$

If there are n discontinuities, and $\Delta\theta$ has the same value on each,

$$\Delta\theta = \frac{\pi}{2n}$$

Then (Fig. 11.2(c))

$$q_f = q_0 + 2c + 2nc\sin\frac{\pi}{2n}$$

The largest value of q_f occurs as n tends to infinity, when

$$q_f = \lim_{n \to \infty}\left[2c + 2nc\sin\frac{\pi}{2n} + q_0\right]$$
$$= c(2 + \pi) + q_0$$
$$= cN_c + q_0N_q$$

where

$$N_c = 2 + \pi = 5\cdot14 \quad \text{and} \quad N_q = 1\cdot0$$

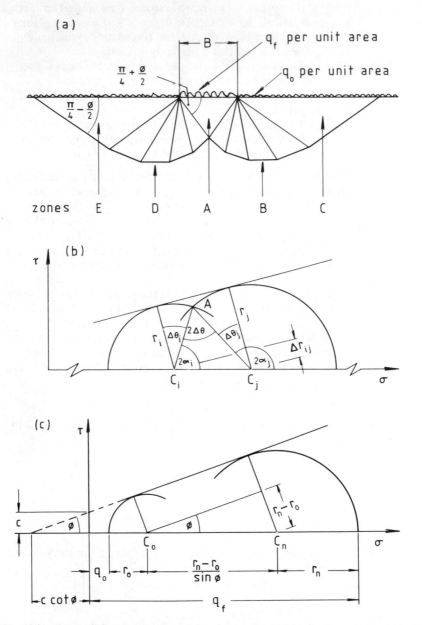

Fig. 11.3   A smooth foundation on weightless frictional soil. (a) The chosen stress field. (b) Mohr circles for stresses at a discontinuity. (c) The complete Mohr stress circles.

(b) *A foundation on frictional soil*. As in the previous case, the stress field (Fig. 11.3(a)) is assumed to consist of radial shear zones between zones of active and passive pressure failure. Let C_i and C_j (Fig. 11.3(b)) be centres of Mohr circles defining the stresses to the left and right of a discontinuity. The rotation of the principal stress direction at this discontinuity is

$$(\alpha_j - \alpha_i) = \tfrac{1}{2}(\Delta\theta_i + \Delta\theta_j) = \Delta\theta$$

Then

$$\Delta r_{ij} = (r_i \sin\Delta\theta_i + r_j \sin\Delta\theta_j)\tan\varphi$$

If $\Delta\theta$ is small, this reduces to

$$\Delta r_{ij} = 2r_i \Delta\theta \tan\varphi$$

and

$$r_j = r_i(1 + 2\Delta\theta\tan\varphi)$$

If there are n discontinuities and $\Delta\theta$ has the same value on each,

$$\Delta\theta = \frac{\pi}{2n}$$

and

$$r_j = r_i\left(1 + \frac{\pi}{n}\tan\varphi\right)$$

Then

$$r_n = r_0\left(1 + \frac{\pi}{n}\tan\varphi\right)^n$$

where r_n and r_0 are the radii of the Mohr stress circles for Zones A and C respectively. The radius r_n has a maximum value as n approaches infinity, when

$$r_n = \lim_{n\to\infty}\left[r_0\left(1 + \frac{\pi}{n}\tan\varphi\right)^n\right]$$

$$= r_0 \exp[\pi\tan\varphi]$$

Then

$$C_0C_n = \frac{r_n - r_0}{\sin\varphi} = \frac{r_0}{\sin\varphi}(\exp[\pi\tan\varphi] - 1)$$

so that

$$q_f = q_0 + r_0 + C_0C_n + r_n$$

$$= q_0 + \frac{r_0}{\sin\varphi}\{\sin\varphi + \exp[\pi\tan\varphi] - 1$$

$$+ \exp(\pi\tan\varphi)\sin\varphi\}$$

But

$$r_0 = (c \cot \varphi + q_0 + r_0) \sin \varphi$$
$$= (c \cot \varphi + q_0) \sin \varphi / (1 - \sin \varphi)$$

Then

$$q_f = q_0 + \frac{c \cot \varphi + q_0}{1 - \sin \varphi} \{\exp[\pi \tan \varphi](1 + \sin \varphi)$$
$$- (1 - \sin \varphi)\}$$

and since

$$(1 + \sin \varphi)/(1 - \sin \varphi) = \tan^2 (\pi/4 + \varphi/2)$$

$$q_f = q_0 + (c \cot \varphi + q_0) \{\exp[\pi \tan \varphi] \tan^2 (\pi/4 + \varphi/2) - 1\}$$
$$= c N_c + q_0 N_q$$

where

$$N_q = \exp[\pi \tan \varphi] \tan^2 (\pi/4 + \varphi/2)$$

and

$$N_c = \cot \varphi (N_q - 1)$$

11.6   *Upper bound solutions for rough foundations on weightless soil:*
(a) *Foundations on frictionless soil.* Consider a foundation
(Fig. 11.4(a)) of width B and infinite length, resting on the
surface of a uniform weightless frictionless soil. The velocity
field below the foundation at failure will be assumed to
consist of five zones, as follows:

(i) Zone A moves vertically as a rigid body with a
velocity v_A equal to that of the foundation.* No slip
takes place at the boundary between the soil and
the foundation, which may be assumed to be
perfectly rough.
(ii) Zones B and D are zones of radial shear, bounded
by the circular arcs LM and LQ.
(iii) Zones C and E move as rigid bodies sliding on MN
and QR.

Energy is dissipated

1.   by sliding on thin deforming layers along JL, KL,
MN, and QR, and
2.   in deforming the radial shear zones B and D.

---

* It might appear that this mechanism is not kinematically admissible, since
the tip of the wedge A must penetrate into the stable soil below Point L. This
is admissible for small movements, however, since the width of the tip of the
wedge is infinitely small.

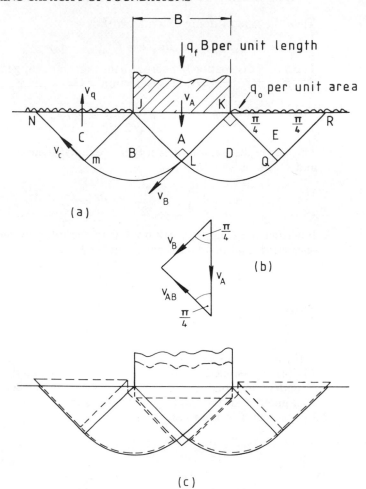

Fig. 11.4   *A rough foundation on weightless frictionless soil: an upper bound solution (after Chen [11.11]).*

Zone A moves down with a velocity v_A. Since the soil below L does not move, Zone B at L must move parallel to the tangent to LM, and therefore normal to JL. From the velocity diagram (Fig. 11.3(b)) it may be seen that v_B (the velocity of B at L) is $v_A/\sqrt{2}$, while v_{AB} (the velocity of B relative to A) is $v_A/\sqrt{2}$ parallel to JL. Zone C moves with a velocity $v_C = v_A/\sqrt{2}$ parallel to MN, and the ground surface JN rises with a velocity $v_A/2$.

Let the mean vertical foundation pressure be q_f, and let the remainder of the ground surface carry a vertical pressure q_0.

Then the rate of doing external work is

$$q_f B v_A - 2q_0 B v_A/2 = B v_A(q_f - q_0)$$

The rate of dissipation of energy on the thin deforming layers JL, KL, MN, and QR is (see Section 8.13)

$$\sum clv\cos\varphi = 4c\frac{B}{\sqrt{2}}\frac{v_A}{\sqrt{2}} = 2cBv_A$$

The rate of dissipation of energy in the radial shear zones B and D is (see Section 8.15)

$$4c\frac{\pi}{2}\frac{v_A}{\sqrt{2}}\frac{B}{\sqrt{2}} = \pi cBv_A$$

Equating the external work done with the internal energy dissipated, and dividing throughout by Bv_A,

$$q_f = c(2 + \pi) + q_0$$
$$= cN_c + q_0 N_q$$

where

$$N_c = 2 + \pi = 5\cdot14 \quad \text{and} \quad N_q = 1\cdot0$$

(b) *A foundation on frictional soil.* The velocity field (Fig. 11.5(a)) is assumed to consist of two zones of radial shear bounded by logarithmic spirals, which move between triangular rigid zones sliding on thin deforming layers. From the geometry of the figure,

$$JL = \frac{B}{2}\sec(\pi/4 + \varphi/2)$$

$$JM = MN = \frac{B}{2}\exp\left[\frac{\pi}{2}\tan\varphi\right]\sec(\pi/4 + \varphi/2)$$

$$JN = B\exp\left[\frac{\pi}{2}\tan\varphi\right]\tan(\pi/4 + \varphi/2)$$

Let Zone A move down with velocity v_A. The direction of motion of Zone B at L must be inclined at φ to the logarithmic spiral LM, and is therefore normal to JL. From the velocity diagram (Fig. 11.5(b)) it may be seen that v_B (the velocity of B at L) is $\frac{1}{2}v_A\sec(\pi/4 + \varphi/2)$, while v_{AB} (the velocity of B relative to A) is $\frac{1}{2}v_A\sec(\pi/4 + \varphi/2)$ inclined at φ to JL. Zone C moves with a velocity

$$v_C = \frac{1}{2}v_A\exp\left[\frac{\pi}{2}\tan\varphi\right]\sec(\pi/4 + \varphi/2)$$

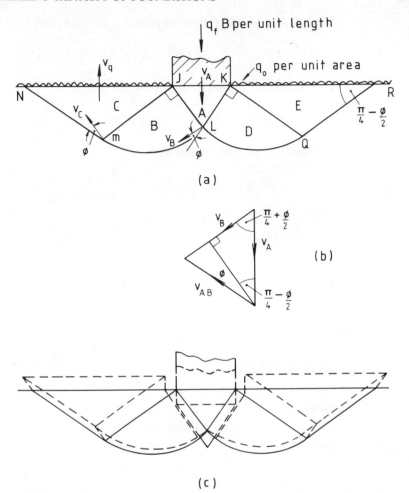

Fig. 11.5 *A rough foundation on weightless frictional soil: an upper bound solution (after Chen [11.11]).*

inclined at φ to MN, and the ground surface JN rises with a velocity

$$v_C \sin(\pi/4 + \varphi/2) = \tfrac{1}{2}v_A \exp\left[\frac{\pi}{2}\tan\varphi\right]\tan(\pi/4 + \varphi/2)$$

The rate of doing external work is

$$q_f B v_A - 2q_0 \frac{B v_A}{2}\left\{\exp\left[\frac{\pi}{2}\tan\varphi\right]\tan(\pi/4 + \varphi/2)\right\}^2$$

$$= B v_A\{q_f - q_0 \exp[\pi\tan\varphi]\tan^2(\pi/4 + \varphi/2)\}$$

The rate of dissipation of energy on JL, KL, MN, and QR is

$$\sum clv \cos\varphi = 2c\frac{B}{2}\frac{v_A}{2}\sec^2(\pi/4 + \varphi/2)(1 + \exp[\pi\tan\varphi])\cos\varphi$$

The dissipation rate in Zones B and D is

$$2c\frac{v_A}{2}\frac{B}{2}\sec^2(\pi/4 + \varphi/2)\cot\varphi\,(\exp[\pi\tan\varphi] - 1)$$

Collecting terms and noting that

$$\sec^2(\pi/4 + \varphi/2) = \frac{2}{(1 - \sin\varphi)}$$

the total dissipation rate is

$$cBv_A\left(\frac{\cot\varphi}{1 - \sin\varphi}\right)$$

$$\times \{\sin\varphi(1 + \exp[\pi\tan\varphi]) + \exp[\pi\tan\varphi] - 1\}$$

$$= cBv_A\cot\varphi\left\{\exp[\pi\tan\varphi]\left(\frac{1 + \sin\varphi}{1 - \sin\varphi}\right) - 1\right\}$$

$$= cBv_A\cot\varphi\,\{\exp[\pi\tan\varphi]\tan^2(\pi/4 + \varphi/2) - 1\}$$

Equating external work done to internal energy dissipated,

$$q_f = cN_c + q_0 N_q$$

where

$$N_q = \exp[\pi\tan\varphi]\tan^2(\pi/4 + \varphi/2)$$

and

$$N_c = \cot\varphi(N_c - 1)$$

11.7 *Validity of the solutions for weightless soil:* The stress fields assumed for smooth foundations (Fig. 11.2 and 11.3) would also be statically admissible if the foundations were rough. Also, the velocity fields assumed for rough foundations (Figs. 11.4 and 11.5) would be kinematically admissible if the foundations were smooth. Thus the lower bound solution for a smooth foundation is also a lower bound if it is rough, while the upper bound solution for a rough foundation is also an upper bound if it is smooth. Since the upper and lower bound solutions are the same, it follows that the solutions are exact *whether the foundations are rough or smooth.* Notice however that this has only been proved for the particular case of a foundation of infinite length on the surface of a weightless soil. Roughness may have an effect in the case of foundations of finite length, or at finite depth below the surface, or on soil having weight. The angle of friction required to prevent all movement between the soil and the foundation is generally small, and in practice all foundations can be assumed to be absolutely rough.

These solutions were first obtained by Prandtl [11.2] for N_c and Reissner [11.13] for N_q.

11.8   Bearing capacity of foundations on soil having weight: No closed solutions have been obtained for the bearing capacity of foundations on soils which have weight, and the approximate methods described in Chapter 8 must be used. As in the case of lateral earth pressure, the computation can be simplified if we assume that we may apply the principle of superposition, and write

$$q_f = cN_c + q_0N_q + \tfrac{1}{2}B\gamma N_\gamma$$

where N_c and N_q have the values obtained above for weightless soil, and N_γ is a coefficient defining the bearing capacity of a soil having weight but no cohesion or surcharge ($c = q_0 = 0$). As before, we must remember that superposition cannot be validly applied when considering the behaviour of a plastic material. The value of q_f is underestimated by this procedure. Terzaghi [11.3] estimated that, for $\varphi = 34°$, the error in the computed value of q_f is about 10%. Other workers have found rather greater discrepancies in some cases, but the procedure is sufficiently accurate for use in practice.

Exact expressions for N_c and N_q were obtained above, but the values of N_γ are still not certainly known. Table 11.1 shows values of N_γ obtained by five methods. Terzaghi [11.3] used the method of limiting equilibrium, assuming that the foundation was rough, and that the rupture surfaces were

TABLE 11.1   *Computed values of N_γ for a strip foundation of infinite length on the surface of a uniform soil*

φ (Degrees)	Terzaghi [11.3]	Meyerhof [11.10]	Sokolovskii [11.1]	Chen [11.11]	Hansen [11.12]
0	0·0	0·0	0·0	0·0	0·0
5	0·05	0·05	0·17	0·46	0·09
10	0·6	0·6	0·6	1·31	0·47
15	1·8	1·8	1·4	2·94	1·42
20	4·9	4·8	3·2	6·20	3·54
25	11·1	10·7	6·9	12·97	8·11
30	24·0	22·9	15·3	27·67	18·08
35	51·8	48·4	35·2	61·49	40·70
40	128·0	116·0	86·5	145·3	95·45

TABLE 11.2   *Bearing capacity coefficients N_c, N_q and N_γ for a strip foundation of infinite length on the surface of a uniform soil.*

φ (Degrees)	N_c Prandtl [11.23]	N_q Reissner [11.13]	N_γ Hansen [11.12]
0	5·14	0·0	0·0
2	5·63	1·20	0·01
4	6·18	1·43	0·05
6	6·81	1·72	0·13
8	7·52	2·06	0·27
10	8·34	2·47	0·47
12	9·28	2·97	0·75
14	10·37	3·59	1·16
16	11·63	4·33	1·72
18	13·10	5·26	2·49
20	14·83	6·40	3·54
22	16·88	7·82	4·96
24	19·32	9·60	6·89
26	22·25	11·85	9·35
28	25·80	14·72	13·13
30	30·14	18·40	18·08
32	35·49	23·18	24·94
34	42·16	29·44	34·53
36	50·58	37·75	48·06
38	61·35	48·93	67·41
40	75·31	64·19	95·45

combinations of logarithmic spirals and straight lines. Meyerhof [11.4, 11.10] has shown that Terzaghi placed unnecessary restrictions on the possible positions of his rupture surfaces, and, using a method similar to Terzaghi's, has obtained rather smaller values. Sokolovskii [11.1] used the slip-line field method to obtain values for smooth foundations. Chen [11.11] used the upper bound method to examine a number of possible failure mechanisms, and has suggested the approximate expression

$$N_\gamma = 2\{1 + \exp[\pi \tan \varphi] \tan^2 (\pi/4 + \varphi/2)\}$$
$$\times \tan \varphi \tan (\pi/4 + \varphi/5)$$

Hansen [11.12] has suggested the approximate expression

$$N_\gamma = 1·80(N_q - 1) \tan \varphi$$

The solutions on which Chen's approximation is based are valid upper bounds. Sokolovskii's solution is probably, but not certainly, a lower bound. The status of the other solutions is not known. Hansen's solution is convenient and is probably as accurate as any available. Table 11.2 gives the values of N_c and N_q from Sections 11.4 to 11.7 above and Hansen's values of N_γ.

There is, in any case, little to be gained by precise determination of N_γ. Inspection of Table 11.2 shows that an increase of $2°$ in φ increases N_γ by about 40%. It is unlikely in practice that φ will be known with greater certainty than this.

EFFECT OF DEPTH AND SHAPE OF FOUNDATION

11.9 Terzaghi's method for shallow foundations: In analysing the previous cases, the foundation has been assumed to be at the surface of the ground. If the foundation is placed at a depth below the surface, the bearing capacity is increased. Provided that the depth (D) below the surface does not exceed the width (B) of the foundation, Terzaghi suggested that a sufficiently accurate solution is obtained by assuming that the soil above the level of the foundation is replaced by

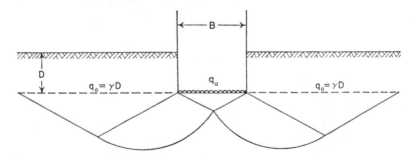

Fig. 11.6   Terzaghi's analysis for shallow foundations.

an equivalent surcharge load $q_0 = \gamma D$ (see Fig. 11.6). This approximation has the effect of ignoring the shearing resistance of that part of the rupture surface which is above foundation level, and the bearing capacity is therefore under-estimated. Provided that $D \leq B$, the error is likely to be small, but the method becomes increasingly conservative as the depth increases.

Because of the complications, few general solutions have been obtained for foundations of finite length. Terzaghi has suggested that the bearing capacity of a square foundation may be written

$$q_f = 1{\cdot}3cN_c + q_0N_q + 0{\cdot}4B\gamma N_\gamma$$

and for a circular foundation

$$q_f = 1{\cdot}3cN_c + q_0N_q + 0{\cdot}3B\gamma N_\gamma$$

where N_c, N_q and N_γ have the values computed for a foundation of infinite length. Others have suggested very similar shape factors.

11.10   Meyerhof's solution for a shallow foundation [11.4]:   For foundations below ground level, Meyerhof analysed the form of failure shown in Fig. 11.7(a). The curved rupture surfaces in the zone of radial shear are assumed to be logarithmic spirals, and all other rupture surfaces are assumed to be plane. Meyerhof's rupture surfaces are more general than Terzaghi's, in that the inclination of the discontinuity *AB* is allowed to have any value between φ and $(\pi/4 + \varphi/2)$ to the horizontal, while the logarithmic spirals in the zone of radial shear may have any centre. The slope of the critical surface obtained in this way is not necessarily continuous at the points C and C', but this is to be expected, since the logarithmic spiral is only an approximation to the real form of the critical surface.

Then, if q_0 is the normal stress on the surface *DA* (called

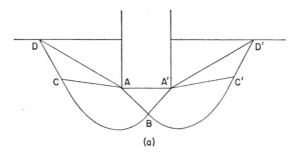

(a)

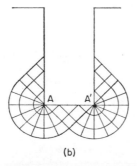   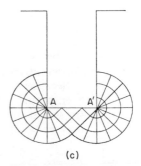

(b)   (c)

Fig. 11.7   Meyerhof's analysis for shallow and deep foundations: (a) Shallow foundations. (b) Deep foundations on frictionless soils—no adhesion to the sides of the foundation. (c) As above but with full adhesion to the sides of the foundation.

the *equivalent free surface*), Meyerhof showed that the bearing capacity may be written in the form

$$q_f = cN_c + q_0N_q + 0 \cdot 5B\gamma N_\gamma$$

This expression is in the same form as Terzaghi's, but the coefficients N_c, N_q, and N_γ are functions of the depth (D) and the shape of the foundation, as well as of the angle of shearing resistance (φ).

Unfortunately, the equivalent free surface cannot be located directly, and the stress q_0 can only be determined by a semi-graphical method which is somewhat cumbersome for routine use.

11.11   *Deep foundations in frictionless soil:*   In the case of deep foundations, the rupture surfaces do not reach the ground surface. In this case, Meyerhof analysed the forms of failure shown in Fig. 11.7(b) and (c), and has computed coefficients N_{cq} for an infinitely long foundation such that

$$q_f = cN_{cq} + \gamma D$$

where N_{cq} depends on both N_c and N_q, although the effect of the latter is small. N_{cq} increases from Prandtl's value of $5 \cdot 14$ at the surface to a maximum at depths exceeding about $2B$. This maximum value depends on the roughness of the sides of the foundation, and is equal to $2\pi + 2 = 8 \cdot 28$ where the shaft is perfectly smooth and $2 \cdot 5\pi + 1 = 8 \cdot 85$ where there is complete adhesion between the soil and the sides of the foundation.

For circular and rectangular foundations, Meyerhof has shown that

$$q_f = \lambda cN_{cq} + \gamma D$$

where $\lambda \simeq 1 \cdot 15$ for circular footings
and $\lambda \simeq (1 + 0 \cdot 15B/L)$ for rectangular foundations.

Gibson [11.5], developing a method of analysis suggested by Bishop, Hill and Mott [11.6], showed that, for a deep circular foundation,

$$N_c = \frac{4}{3}\left(\log_e \frac{E}{c} + 1\right) + 1$$

where E is the elastic modulus of the soil.

In deriving this expression, it is assumed that the rupture zone is spherical, and is entirely surrounded by soil in the elastic state. Gibson assumed E/c to vary between 50 and 200, giving N_c in the range from $7 \cdot 6$ to $9 \cdot 4$. The rather larger

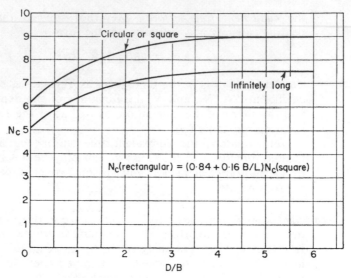

After Skempton [11.7]

Fig. 11.8 Skempton's values for N_c.

values of E/c quoted in Section 12.15 would give marginally larger values for N_c.

Based on a critical examination of this analysis, and of the methods of Meyerhof and others, and on the results of a number of model tests, Skempton [11.7] suggested that the bearing capacity may be taken to be

$$q_f = cN_c + q_0$$

where N_c has the values given in Fig. 11.8. Alternatively, N_c may be taken to have the approximate value

$$N_c \simeq 5\left(1 + \frac{B}{5L}\right)\left(1 + \frac{D}{5B}\right) \quad \text{if} \quad D \le 2{\cdot}5B$$

or

$$N_c \simeq 7{\cdot}5\left(1 + \frac{B}{5L}\right) \quad\quad\quad \text{if} \quad D \ge 2{\cdot}5B$$

These values of N_c compare well with values inferred from known failures, although N_c may be expected to vary somewhat in different soils as a result of the variation in the value of E/c. Bjerrum has suggested slightly greater values for deep foundations in certain Norwegian clays [11.8].

11.12 Deep foundations in cohesionless soils: Meyerhof has computed resultant bearing capacity factors $N_{\gamma q}$ such that

$$q_f = 0{\cdot}5B\gamma N_{\gamma q}$$

where $N_{\gamma q}$ depends on N_γ and N_q, on the angle of shearing resistance (φ), on the roughness of the sides of the foundation, and on the lateral earth pressure on the sides of the foundation ($K_s \gamma D$).

Meyerhof's values of q_f are large for deep foundations, and, being based on an assumed rupture surface, probably represent an upper limit for the bearing capacity. Moreover, the settlements which would occur before the full bearing resistance was developed would generally be intolerably large. The allowable bearing pressure on a deep foundation in cohesionless soil is almost always limited by the allowable settlement.

ALLOWABLE BEARING PRESSURE

11.13 Total and effective stress calculations: The expressions given above have been derived solely from the conditions of equilibrium and from the Mohr–Coulomb failure condition. They are therefore equally valid if expressed in terms of effective stress. For example, in terms of total stress, Terzaghi's expression for the bearing capacity of a square foundation is

$$q_f = 1 \cdot 3 c N_c + q_0 N_q + 0 \cdot 4 B \gamma N_\gamma$$

Similarly, in terms of effective stress,

$$q'_f = 1 \cdot 3 c' N_c + q'_0 N_q + 0 \cdot 4 B Z N_\gamma$$

where q'_f is the effective stress on the surface AA' at failure

q'_0 is the effective surcharge at foundation level

N_c, N_q, and N_γ are the bearing capacity coefficients corresponding to the value of φ'

Z is the vertical component of the effective body force per unit volume (see Section 8.6).

Above the free surface, which is assumed to be coincident with the water table, $Z = \gamma$. If there is no flow, $Z = (\gamma - \gamma_w)$ below the free surface.

Example 11.1

A foundation 2·0 m square is installed 1·2 m below the surface of a uniform sandy gravel having a density (γ) of 19·2 kN/m³ above the water table, and a submerged density ($\gamma - \gamma_w$) of 10·1 kN/m³. The shear strength parameters with respect to effective stress are

$$c' = 0 \qquad \varphi' = 30°$$

For $\varphi' = 30°$, Table 11.2 gives $N_q = 18\cdot4$ and $N_\gamma = 18\cdot1$.

(a)   If the water table is well below the base of the foundation (so that the whole of the plastic zones are above the water table),

$$q'_0 = q_0 = 1\cdot2 \times 19\cdot2 = 23\cdot0\,\text{kN/m}^2; \quad Z = \gamma = 19\cdot2\,\text{kN/m}^3$$

then the gross ultimate bearing capacity, in terms of effective stress, is

$$q'_f = q_f = 23\cdot0 \times 18\cdot4 + 0\cdot4 \times 2\cdot0 \times 19\cdot2 \times 18\cdot1$$
$$= 701\,\text{kN/m}^2$$

(b)   If the water table rises to the level of the base of the foundation,

$$q'_0 = 23\cdot0\,\text{kN/m}^2; \quad Z = 10\cdot1\,\text{kN/m}^3$$

then

$$q'_f = 23\cdot0 \times 18\cdot4 + 0\cdot4 \times 2\cdot0 \times 10\cdot1 \times 18\cdot1$$
$$= 569\,\text{kN/m}^2$$

(c)   If the water table rises to ground level,

$$q'_0 = 1\cdot2 \times 10\cdot1 = 12\cdot1\,\text{kN/m}^2; \quad Z = 10\cdot1\,\text{kN/m}^3$$

then

$$q'_f = 12\cdot1 \times 18\cdot4 + 0\cdot4 \times 2\cdot0 \times 10\cdot1 \times 18\cdot1$$
$$= 369\,\text{kN/m}^2$$

This illustrates the considerable reduction in the bearing capacity which occurs if a foundation is flooded by a rising water table.

11.14   *The effect of upward flow of water:*   The bearing capacity is further reduced if there is upward flow of water in the soil. Where there is an upward hydraulic gradient (i_z), the effective vertical body force per unit volume is

$$Z = (\gamma - \gamma_w) - i_z\gamma_w$$

Example 11.2
If, in the previous example, the water table had been maintained at ground level, but there had been an upward hydraulic gradient in the soil of $0\cdot5$, then

$$Z = 10\cdot1 - 0\cdot5 \times 9\cdot81 = 5\cdot2\,\text{kN/m}^3$$
$$q'_0 = 5\cdot2 \times 1\cdot2 = 6\cdot2\,\text{kN/m}^2$$

Then

$$q'_f = 6\cdot2 \times 18\cdot4 + 0\cdot4 \times 2\cdot0 \times 5\cdot2 \times 18\cdot1$$
$$= 189\,\text{kN/m}^2$$

This illustrates the loss of bearing capacity which results from upward flow, even where—as in this case—the hydraulic gradient is considerably less than the critical value (i_c).

11.15 Net bearing pressure and the load factor: The ultimate bearing capacity (q_f) is the gross pressure at foundation level at failure. However, a uniform pressure of q_0 over the whole surface of the ground gives rise to no shear stress. The shear stresses in the soil are governed by the net difference (q_n) between the foundation pressure and the surcharge. Then, at failure,

$$q_n = q_f - q_0$$

The net allowable foundation pressure is determined by dividing the net ultimate bearing capacity by a suitable load factor (F). The value of this load factor is a matter for the engineer to decide in each case, but it is usual to ensure that F is at least 3·0. Then the gross allowable foundation pressure is

$$q_a = \frac{q_f - q_0}{F} + q_0$$

But

$$q'_f - q'_0 = (q_f - u) - (q_0 - u)$$
$$= q_f - q_0$$

Then, in terms of effective stress,

$$q_a = \frac{q'_f - q'_0}{F} + q_0$$

Example 11.3
(a) In Example 11.1(a), where the water table is well below the base of the foundation,

$$q'_0 = q_0 = 23\cdot0 \text{ kN/m}^2$$

Then, if $F = 3\cdot0$,

$$q_a = \frac{701 - 23\cdot0}{3\cdot0} + 23\cdot0$$

$$= 249 \text{ kN/m}^2$$

(b) If the water table rises to the ground surface,

$$q'_0 = 12\cdot1 \text{ kN/m}^2 \qquad q_0 = 23\cdot9 \text{ kN/m}^2$$

and

$$q_a = \frac{369 - 12\cdot1}{3\cdot0} + 23\cdot9$$

$$= 143 \text{ kN/m}^2$$

APPLICATIONS TO PRACTICAL PROBLEMS

11.16   *Bearing capacity of foundations on clay soils:*   The application of structural loads to the ground results in excess pore pressures within the soil. The rate at which these excess pore pressures are dissipated in clay soils is generally slow compared with the rate of application of the loads. Clay soils supporting foundations are therefore usually almost undrained at the end of the construction period. In the case of saturated clays, the stability at this moment may be computed by a total stress analysis, based on the undrained shear strength of the clay. Since the structural loads increase the principal stresses in the soil, the pore pressures are increased during loading, and subsequently reduce. Thus the effective stresses have their least value at the end of the construction period, and increase as the soil consolidates. The foundation therefore becomes most stable with time, and the long term stability need not generally be considered, provided that the foundation is shown to be safe initially. If the construction period is unusually long, or if the load is applied in stages over a long period, some significant dissipation of the excess pore pressure may take place before loading is complete. A total stress analysis based on the undrained condition may then be unduly conservative. An analysis must be made in terms of effective stress, taking account of the dissipation of the pore pressure during loading.

11.17   *Bearing capacity of foundations on sands and gravels:*   These soils have high permeabilities, and are almost always fully drained. The bearing capacity is computed in terms of effective stress, the pore pressures in the soil being determined either from a flow net, or from observations of the water levels in borings on the site.

The principal difficulty in practice lies in determining the angle of shearing resistance of the soil, since it is not generally possible to obtain undisturbed samples for testing. It is usually necessary to make some empirical estimate, based on the results of penetration tests.

The bearing capacity of sands and gravels increases rapidly with increasing size of the foundation. Allowable loads on large foundations are therefore limited only by the allowable settlements. The bearing capacity is seldom critical, except in the case of small footings on saturated soils.

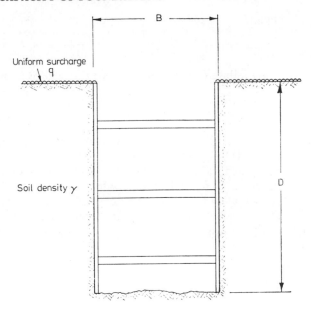

Fig. 11.9 Base heave in excavations.

11.18 Base heave in excavations: Where deep excavations are made in soft clays, shear failure may occur in the soil at the base, resulting in an inflow of clay into the excavation. This is, in effect, a foundation failure in reverse, the shear stresses in the soil being caused by the weight of the material removed from the excavation.

Since the problem arises in temporary excavations, a total stress analysis based on the undrained strength is relevant. Then (Fig. 11.9)

$$\gamma D + q = \frac{c_u N_c}{F}$$

$$F = \frac{c_u N_c}{\gamma D + q}$$

Example 11.4

An excavation, 5 m in diameter, is to be made to a depth of 4 m in soft clay having a density of 19·0 kN/m³ and an undrained shear strength of 16·5 kN/m². The ground surrounding the excavation carries a surcharge load of 10 kN/m². From Fig. 11.8

$$\frac{D}{B} = 0.8 \quad \text{and} \quad N_c = 7.5$$

Then

$$F = \frac{7 \cdot 5 \times 16 \cdot 5}{19 \cdot 0 \times 4 + 10} = 1 \cdot 44$$

At failure (*i.e.* where $F = 1 \cdot 0$) the critical depth is

$$D_c = \frac{c_u N_c - q}{\gamma}$$

Bjerrum and Eide have examined a number of cases of base failure [11.8], and have shown good agreement between the calculated value of D_c and the depth at which failure was observed. The calculated value appears to be slightly conservative for very deep excavations, but they state that this may have been due to the effect of a weathered crust near the surface of the clay in some of the cases they examined.

REFERENCES

11.1   SOKOLOVSKII, V. V. 1965. *Statics of granular media.* Pergamon (London).
11.2   PRANDTL, L. 1920. Uber die Härte plastischer Körper. *Nachr. Kgl. Ges. Wiss. Göttingen Math. Phys. Kl.*
11.3   TERZAGHI, K. 1943. *Theoretical soil mechanics.* Wiley (New York).
11.4   MEYERHOF, G. G. 1951. The ultimate bearing capacity of foundations. *Géotechnique,* 2.
11.5   GIBSON, R. E. 1950. Discussion on 'The bearing capacity of screwed piles and screwcrete cylinders' (Wilson). *Jour. Inst. Civil Eng.,* 34.
11.6   BISHOP, R. F., HILL, R. and MOTT, N. F. 1945. The theory of indentation and hardness tests. *Proc. Phys. Soc.,* 57.
11.7   SKEMPTON, A. W. 1951. The bearing capacity of clays. *Proc. Building Res. Congress.*
11.8   BJERRUM, L. and EIDE, O. 1956. The stability of strutted excavations in clay. *Géotechnique,* 6.
11.9   British Standards Institution. *Code of Practice for Foundations.* CP 2004: 1972.
11.10 MEYERHOF, G. G. 1948. An investigation of the bearing capacity of shallow footings in dry sand. *Proc. 2nd Int. Conf. on Soil Mech. and Found. Eng.*
11.11 CHEN, W-F. 1975. *Limit analysis and soil plasticity.* Elsevier (Amsterdam).
11.12 HANSEN, J. B. 1961. A general formula for bearing capacity. *Danish Geotech. Inst. Bull.,* 11.
11.13 REISSNER, H. 1924. Zum Erddruckproblem. *Proc. 1st Int. Conf. Appl. Mech. (Delft).*

CHAPTER 12

Settlement of foundations

SETTLEMENT ANALYSIS

12.1   Stability and deformation: The previous three chapters have been mainly concerned with the ultimate collapse of structures, as a result of shear stresses which exceed the shear strength of the soil. In some cases, however, large soil deformations may occur before this state of ultimate collapse is reached, and such deformations may lead to unacceptable movements and distortion of the superimposed structures. In particular, considerable foundation settlements may be caused by loads which are much less than those which are required to cause complete collapse.

In designing foundations, therefore, it is generally not enough to ensure that the soil has adequate bearing capacity. It is also necessary to estimate the likely settlements under the effect of the working loads, and to ensure that these settlements are not so large as to endanger the structures above.

12.2   Distortion and consolidation: Settlement of a foundation results from

(a)   distortion of the underlying soil mass, in consequence of the shear stresses imposed by the external loads, and

(b)   volume changes which occur as the soil is consolidated.

Distortion, which is partly elastic and partly plastic, takes place relatively rapidly, and is substantially complete within the period of construction. In soils of very low permeability, however, the consolidation settlement may continue throughout the working life of the structure. In assessing the effects of consolidation, therefore, it is generally necessary to examine both the extent of the settlement and the rate at which it takes place.

Because of the non-linear stress/strain properties of soils, and because of the complex geometry of most practical problems, it is generally necessary to make drastic simplifications before even an approximate estimate of settlement is possible. In estimating settlements due to distortion, the

313

soil is usually assumed to behave as a uniform isotropic elastic solid conforming to Hooke's law. In computing consolidation settlements, it is usual to assume that

(a) the stress distribution in the soil is that which would occur in a uniform elastic material conforming to Hooke's law, and

(b) the consolidation is one-dimensional, in a vertical direction only. The volume change is assumed to be dependent only on the change in the vertical component of normal effective stress (σ'_z).

Before considering the settlements of real foundations, it is therefore necessary to determine the stresses and deformations in this hypothetical elastic material under different configurations of load.

STRESS IN A SEMI-INFINITE ELASTIC MEDIUM LOADED AT THE SURFACE

12.3 *Stress resulting from a point load normal to the surface:* Boussinesq determined the stress induced at any point within a semi-infinite elastic medium by a single point load (Q) normal to the surface [12.1]. In polar co-ordinates (Fig. 12.1), the components of this stress are

$$\sigma_z = \frac{Q}{z^2}\frac{3}{2\pi}\frac{z^5}{R^5} = \frac{Q}{z^2}\frac{3}{2\pi}\cos^5\psi$$

$$\sigma_r = \frac{Q}{z^2}\left(\frac{3z^3 r^2}{2\pi R^5} - \frac{(1-2v)z^2}{R(R+z)}\right)$$

$$\sigma_\theta = \frac{Q}{z^2}\frac{(2v-1)}{2\pi}\left(\frac{z^3}{R^3} - \frac{z^2}{R(R+z)}\right)$$

$$\tau_{rz} = \frac{Q}{z^2}\frac{3}{2\pi}\frac{z^4 r}{R^5} \qquad \tau_{\theta r} = \tau_{\theta z} = 0$$

where the significance of z, r, R, and ψ are as shown in Fig. 12.1.

Only the vertical normal stress component (σ_z) and the shear stress component (τ_{rz}) are independent of Poisson's ratio (v). If $v = 0.5$ (*i.e.* if there is no volume change), σ_θ is everywhere zero.

The vertical normal stress component (σ_z) may be written in the form

$$\sigma_z = \frac{Q}{z^2}I_\sigma$$

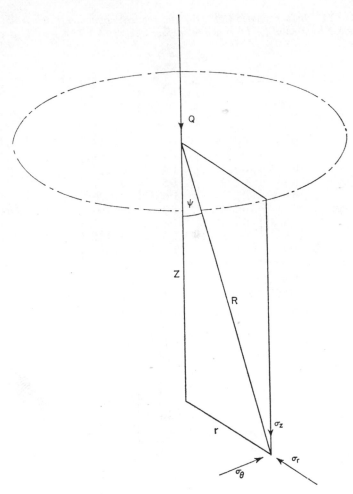

Fig. 12.1 Stress beneath a point load at the surface of a semi-infinite elastic solid (Boussinesq's solution).

where

$$I_\sigma = \frac{3}{2\pi} \frac{1}{[1 + (r/z)^2]^{5/2}}$$

Values of I_σ are given in Table 12.1.

12.4 *Stress below the centre of a uniformly-loaded circular area at the surface:* Where the load is distributed over the surface, the resultant stress may be determined by integrating Boussinesq's expressions over the relevant area. As an example of this, consider the vertical component of stress

TABLE 12.1 *Influence coefficients for the vertical components of normal stress* (σ_z) *induced by a point load* (Q) *at the surface of a semi-infinite elastic solid*

$$\sigma_z = \frac{Q}{z^2} I_\sigma$$

r/z	I_σ	r/z	I_σ	r/z	I_σ
0	0·4775				
0·1	0·4657	1·1	0·0658	2·1	0·0070
0·2	0·4329	1·2	0·0513	2·2	0·0058
0·3	0·3849	1·3	0·0402	2·3	0·0048
0·4	0·3294	1·4	0·0317	2·4	0·0040
0·5	0·2733	1·5	0·0251	2·5	0·0034
0·6	0·2214	1·6	0·0200	3·0	0·0015
0·7	0·1762	1·7	0·0160	4·0	0·0004
0·8	0·1386	1·8	0·0129	5·0	0·0001
0·9	0·1083	1·9	0·0105	10·0	0·0000
1·0	0·0844	2·0	0·0085		

below the centre of a uniformly-loaded circular area of radius a. The load on the element of area shown in Fig. 12.2 is $qr\,d\theta\,dr$. At a depth z below the centre of the loaded area,

$$
\begin{aligned}
\sigma_z &= \int_0^{2\pi} \int_0^a \frac{qr}{z^2} \frac{3}{2\pi} \frac{z^5}{R^5} d\theta\,dr \\
&= \frac{3qz^3}{2\pi} \int_0^{2\pi} \int_0^a \frac{r}{R^5} d\theta\,dr \\
&= \frac{3qz^3}{2\pi} \int_0^{2\pi} \int_0^a \frac{r\,d\theta\,dr}{(r^2 + z^2)^{5/2}} \\
&= 3qz^3 \int_0^a \frac{r\,dr}{(r^2 + z^2)^{5/2}} \\
&= q\left(1 - \frac{1}{[1 + (a/z)^2]^{3/2}}\right) \\
&= qI_\sigma
\end{aligned}
$$

where

$$I_\sigma = 1 - \frac{1}{[1 + (a/z)^2]^{3/2}}$$

Values of I_σ for this case are given in Table 12.2.

Example 12.1

Compare the vertical normal stress components at points vertically below

(a) a point load of 500 kN at the surface, and

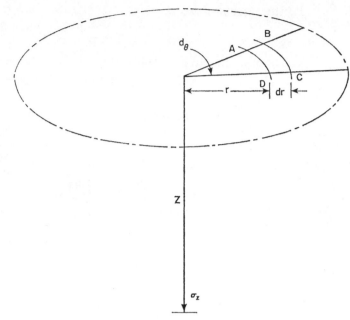

Fig. 12.2 Stress below the centre of a uniformly-loaded circular area at the surface.

    (b)   the same load spread over a circular area of radius 1·0 m.

For the point load,

$$\sigma_z = \frac{Q}{z^2} I_\sigma = \frac{Q}{z^2} \times 0.478$$

TABLE 12.2 *Influence coefficients for the vertical component of normal stress* (σ_z) *below the centre of a uniformly-loaded circular area at the surface.*

$$\sigma_z = q I_\sigma$$

z/a	I_σ	z/a	I_σ	z/a	I_σ
0	1·000				
0·1	0·999	1·1	0·595	2·1	0·264
0·2	0·992	1·2	0·547	2·2	0·245
0·3	0·970	1·3	0·502	2·3	0·229
0·4	0·949	1·4	0·461	2·4	0·214
0·5	0·911	1·5	0·424	2·5	0·200
0·6	0·864	1·6	0·390	3·0	0·146
0·7	0·818	1·7	0·360	4·0	0·087
0·8	0·756	1·8	0·332	5·0	0·057
0·9	0·701	1·9	0·307	10·0	0·015
1·0	0·646	2·0	0·284		

For the circular load,

$$\sigma_z = \frac{Q}{\pi a^2} I_\sigma$$

Values of σ_z are given in Table 12.3

TABLE 12.3 (*Example* 12.1)

z (m)	Point load		Circular load	
	z^2	σ_z (kN/m²)	I_σ	σ_z (kN/m²)
0	0	∞	1·000	159
0·5	0·25	956	0·911	145
1·0	1·0	239	0·646	103
1·5	2·25	106	0·424	68
2·0	4·0	60	0·284	46
2·5	6·25	38	0·200	32
3·0	9·0	27	0·146	23
4·0	16·0	15	0·087	14
5·0	25·0	9·5	0·057	9·0

12.5 *Stress below a uniformly-loaded strip of infinite length:* In this case, it may be shown that the principal stresses in the elastic medium are

$$\sigma_1 = \frac{q}{\pi}(\beta + \sin\beta)$$

$$\sigma_3 = \frac{q}{\pi}(\beta - \sin\beta)$$

$$\tau_{\max} = \frac{q}{\pi}\sin\beta$$

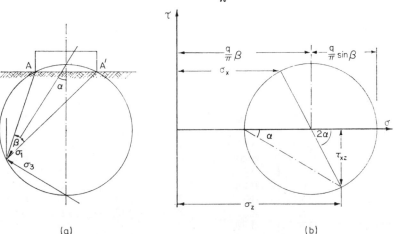

(a)   (b)

Fig. 12.3   *Stress beneath a uniform load of infinite length at the surface.*

where β is the angle subtended by the base of the foundation, as shown in Fig. 12.3(a). All the stress components are independent of the elastic constants in this case. Since the angle β is the same for any point on a circle drawn through A and A', all such circles are lines of constant principal stress. It may also be shown that the principal stresses act in the directions shown in Fig. 12.3(a).

Consider the Mohr circle diagram shown in Fig. 12.3(b). If α is the angle between the vertical and the direction of the major principal stress (σ_1),

$$\tau_{xz} = \tau_{max} \sin 2\alpha = \frac{q}{\pi} \sin \beta \sin 2\alpha$$

$$\sigma_z = \frac{q}{\pi}(\beta + \sin \beta \cos 2\alpha)$$

$$\sigma_x = \frac{q}{\pi}(\beta - \sin \beta \cos 2\alpha)$$

Values of β/π and $(\sin \beta)/\pi$ are given in Table 12.4.

TABLE 12.4 *Influence coefficients for the vertical component of normal stress* (σ_z) *induced by a uniform load of constant width and infinite length*

$$\sigma_z = \frac{q}{\pi}(\beta + \sin \beta \cos 2\alpha)$$

$\beta^{(\circ)}$	$\dfrac{\beta}{\pi}$	$\dfrac{\sin \beta}{\pi}$	$\beta^{(\circ)}$	$\dfrac{\beta}{\pi}$	$\dfrac{\sin \beta}{\pi}$
0	0	0			
5	0·028	0·028	95	0·528	0·317
10	0·056	0·055	100	0·556	0·314
15	0·083	0·082	105	0·583	0·308
20	0·111	0·109	110	0·611	0·299
25	0·139	0·135	115	0·639	0·289
30	0·167	0·159	120	0·667	0·276
35	0·194	0·183	125	0·694	0·261
40	0·222	0·205	130	0·722	0·244
45	0·250	0·225	135	0·750	0·225
50	0·278	0·244	140	0·778	0·205
55	0·306	0·261	145	0·806	0·183
60	0·333	0·276	150	0·833	0·159
65	0·361	0·289	155	0·861	0·135
70	0·389	0·299	160	0·889	0·109
75	0·417	0·308	165	0·917	0·082
80	0·444	0·314	170	0·944	0·055
85	0·472	0·317	175	0·972	0·028
90	0·500	0·318	180	1·000	0

12.6 Stress beneath a rectangular loaded area: Newmark [12.2] has shown that the vertical component of stress at a depth z below the corner of a uniformly-loaded rectangular area, of length L and breadth B, is

$$\sigma_z = qI_\sigma$$

where

$$I_\sigma = \frac{1}{4\pi}\left(\frac{2mn(m^2 + n^2 + 1)^{\frac{1}{2}}}{m^2 + n^2 + m^2n^2 + 1} \cdot \frac{m^2 + n^2 + 2}{m^2 + n^2 + 1}\right.$$
$$\left. + \tan^{-1}\frac{2mn(m^2 + n^2 + 1)^{\frac{1}{2}}}{m^2 + n^2 - m^2n^2 + 1}\right)$$

and

$$m = \frac{B}{z}; \quad n = \frac{L}{z}$$

The coefficients m and n are interchangeable.

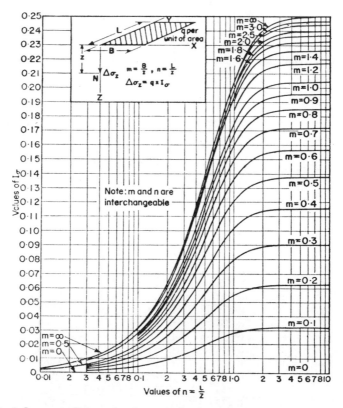

Fig. 12.4 *Influence coefficients for the vertical component of stress below the corner of a uniformly-loaded rectangular area.*

Fadum [12.3] has plotted curves (from which Fig. 12.4 has been prepared) from which I_σ may be obtained for this case.

Example 12.2

A load of 500 kN is uniformly distributed over a rectangle 1·0 m by 1·5 m. Determine the vertical component of stress at a depth of 2 m below one corner.

$$m = \frac{1\cdot0}{2\cdot0} = 0\cdot5 \qquad n = \frac{1\cdot5}{2\cdot0} = 0\cdot75$$

Therefore

$$I_\sigma = 0\cdot110 \text{ (from Fig. 12.4)}$$

$$\sigma_z = \frac{500}{1\cdot5 \times 1\cdot0} \times 0\cdot110 = 36\cdot8 \text{ kN/m}^2$$

Where the stress is required below some point other than a corner, or where the loaded area is not rectangular, the stress may be determined by taking the algebraic sum of the effects of a suitable combination of rectangular areas.

Example 12.3

For the same load as in the previous example, determine the stress at a depth of 2·0 m

(a)   below the centre (C), and
(b)   below the point marked G in Fig. 12.5.

(a)   The loaded area may be divided into four rectangles similar to $ACDE$. For each rectangle,

$$m = \frac{0\cdot5}{2\cdot0} = 0\cdot25 \qquad n = \frac{0\cdot75}{2\cdot0} = 0\cdot375$$

Then

$$I_\sigma = 0\cdot038 \text{ (from Fig. 12.4)}$$

$$\sigma_z = 4 \times \frac{500}{1\cdot5 \times 1\cdot0} \times 0\cdot038 = 51 \text{ kN/m}^2$$

(b)   The loaded area is twice the area $GHJF$ less twice the area $GAEF$.

For $GHJF$,

$$m = 1\cdot0 \qquad n = 0\cdot375 \quad \text{and} \quad I_\sigma = 0\cdot098$$

For $GAEF$,

$$m = 0\cdot5 \qquad n = 0\cdot375 \quad \text{and} \quad I_\sigma = 0\cdot068$$

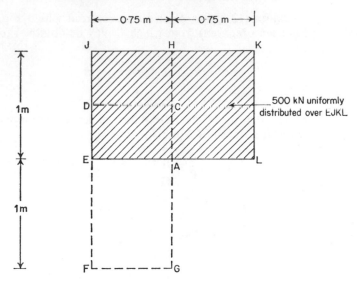

Fig. 12.5   Example 12.3.

Then

$$\sigma_z = \frac{500}{1 \cdot 5 \times 1 \cdot 0} \times (2 \times 0 \cdot 098 - 2 \times 0 \cdot 068)$$

$$= 20 \text{ kN/m}^2$$

12.7   *Newmark's influence charts:*   Newmark has prepared charts, similar to Fig. 12.6(b), on which the integration of Boussinesq's expression for σ_z may be conveniently carried out in cases where the loaded area is of irregular shape. Suppose that σ_z is required at some point (A) at a depth z below the surface. An outline plan of the loaded area is drawn to such a scale that the line OQ (Fig. 12.6(b)) represents z to the same scale. This outline is placed over the chart, so that the centre of the chart coincides with the point on the plan immediately over A. The 'squares' on the chart which are covered by the plan are counted. Then

$$\sigma_z = q \times \text{(number of 'squares')} \times I$$

where I is a scale factor for the particular chart being used.

Example 12.4

Determine the value of σ_z at a depth of 2 m below the loaded area shown in Fig. 12.6(a). The plan of the loaded area is drawn on Fig. 12.6(b) to such a scale that $OQ = 2$ m.

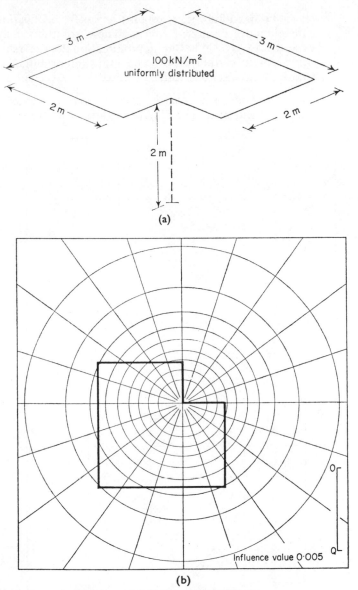

(a)

(b)

Fig. 12.6   Example 12.4 (Newmark's influence charts [12.4]).

The number of squares covered by the plan is 83·3. Then

$$\sigma_z = 100 \times 83\cdot3 \times 0\cdot005$$
$$= 41\cdot6 \text{ kN/m}^2$$

12.8   Other elastic models: The stress distribution has been analysed in
a number of cases where loads are applied to the surfaces of

elastic bodies differing in certain respects from Boussinesq's simple model discussed above. These solutions may be expected to give a better approximation to the real soil behaviour in certain cases. The more important of these solutions are discussed briefly below.

(a) *Stresses in an elastic layer of finite thickness.* Biot [12.5] and Burmister [12.6] have analysed the stresses in an

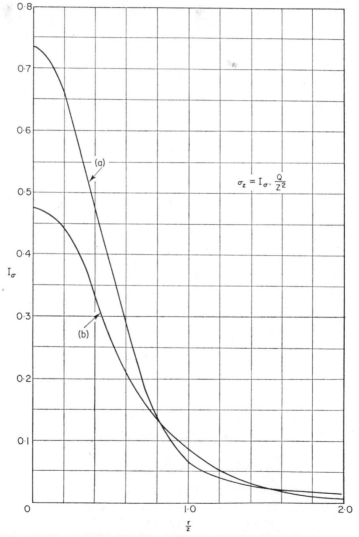

$$\sigma_z = I_\sigma \cdot \frac{Q}{z^2}$$

Fig. 12.7   Influence coefficients for the vertical component of stress beneath a point load at the surface. (a) At the base of an elastic layer of thickness z, resting on a rigid base. (b) At the same depth in a semi-infinite elastic body.

elastic layer of finite thickness resting on a rigid base, and Burmister has prepared charts from which influence coefficients may be obtained. The presence of the rigid base at a finite depth increases the vertical component of stress directly below the load, and reduces the lateral spread of the load through the elastic material. Figure 12.7 compares the vertical component of stress at the surface of the rigid layer with the vertical component at the same depth in a semi-infinite elastic solid, caused in each case by a single point load at the surface. The stress vertically under the load is increased by 53 %.

(b) *Westergaard's analysis* [*12.7*]. Westergaard analysed the stresses in an elastic material in which horizontal extension is prevented by the inclusion of a large number of thin inextensible sheets. This model was thought likely to give a better approximation to the behaviour of certain stratified or 'varved' silts and clays, in which thin bands of coarser material occur at regular vertical intervals. Westergaard's solution gives smaller values than Boussinesq's model for the vertical component of stress vertically below a point load.

Newmark and Fadum have published charts, similar in form to Fig. 12.6 and Fig. 12.4, from which the vertical stress components may be computed for Westergaard's model.

(c) *Stresses in two-layer and three-layer elastic systems.* Acum and Fox [12.8] have prepared tables from which the stresses may be computed in an elastic body consisting of two or three horizontal layers having different elastic constants. The tables were prepared for the analysis of stresses in road pavements, and they are therefore based on a uniformly-loaded circular area at the surface.

(d) *Inhomogeneous soils.* Gibson [12.18] has examined the stresses and displacements caused by a distributed load at the surface of an inhomogeneous elastic half-space, where E varies with depth z according to the law

$$E(z) = E(0) + mz$$

where $E(z)$ and $E(0)$ are the values of E at depth z and at the surface ($z = 0$) respectively

m is a constant.

This is of particular interest, because recent analysis of some observed displacements of structures (for example [12.21], [12.22]) has shown that they are most closely matched by values of E increasing nearly linearly with depth.

For the particular case of an incompressible material ($v = 0.5$), and where the modulus is zero at the surface [$E(0) = 0$], Gibson obtained the simple and rather surprising result that the settlement within the loaded area is directly proportional to the contact pressure, and that the surface settlement outside the loaded area is zero. This is identical with the model proposed by Winkler and frequently used to predict moments and shear forces in foundation beams and rafts [12.34].

12.9   *Loads applied at a depth below the surface:* In all the models discussed above, the load has been assumed to be applied at the surface. In practice, most structures are founded below ground level—sometimes at considerable depths. Mindlin [12.9] has analysed the stresses in a semi-infinite elastic solid loaded below the surface. The solution is too complex to be fully discussed here. The stresses in Mindlin's model decrease with increasing depth of the load below the surface. Where $v = 0.5$ (*i.e.* where there is no volume change) the stress imposed by a load at infinite depth is equal to half that imposed by the same load at the surface.

12.10   *Actual stress distribution in soils:* Because of the difficulty of making accurate measurements of soil stress, comparisons of computed and measured stresses have only been made in a very few cases. Such evidence as there is suggests that Boussinesq's model provides a reasonably good approximation to the real stress distribution, even where there is considerable variation in the elastic constants in different parts of the soil mass. The other solutions discussed above may give a rather closer approximation in suitable cases.

SURFACE DISPLACEMENT OF LOADED ELASTIC SOLIDS

12.11   *Vertical displacement of the surface due to surface loads:* Boussinesq showed that, where a point load is applied at the surface of a semi-infinite solid, the settlement of the surface at a radius r from the point of application is

$$\rho = \frac{Q(1 - v^2)}{\pi E r}$$

The settlement caused by a distributed load may be determined by integration of this expression over the relevant area. For example, consider the circular loaded area of radius a shown in Fig. 12.2. The load on the element $ABCD$

is $qr\,d\theta\,dr$. Then the total settlement of the centre of the loaded area is

$$\rho = \int_0^{2\pi} \int_0^a \frac{qr(1 - v^2)}{\pi Er}\, d\theta\, dr$$

$$= \frac{q(1 - v^2)}{\pi E} \int_0^{2\pi} \int_0^a d\theta\, dr$$

$$= \frac{2q(1 - v^2)a}{E}$$

The settlement of a corner of a rectangular loaded area may be written as

$$\rho = \frac{qB(1 - v^2)}{E} I_\rho$$

where B is the breadth of the loaded area
and I_ρ is an influence coefficient dependent only on the ratio (L/B) of the sides of the rectangle.
Values of I_ρ are given in Table 12.5.

TABLE 12.5 *Influence coefficients for vertical displacement of the corner of a uniformly-loaded rectangular area on the surface of a semi-infinite elastic solid.*

$$\rho = \frac{qB(1 - v^2)}{E} I_\rho$$

$\dfrac{L}{B}$	1·0	1·5	2·0	2·5	3·0	4·0	5·0
I_ρ	0·56	0·68	0·76	0·84	0·89	0·98	1·05

Where the elastic layer is of finite thickness, the influence factor I_ρ may be obtained from Steinbrenner's curves (Fig. 12.8).

$$I_\rho = F_1 + F_2(1 - 2v)/(1 - v)$$

12.12 Settlement of loads applied below the surface: Mindlin has determined expressions for the settlement of loads applied below the surface of a semi-infinite elastic solid. These expressions are too cumbersome for general use, but Fox has used Mindlin's solution to compare the settlements of loads at the surface with those of loads applied at various depths below the surface. Fox's coefficients, which may be obtained from Fig. 12.9, may be used in estimating the settlements of structures founded below ground level [12.11].

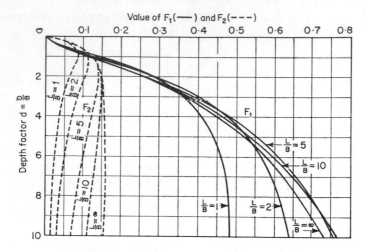

Fig. 12.8 Influence coefficients for the settlement of one corner of a uniformly-loaded rectangular area on the surface of an elastic layer of thickness D. (After Steinbrenner [12.10].)

SETTLEMENT OF STRUCTURES

12.13 *Settlement of structures on clay soils:* Ground movements resulting from the construction and loading of foundations may be divided into three parts:

    (a) 'Heave' or uplift, which occurs during excavation to foundation level.

    (b) Immediate settlement, which occurs during construction and loading.

    (c) Consolidation settlement.

12.14 *'Heave' or uplift during excavation:* This is caused by relief of stress in the soil, as the weight of overburden is removed. The effect is largely elastic, so that the net uplift is practically reduced to zero when a foundation pressure equal to that of the original overburden has been applied. Where necessary, the heave may be calculated, using the elastic theory discussed above [12.12].

12.15 *Net immediate settlement:* This is primarily the result of distortion of the ground due to the net increase in the foundation pressure (q_n). An estimate of the extent of this settlement may be made, using elastic theory. For clay soils, Poisson's ratio (v) may generally be taken to be 0·5.

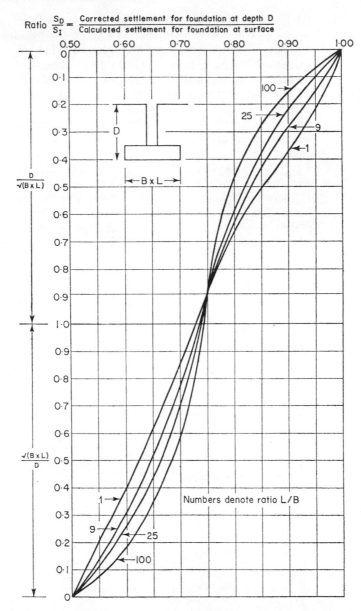

Fig. 12.9 *Correction factors for the settlements of loads applied below the surface. (After Fox [12.11].)*

Values of the undrained modulus (E_u) obtained from laboratory tests are generally subject to considerable errors, and more reliable estimates of E_u may be obtained from *in situ* tests or from observation of structures. For example,

Burland *et al.* [12.19] and Marsland [12.20] have determined values of E_u/c_u from plate bearing tests in the London clay which vary between 600 and 800. These are far larger than values which have been obtained from laboratory tests, which are typically about 150. However, Marsland has shown that the plate test values of E_u are considerably reduced by delays in testing. About ten hours after excavation, the modulus had been reduced by half. By analysing the measured displacements of structures, using the finite element method, it is possible to estimate the values of E_u (Cole and Burland [12.21], Hooper [12.22]). Such analyses for structures in the London clay have shown that E_u increases with depth, and that E_u/c_u is in the range from 200 to 500—that is, somewhere between the laboratory and plate test results. The results of this type of back-analysis, where available, will generally yield the most reliable values of E_u.

Example 12.5 (Fig. 12.10)

A flexible foundation 3 m square is to carry a uniformly-distributed load of 2 500 kN, and will be founded at a

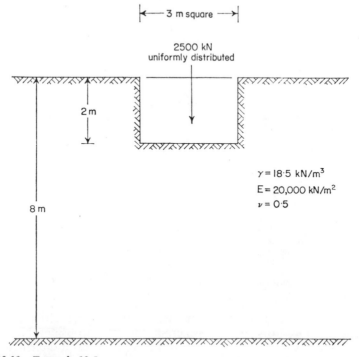

Fig. 12.10   Example 12.5.

depth of 2 m below the surface of a clay whose mean density is 18.5 kN/m^3. Triaxial compression tests indicate that E has a mean value of 20 000 kN/m^2. The clay stratum rests, at a depth of 8 m below the ground surface, on a thick stratum of dense sandy gravel which may be taken to be rigid.

The increase in vertical stress (q_n) at foundation level is

$$2\,500/9 - 18.5 \times 2 = 241 \text{ kN/m}^2$$

The settlement may be computed from Steinbrenner's curves (Fig. 12.8). Since $v = 0.5$, $I_\rho = F_1$.

(a) *At a corner of the foundation,*

$$\frac{B}{L} = 1 \quad \text{and} \quad \frac{D}{B} = 2 \quad \text{so that} \quad F_1 = 0.29$$

Therefore

$$\rho = \frac{241 \times 3 \times (1 - 0.25)}{20\,000} \times 0.29 \text{ m}$$

$$= 8 \text{ mm}$$

(b) *At the centre of the foundation* (considered as four equal square loaded areas of side 1.5 m),

$$\frac{B}{L} = 1 \quad \text{and} \quad \frac{D}{B} = 4 \quad \text{so that} \quad F_1 = 0.41$$

Therefore

$$\rho = \frac{241 \times 1.5 \times (1 - 0.25)}{20\,000} \times 0.41 \times 4 \text{ m}$$

$$= 22 \text{ mm}$$

12.16   Consolidation settlement: In computing the consolidation settlement of a foundation, the following procedure is adopted:

(a)   The consolidating stratum is divided into a convenient number of horizontal strips (Fig. 12.11).

(b)   The mean value of the effective overburden pressure (σ'_i) is calculated for each strip. For practical purposes, the mean value of σ'_i may be taken to be the value at the mid-depth of the strip.

(c)   The mean value of the net increase in vertical stress ($\Delta\sigma'_z$) is calculated for each strip. Provided that the depth of each strip is small compared with the width of the foundation, the mean value of $\Delta\sigma'_z$ may be taken to be the value at the mid-depth of the strip. In computing $\Delta\sigma'_z$, it is generally

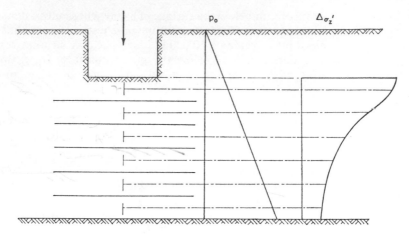

Fig. 12.11   Calculation of consolidation settlement.

assumed that the foundation is completely flexible, and that the soil is linearly elastic. The value of $\Delta\sigma'_z$ may then be calculated by the methods of Sections 12.3 to 12.10.

(d) The reduction (ΔH) in the thickness of each strip, as a result of this increase in vertical stress, is then computed.

(e) The consolidation settlement of any point in the foundation is assumed to be the sum of the values of ΔH for all strips vertically below the point concerned.

$$\rho = \Sigma\, \Delta H$$

For normally consolidated clays, the value of ΔH may be determined from the Compression Index (C_c) (see Section 5.8)

$$\Delta H = \frac{1}{1 + e_i}\, H_i C_c \log_{10} \frac{\sigma'_i + \Delta\sigma'_z}{\sigma'_i}$$

where H_i and e_i are the initial values of H and e respectively.

Where the results of the oedometer test are reported in the form of a relationship between the void ratio (e) and the effective stress (σ'_z), the change in thickness (ΔH) may be calculated from the expression

$$\Delta H = \frac{1}{1 + e_i}\, H_i(e_i - e_f)$$

where e_i is the initial value of e, corresponding to $\sigma'_z = \sigma'_i$
e_f is the final value of e, corresponding to $\sigma'_z = (\sigma'_i + \Delta\sigma'_z)$.

Where the results of the oedometer test are reported in terms of the coefficient of volume change (m_v), ΔH is given by

$$\Delta H = m_v \times \Delta\sigma'_z \times H$$

where m_v is the mean value of the coefficient of volume change over the range of stress from σ'_i to $(\sigma'_i + \Delta\sigma'_z)$.

12.17 Consolidation settlement in overconsolidated clays: In the oedometer test, the condition of one dimensional consolidation is enforced, since the steel ring practically prevents any lateral strains. In applying the results of the test to the prediction of settlements in structures, it is implicitly assumed that the condition of one dimensional consolidation also applies to the soil beneath the foundations. In fact, however, there is far less restraint of lateral movement in the ground, and the shear strains are therefore much greater than in the test samples. In overconsolidated clays, the pore pressures set up during loading, and the volume changes which take place during consolidation, are greatly reduced by increasing shear strains, as these soils are dilatant—that is, they tend to expand as a result of shear strains. As a result, the volume changes in such soils in the ground are generally much smaller than would

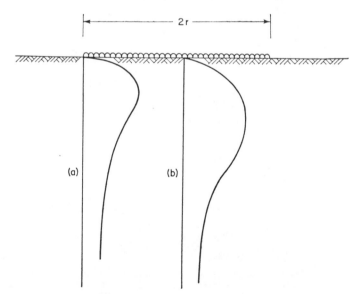

After Jurgenson [12.17]

Fig. 12.12 Variation of maximum shear stress with depth below a uniformly-loaded circular area.

be predicted from the results of the oedometer test alone, and the effect is most marked in those parts of the soil mass in which the shear stresses are greatest. This is the region immediately below the foundation (Fig. 12.12) in which the major part of the consolidation settlement takes place. Skempton and Bjerrum [12.16] have examined this effect, and have suggested values of a coefficient μ such that

$$\mu = \frac{\rho_c}{\rho_{\text{oed}}}$$

where ρ_c is the actual consolidation settlement

ρ_{oed} is the consolidation settlement predicted (as above) from the results of oedometer tests.

The coefficient μ is a function of the pore pressure parameter A, and of the shape of the foundation. The values of μ as modified by R. F. Scott [12.35] are shown in Fig. 12.13.

12.18   Rate of consolidation: The time factor (T_v) was defined in Section 5.11 as

$$T_v = \frac{c_v t}{d^2}$$

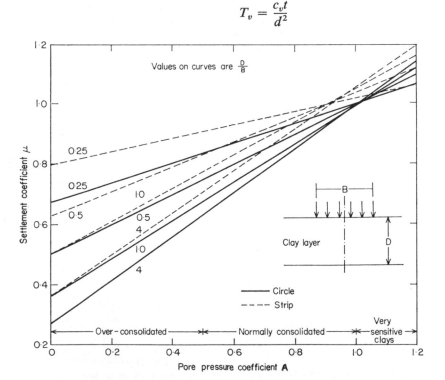

Fig. 12.13   *Correction factors for consolidation settlement in overconsolidated clays.*

where t is the time

   c_v is the coefficient of consolidation

   d is the length of the drainage path.

The relationship between the mean degree of consolidation ($\bar{U}$) and T_v depends on the boundary conditions, and the relationship for the conditions of the oedometer test was given in Table 5.1. In the ground, the boundary conditions may differ from those in the test, since

(a)   the consolidating layer may be free to drain in one direction only, and

(b)   the stress increment (and therefore the initial excess pore pressure) will generally vary with depth below the foundation.

Table 12.6 shows the relationship between $\bar{U}$ and T_v for several different conditions. In estimating the rate of consolidation of a stratum, the value of T_v corresponding to any specified value of $\bar{U}$ is selected from the most appropriate set of boundary conditions. The time required for this degree of consolidation is then

$$t = \frac{T_v d^2}{c_v}$$

TABLE 12.6   *Relationship between $\bar{U}$ and T_v for three cases of initial excess pore pressure distribution.*

$\bar{U}$	Permeable / Permeable T_v	Permeable / Impermeable T_v	Impermeable / Permeable T_v
0	0	0	0
0·1	0·008	0·003	0·047
0·2	0·031	0·009	0·100
0·3	0·071	0·024	0·158
0·4	0·126	0·048	0·221
0·5	0·196	0·092	0·294
0·6	0·287	0·160	0·383
0·7	0·403	0·271	0·500
0·8	0·567	0·440	0·665
0·9	0·848	0·720	0·940
1·0	∞	∞	∞

12.19 *Other methods of analysis:* A number of analytical methods have been suggested, based on consideration of the stress path followed during consolidation (Lambe [12.23], Simons and Som [12.24]) or on applications of elastic theory (Davis and Poulos [12.25]). With the exception of Simons and Som, all these methods require extensive special testing which generally rules them out for routine use. They may, however, be useful in special circumstances.

12.20 *Accuracy of settlement predictions:* Under exceptionally favourable circumstances, predicted settlements may be expected to agree with the actual values to within 20%, but in many cases errors of 50% or more may be found. It is important that the results of settlement calculations should not be reported in such a manner as to imply a degree of precision which the accuracy of the data does not warrant.

In attempting to predict the settlements of structures on lightly overconsolidated soils, it is important to determine the preconsolidation pressure (σ'_c), which may be the result of removal of overburden, fluctuations of the ground water level, electrochemical changes at the particle contacts, or delayed consolidation of the kind described in Section 5.19. Where the stress increment does not exceed $(\sigma'_c - \sigma'_i)$, the settlements may be very small, particularly if the overconsolidation is the result of stress changes such as those resulting from the removal of overburden or changes in ground water level.

Computed rates of settlement, determined from the results of oedometer tests, are frequently much too slow. There are two principal reasons for this. Firstly, there is often a continuous network or 'fabric' of silt or sand inclusions in fissures or bedding planes (Rowe [12.26]) and these can greatly reduce the effective drainage path in the field, although the oedometer sample is too small to be affected. Secondly, some laminated soils have a high permeability in a horizontal direction. The oedometer test, in which the flow is vertical, does not reflect this. Much more reliable values of c_v can be obtained either by using a large hydraulic oedometer, or by measuring the permeability *in situ* (Section 4.16). A three-dimensional analysis may also be necessary in some cases (Murray [12.27]).

12.21 *Soil–structure interaction:* In the methods of analysis described in the preceding sections, it is generally assumed that the distribution of the load over the foundation soil is known, and that no redistribution results from stiffness of the structure. Few real structures will behave in this way, and estimates of differential settlement by these methods are often substantially too large. If more reliable predictions of differential settlement are required, the analysis must take account of the interaction between the soil and the structure. Although the mean total settlement depends mainly on the soil properties, the differential settlement is often predominantly determined by the structural stiffness. It is therefore usually sufficient to use fairly simple models to represent the soil behaviour. The soil may be represented by a series of springs (the Winkler model), or by a semi-infinite elastic half-space. A review of this problem is given in ref. [12.36]. Summaries of available methods and solutions are given by Hooper [12.37], and Poulos and Davis [12.38]. The related problem of the tolerable settlement of structures is discussed by Burland and Wroth [12.39].

12.22   *Settlement of structures on granular soils:* The settlement of structures on these soils is generally substantially complete at the end of construction, although there is some evidence of continuing movements in some cases (Schmertmann [12.28]). Since undisturbed samples can seldom be obtained, the settlements have generally to be predicted from the results of empirical tests on the soil *in situ*.

The procedure for *plate bearing tests* is described in Section 15.21 below. In applying the test results to estimate the settlement of a foundation, it is necessary to allow for the different sizes of the foundation and of the plate. Terzaghi and Peck [12.13] suggested the following approximate relationship:

$$\rho_B = \rho_1 \left(\frac{2B}{B+1} \right)^2$$

where ρ_1 is the settlement of a test plate 1 ft (0·3 m) square
ρ_B is the settlement of a foundation carrying the same intensity of loading
B is the breadth of the foundation measured in feet.

This expression implies that the ratio of the settlement of the foundation to that of the plate will never be more than 4. More recently, evidence has accumulated to indicate that this expression may under-estimate the settlement of a foundation, and settlement ratios of 20 or more have been recorded. Bjerrum and Eggestad [12.29] have detected a tendency for the higher settlement ratios to occur on more loosely packed soils, but this has not been confirmed in other cases. At the moment, there does not seem to be any reliable way of predicting foundation settlements from the results of small plate tests.

The *Standard Penetration Test* is described in Section 15.19, but some correction of the observed values is generally necessary before they are used in settlement predictions. In saturated very fine or silty sands, where the permeability is relatively low, it is possible for a considerable drop in pore pressure to occur near the base of the spoon. This results in an increase in effective stress and therefore in the penetration resistance, and the following correction is commonly used to allow for this:

$$N_{cor} = \tfrac{1}{2}(N - 15) + 15$$

where N_{cor} is the corrected value of N.

There is also evidence that the reduced confining pressure near the ground surface results in too small a value of N, although there is some disagreement as to the extent of the allowance which should be made for this (Sutherland [12.30], Tomlinson [12.31]).

Terzaghi and Peck [12.13] suggested the correlation between allowable bearing pressure and N values, shown in Fig. 12.14. This correlation was originally suggested, on the basis of rather limited information, as a means of selecting a safe allowable pressure which would not cause settlements exceeding 1 in (25 mm). As a means of predicting settlements, this correlation is very conservative. Terzaghi and Peck also proposed that the allowable bearing pressure should be halved if the ground water level were at the base of the foundation. This stipulation is now generally agreed to be too conservative, and the position of the ground water is often ignored, on the grounds that it is already reflected in the measured value of N.

Meyerhof [12.32] has proposed that the allowable bearing pressures in Fig. 12.14 should be increased by 50% and that the position of the ground water level should be ignored. This still appears to be slightly conservative, and Meyerhof quotes predicted settlements from 1·2 to 4 times the observed values.

Settlements may also be estimated from the results of tests with the '*Dutch*' *Cone Penetrometer*. Terzaghi's expression for settlement (Section 5.8) may be written in the form

$$\Delta H = \frac{H}{C} \log_e \frac{\sigma'_i + \Delta\sigma'_z}{\sigma'_i}$$

where C is a coefficient defining the compressibility of the soil. Buissman [12.14] suggested the empirical relation

$$C = 1·5 \frac{C_{kd}}{\sigma'_i}$$

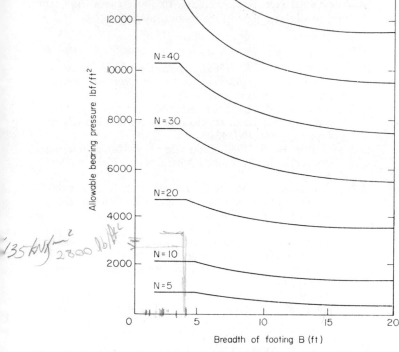

Fig. 12.14   Relationship between Standard Penetration Test 'N' value and allowable bearing capacity. (After Terzaghi and Peck [12.13].)

where C_{kd} is the cone resistance. In the procedure suggested by De Beer and Martens [12.15], the granular stratum is divided into a number of horizontal layers. For each layer, the mean values of C_{kd}, σ'_i and $\Delta\sigma'_z$ are determined. Hence, the reduction in the thickness (ΔH) of each layer may be determined. Then

$$\rho = \Sigma \Delta H$$

Field measurements indicate that the average ratio of predicted to observed settlements is about 2, and several modifications to the method have been suggested which have the effect of increasing C to about $2{\cdot}0 C_{kd}/\sigma'_i$ or more.

Since the Standard Penetration Test and the 'Dutch' Cone Test are, in effect, measuring ultimate penetration resistance, they do not distinguish between normally consolidated and overconsolidated soils at the same relative density. Settlement of a foundation on an overconsolidated soil is likely to be considerably less than that predicted by the Buissman–De Beer method, which was intended for normally consolidated soils (De Beer [12.33]). In most cases, however, the preconsolidation pressure is unknown.

REFERENCES

12.1 BOUSSINESQ, J. V. 1885. *Applications des potentials à l'étude de l'équilibre et du mouvement des solides élastiques.* Gautier-Villars (Paris).

12.2 NEWMARK, N. M. 1935. Simplified computations for vertical pressure in elastic foundations. *Univ. Illinois Eng. Exp. Sta. Bul.* 429.

12.3 FADUM, R. E. 1948. Influence values for estimating stresses in elastic foundations. *2nd Int. Conf. on Soil Mech. and Found. Eng.*

12.4 NEWMARK, N. M. 1942. Influence charts for computation of stresses in elastic foundations. *Univ. Illinois Eng. Exp. Sta. Bul.* 338.

12.5 BIOT, M. A. 1935. Effect of certain discontinuities on the pressure distribution in a loaded soil. *Physics,* **6.**

12.6 BURMISTER. 1945. The general theory of stresses and displacement in layered soil systems. *Journal of Appl. Physics,* **16.**

12.7 WESTERGAARD, H. M. 1939. A problem in elasticity suggested by a problem in soil mechanics. In *Contributions to mechanics of solids.* Macmillan (New York).

12.8 ACUM, W. A. E. and FOX, L. 1951. Computation of load stresses in a three-layer elastic system. *Géotechnique,* **2.**

12.9 MINDLIN, R. D. 1936. Force at a point in the interior of a semi-infinite solid. *Physics,* **7.**

12.10 STEINBRENNER, W. 1934. Tafeln zur Setzengberechnung. *Die Strasse,* **1.**

12.11 FOX, E. N. 1948. The mean elastic settlement of a uniformly loaded area at a depth below the ground surface. *Proc. 2nd Int. Conf. on Soil Mech. and Found. Eng.*

12.12 SEROTA, S. and JENNINGS, R. A. J. 1959. The elastic heave in the bottom of excavations. *Géotechnique,* **9.**

12.13 TERZAGHI, K. and PECK, R. B. 1967. *Soil mechanics in engineering practice.* Wiley (New York). 2nd Ed.

12.14 BUISSMAN, A. K. 1943. *Grondmechanica,* **2.**

12.15 DE BEER, E. and MARTENS, A. 1957. Method of computation of an upper limit for the influence of heterogeneity of sand layers in the settlement of bridges. *Proc. 4th Int. Conf. on Soil Mech. and Found. Eng.*

12.16 SKEMPTON, A. W. and BJERRUM, L. 1957. A contribution to the settlement analysis of foundations on clay. *Géotechnique,* **7.**

12.17 JURGENSON, L. 1940. The application of theories of elasticity and plasticity to foundation problems. In *Contributions to soil mechanics 1925–1940*. Boston Soc. Civil Eng.

12.18 GIBSON, R. E. 1967. Some results concerning displacements and stresses in a non-homogeneous elastic half-space. *Géotechnique*, **17**.

12.19 BURLAND, J. B., BUTLER, F. G. and DUNICAN, P. 1966. The behaviour and design of large diameter bored piles in stiff clay. *Proc. Symp. on Large Bored Piles*. Inst. Civ. Eng.

12.20 MARSLAND, A. 1971. The shear strength of stiff fissured clays. *Proc. Roscoe Memorial Symp.* Cambridge.

12.21 COLE, K. W. and BURLAND, J. B. 1972. Observations of retaining wall movements associated with large excavations. *Proc. 5th Europ. Conf. on Soil Mech. and Found. Eng.*

12.22 HOOPER, J. A. 1973. Observations on the behaviour of a piled raft foundation on London clay. *Proc. Inst. Civ. Eng.*

12.23 LAMBE, T. W. 1964. Methods of estimating settlement. *J. Soil Mech. Div. ASCE*, **90**.

12.24 SIMONS, N. and SOM, N. N. 1970. *Settlement of structures on clay*. CIRIA Report No. 22.

12.25 DAVIS, E. H. and POULOS, H. G. 1968. Settlement predictions under three-dimensional conditions. *Géotechnique*, **22**.

12.26 ROWE, P. W. 1972. The relevance of soil fabric to site investigation practice. *Géotechnique*, **22**.

12.27 MURRAY, R. T. 1971. *Embankments constructed on soft foundations: settlement study at Avonmouth*. Transport and Road Research Laboratory Report LR419.

12.28 SCHMERTMANN, J. H. 1970. Static cone to compute static settlement over sand. *J. Soil Mech. Div. ASCE*, **96**.

12.29 BJERRUM, L. and EGGESTAD, A. 1963. Interpretation of loading tests on sand. *Proc. 3rd Europ. Conf. on Soil Mech. and Found. Eng.*

12.30 SUTHERLAND, H. B. 1963. The use of *in situ* tests to estimate the allowable bearing pressure of cohesionless soils. *Struct. Eng.*, **41**.

12.31 TOMLINSON, M. J. 1969. *Foundation design and construction*. Pitman (London). 2nd Ed.

12.32 MEYERHOF, G. G. 1965. Shallow foundations. *J. Soil Mech. Div. ASCE*, **91**.

12.33 DE BEER, E. 1965. Bearing capacity and settlement of shallow foundations on sand. *Proc. Symp. on Bearing Capacity and Settlement of Foundations*. Duke Univ.

12.34 WINKLER, E. 1867. *Die Lehre von Elastizität und Festigkeit*. Prague.

12.35 SCOTT, R. F. 1963. *Principles of soil mechanics*. Addison-Wesley (London).

12.36   LITTLEJOHN, G. S. and MACLEOD, I. A. (Eds.) 1974. *Report on structure soil interaction in relation to buildings.* Inst. Struct. Engrs.

12.37   HOOPER, J. A. 1978. Foundation interaction analysis. In *Developments in soil mechanics.* (Ed. SCOTT, C. R.) Applied Science Publ. (London).

12.38   POULOS, H. G. and DAVIS, E. H. 1974. *Elastic solutions for soil and rock mechanics.* Wiley (New York).

12.39   BURLAND, J. B. and WROTH, C. P. 1975. Allowable and differential settlement of structures including damage and soil structure interaction. *Proc. Conf. on Settlement of Structures.* Pentech Press (London).

CHAPTER 13

Piled foundations

TYPES OF PILED FOUNDATION

13.1   Purpose of piled foundations: The engineer's principal object in using a piled foundation is usually to transfer the structural loads to a firm stratum at some depth below the base of the structure. Piled foundations are commonly required in the following circumstances:

(a) Where the soil below the structure has insufficient strength to support the structural loads on shallow foundations.

(b) Where the compressibility of the soil below the structure is so large that excessive settlements would occur if shallow foundations were used.

(c) Where variations in the compressibility of the soil, or non-uniform distribution of the structural loads, would lead to excessive differential settlements of shallow foundations.

(d) Where the foundations may have to resist lateral or uplift forces.

(e) Where excavation to a firm stratum would prove difficult or expensive (*e.g.* in soft water-bearing alluvial deposits, or for foundations under water).

13.2   Types of pile: In selecting a suitable type of pile, the engineer must take account of the ground and other conditions of the particular site, and of the relative cost of different types of pile and of their installation in the special circumstances of that site. In assessing the bearing capacity of piled foundations, an important factor is the effect of the method of installation on the soil behaviour. From this point of view, piles may be conveniently divided into the following groups:

(a) Displacement piles.
    (i) Driven piles (prefabricated or cast-in-place).
    (ii) Jacked piles.
    (iii) Screw piles.

(b) Non-displacement piles (bored piles).

13.3 Driven piles: Driven piles are commonly constructed of timber, steel or concrete.

Timber piles are generally used only for temporary works in Britain, although they may be used for permanent structures in countries where timber is plentiful and cheap. They are easily cut, and are fairly easily extended if required. Timber offers good resistance to decay if permanently submerged. Timber piles from Roman and medieval foundations are often found to be still in excellent condition where they have always been below the level of the ground water. Above the water table, the piles should be protected by pressure creosoting, or by some similar process, if they are to be used in permanent works, but it is generally better to cap them below water level, or to cut them off and to extend them in concrete. In sea water, timber piles are also liable to attack by marine borers.

Steel piles may be tubular, or box- or H-section. They are easily handled and driven, and may be readily cut or extended if required. However, they are liable to corrosion, and, where they are used in permanent marine structures, some form of cathodic protection is generally required to prevent electrolytic attack by sea water. Tubular and box-section piles may be driven open-ended, or the end may be closed with a steel plate or with a plug of gravel or concrete.

Precast concrete piles may be normally reinforced or prestressed. They are best used where large numbers are required, so that a casting yard on site becomes economical. They are not easily cut or extended (particularly in the case of prestressed piles). Owing to its brittle nature, precast concrete is very liable to damage during heavy driving, particularly if the section has been cracked by careless handling. This damage may take place below ground level, and is often difficult to detect. Longitudinal steel reinforcement (or prestress) is required to resist the bending stresses during handling and pitching. The stresses occurring during subsequent service are usually small by comparison. Heavy transverse reinforcement is also required in the head and toe of the pile to resist the driving stresses.

Driven cast-in-place piles are installed by first driving a hollow steel or concrete casing into the ground. The hole thus formed is then filled with wet concrete. The casing may be left in place to form part of the finished pile, or may be withdrawn for reuse as the wet concrete is placed. The wet concrete may be rammed as the casing is withdrawn, to ensure firm compaction against the soil, or to form an enlarged base to the pile, which greatly increases the bearing capacity in suitable soils. These piles have some of the advantages of driven piles (*e.g.* the compaction of loose strata as the casing is driven) and of cast-in-place piles (*e.g.* no risk of damage to the concrete core during driving). However, careless installation may result in 'necking'—that is, inward bulging of the soil which displaces the wet concrete in the shaft of the pile as the casing is withdrawn. This reduces the effective cross-section of the pile, and greatly reduces the load carrying capacity. It is not easy to detect. It is usually caused by either over-ramming the wet concrete or by withdrawing the casing too fast.

13.4 Jacked piles: These are commonly used for underpinning existing structures, where lack of headroom often precludes the use of a hammer, and where the structure itself provides a convenient reaction against which the jack may be operated. The piles are generally built up from a series of short precast sections which may be inserted singly as jacking proceeds. Since the load on the jack is easily measured, each pile is effectively tested to failure during installation, and a fairly low load factor is acceptable. However, if the structure itself is being used to provide a reaction for the jack, some care is needed to ensure that the jacking loads are not so great as to cause damage. It may be necessary to limit the jacking loads to little more than the working loads on the piles. A load factor of about 1·5 is quite commonly used.

13.5 Screw piles: Screw piles are simple steel or concrete cylinders, to which one or more sets of helical blades are attached. The pile is forced into the ground by rotating the blades with a capstan attached to the head of the pile. These piles are most effective where it is necessary to penetrate soft or loose alluvial deposits to reach a hard stratum below. Owing to the large diameter of the blades, these piles offer good resistance to uplift forces.

13.6 Bored cast-in-place piles: The hole for the shaft of the pile is first formed by any convenient drilling method, and may be lined if necessary to support the sides. In stiff clays, the bottom of the shaft may be enlarged by under-reaming, to give a greatly increased end bearing capacity to the pile. This method of installation is generally free from vibration, and allows the hole to be inspected before the concrete is placed. However, the drilling process may loosen the soil against the sides of the shaft or at the base. Clay soils may be softened by absorption of water from the atmosphere— if the hole is left open for too long—or from excess water in the wet concrete. This reduces the bearing capacity of the pile, particularly that due to adhesion between the soil and the shaft. Necking may also occur if the casing is withdrawn too fast.

13.7   Piles to resist uplift: Unless the pile has an enlarged base, or is anchored in sound rock, the only resistance to uplift forces is that developed by shaft friction and adhesion. Where a stratum of sound rock exists at a moderate depth, the bases of the piles may be anchored into it, by grouting in dowel bars or anchorages attached to high tensile steel tendons. The latter may be prestressed if required. Concrete piles must have sufficient longitudinal reinforcement to resist the whole of the uplift forces.

13.8   Piles to resist horizontal forces: Although vertical piles can offer considerable resistance to horizontal forces, the horizontal displacements which occur before this resistance is fully developed are usually quite unacceptable. Horizontal forces are therefore usually provided for by driving inclined or 'raking' piles. The angle of rake is generally limited by the method of installation. Drop hammers and single acting steam hammers cannot be operated effectively at a rake of more than about 15°, while the limit with a double acting hammer is about 40°. Bored piles can be installed at almost any angle, but some difficulty may be experienced in keeping the boring straight.

BEARING CAPACITY OF A SINGLE PILE

13.9   Method of assessment: The bearing capacity of a single pile may be estimated in one of three ways:

(a)   *By test loading.* This is the most reliable method, but it is also the most expensive, and it requires extensive access to the site at an early stage if the results are to be used in design. Moreover, the test results cannot be easily extrapolated to cover changes in the size or type of pile to be used.
(b)   *By static formulae.* The bearing capacity of the pile base, and the friction and adhesion on the pile shaft, are assessed from a knowledge of the soil properties. This is clearly the most logical method, but it requires a very full site investigation if the results are to be reliable. It is also necessary to take account of the effects which the method of installation may have on the soil behaviour.

(c) *By dynamic formulae.* In using these formulae, an attempt is made to assess the load bearing capacity of the pile from measurements of the penetration of the pile caused by hammer blows of known energy. While these measurements are related to the dynamic resistance to driving, all attempts to formulate a general correlation between the static and dynamic resistance have proved to be, at best, highly unreliable. In highly permeable coarse sands and gravels, it should be possible to estimate the static load bearing capacity of a pile to within about 40% by this means. While this error may seem large, it must be remembered that, in these soils, static formulae must often be based on the results of dynamic penetration tests, which are hardly more reliable. In fine sands, and in finer-grained soils, much greater errors may result because of the very different pore pressures which occur under static and dynamic loads in these soils. The approach is of most value where a large number of piles are to be driven in one series of strata on a single site. Here it may be possible to obtain a reliable correlation between the dynamic penetration resistance and the results of static load tests. This correlation may then be used to check the load bearing capacity of similar piles similarly driven on the same site.

13.10   Test loading procedure: A full-size pile, of the type to be used in the final structure, is installed in the ground. The pile is loaded by means of a hydraulic jack on the head. The reaction for the jack is provided either by kentledge supported on a frame over the head of the pile, or by the resistance to uplift of a pair of piles installed on either side of the test pile. In the latter case, the uplift piles must not be installed so close to the test pile that the stresses induced in the soil by the uplift forces could affect the displacement or load carrying capacity of the test pile. The test may be carried out either

(a)   by the Maintained Load method, or
(b)   by the Constant Penetration Rate method.

In the Maintained Load method, the load is applied to the pile in discrete increments, the load being maintained constant at each increment until settlement of the pile has substantially ceased. A curve may then be prepared showing the relationship between load and settlement. Where it is necessary to estimate the settlement of piles under working loads, this is the most reliable procedure. However, where the ultimate load is to be determined, the results are less satisfactory. As failure approaches, it becomes increasingly difficult to decide at what stage the settlement should be assumed to have substantially ceased.

A more satisfactory procedure is provided by the Constant Penetration Rate test, developed by Whitaker and Cooke [13.1]. In this test, the pile is jacked into the ground at a constant rate, and a continuous record of the load is taken. Figure 13.1 shows typical curves obtained from this test. The results of the test do not seem to be very sensitive to the rate of penetration.

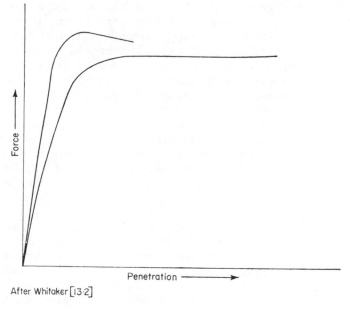

After Whitaker [13·2]

Fig. 13.1   Constant rate of penetration test. Typical force/penetration curves for friction piles.

13.11   Definition of the load at failure: Civil Engineering Code of Practice No. 4 (1954) defined the load at failure as 'the load at which the rate of settlement continues undiminished without further increment of load'. This definition is not generally very satisfactory. In many cases, it is doubtful if any such load could be determined, since the load bearing capacity of the pile increases continuously with increasing penetration (Fig. 13.2). For piles whose resistance to penetration depends mainly on shaft friction and adhesion, the penetration required to develop the full resistance is small, and the curve commonly shows a clear maximum value of the load which may be taken to be the load at failure. The full mobilisation of the end bearing resistance

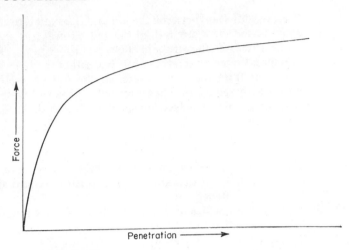

Force

Penetration ⟶

After Whitaker [13·2]

Fig. 13.2   Constant rate of penetration pile tests. Typical force/penetration curve for an end-bearing pile.

requires a considerably greater penetration and the point at which this is fully developed is less easy to determine. The load at failure is commonly defined as 'the load corresponding to a settlement equal to $\frac{1}{10}$ of the pile diameter'. This definition was first suggested by Terzaghi, and Whitaker [13.2] has suggested that it may be applied to the results of the Constant Penetration Rate test, although it probably represents a lower limit to the penetration necessary to develop the full end resistance.

13.12   Static formulae—clay soils: It is usual to assume that the soil surrounding the pile is undrained at the time the pile is loaded, and to base the design on the undrained test results. However, there is considerable evidence that the rate of drainage in the soil immediately adjacent to the pile shaft is fairly rapid, and considerable dissipation has taken place by the time the pile is fully loaded. It would be more logical, therefore, to design in terms of effective stress. However, this requires a knowledge of the effective lateral earth pressure on the shaft, and even if the original value of K_0 is known, it is not easy to say what it may be after installation of the pile. We are therefore obliged to use the undrained strength, corrected where necessary by empirical coefficients. Burland [13.11] has suggested a possible approach to this problem.

The ultimate load bearing capacity of a pile is the sum of the end bearing capacity and the shaft adhesion. For bored piles, the adhesion between the soil and the shaft is practically equal to the shear strength of the soil. However, the boring process releases the stress in the soil, and reduces the pore pressure. Water is therefore absorbed either from the atmosphere, or from the excess water in the wet concrete. This may greatly reduce the shear strength of the soil close to the shaft of the pile. The bearing capacity due to shaft adhesion may therefore be written

Bearing Capacity (shaft Adhesion) $= A_s \alpha \bar{c}$

where A_s is the surface area of the shaft

$\bar{c}$ is the mean value of the shear strength of the unsoftened soil

α is a coefficient expressing the loss of strength due to softening.

Skempton [13.3] has shown that, for bored cast-in-place piles in the London clay, α is generally in the range from 0·3 to 0·6, and has suggested a value of 0·45 in normal circumstances. Tomlinson [13.4] has pointed out that the lapse of time between boring and placing the concrete may affect the adhesion. He suggests that α should be taken to be 0·3 wherever there may be considerable delay before concrete is placed.

The published evidence for the values of α in other clays is inconclusive, the small number of test results indicating values of α varying from 0·2 to 1·2. In the absence of previous experience on a particular site, Tomlinson suggests taking $\alpha = 0·3$ as a preliminary guide, and confirming the pile design by means of load tests to failure.

The adhesion factor α for a driven pile is affected by the disturbance and remoulding of the clay during driving. For piles driven in soft clays, α is nearly unity, but in stiff and very stiff clays α varies considerably over the embedded length of the pile. For a considerable depth below the ground surface, there is generally little or no adhesion, and almost no transfer of load from the pile to the soil. Below this, the pile is covered with a dry skin of remoulded clay firmly adhering to the pile surface. If failure occurs, it takes the form of sliding between this skin and the surrounding clay. This develops a shearing resistance which may nearly equal the full undrained strength of the undisturbed soil. Where a pile is driven into stiff clay through an overlying stratum, some of the overlying soil is dragged down with the pile. If the overlying soil is soft clay, this may substantially reduce the adhesion in the upper part of the stiff clay below. If, on the other hand, the overlying soil is sand, the adhesion may be increased [13.6].

Where the pile passes through compressible fill, or soft

consolidating sediments, the settlement of the soil surrounding the pile may cause a downward force on the shaft which must be deducted from the bearing capacity. Where the base of the pile is enlarged, there may also be some downward force on the lower part of the shaft caused by the weight of the soil above the bell, which may be partly unsupported as the pile settles under load. It is usual to make no allowance for shaft adhesion over a length equal to twice the shaft diameter above the top of the bell.

For deep circular foundations in clay soils, the bearing capacity coefficient $N_c = 9.0$ (see Section 11.11, above). Then, assuming that the weight of the pile is approximately equal to the weight of soil which it has displaced, the end bearing capacity of the pile is

$$Q_b = 9.0c_u A_b$$

where A_b is the base area of the pile

$\quad c_u$ is the undrained shear strength of the soil below the base.

In stiff fissured clays, the average results of conventional tests on small samples may over-estimate the shear strength of the soil in the mass. Whitaker and Cooke [13.5] have shown that the value of c_u to be used in the expression above is near the lower end of the range of observed test results.

13.13   *Static formulae—granular soils:*   For penetrations of between 5 and 20 diameters, the resistance of a pile driven into cohesionless soil is given by

$$Q = Q_b + Q_s = q_0 N_q A_b + \tfrac{1}{2}K_s \tan \delta A_s$$

where q_0 is the effective overburden pressure at the base of the pile.

$\quad N_q$ is a bearing capacity coefficient depending on φ' and the ratio of penetration depth to diameter.

$\quad K_s$ is a coefficient of lateral earth pressure, depending on the relative density of the soil (RD).

$\quad \delta$ is the angle of friction between the soil and the pile.

Values of N_q for deep foundations have been derived by Meyerhof [13.7] and Berezantsev *et al.* [13.10] amongst others. Meyerhof's values are large, and probably represent an upper limit to the end bearing capacity. The settlement which would occur before the full resistance was developed would generally be intolerably large. The values given by Berezantsev *et al.* have been shown [13.12] to correspond more nearly with practical limits of settlement. Values of K_s

have been given by Broms [13.13]. The value of δ depends on the nature of the pile surface, and varies from $0 \cdot 5\varphi'$ for clean steel to φ' for concrete cast against the soil.

Because of the difficulty of obtaining undisturbed samples of cohesive soils, values of RD and φ' must generally be estimated from the results of Standard Penetration Tests (Section 15.19). Alternatively, the values of Q_b and Q_s may be estimated directly from the results of Dutch Cone Penetration Tests (Section 15.20).

The principal difficulty in assessing the bearing capacity of bored piles in granular soil lies in determining the effect of the method of installation on the strength of the soil. Some loss of strength results from disturbance of the soil by the boring tools and during extraction of the casing. Further disturbance may occur if water is allowed to flow rapidly into the boring. However, if the wet concrete can be firmly rammed into the base of the pile, the loosened soil may, to some extent at least, be recompacted.

BEARING CAPACITY OF GROUPS OF PILES

13.14 Pile groups in cohesive soils: Except in the case of end bearing piles driven to rock, the bearing capacity of a group of piles is not generally equal to the sum of the bearing capacities of individual piles considered separately. In the case of piles in clay soils, the bearing capacity of the group may be considerably reduced if the piles are close together.

If the piles are very closely placed, failure may occur as a result of shearing of the soil along the perimeter of the whole group, the soil between the piles being carried down with them. The resistance to such a 'block failure' is

Resistance to failure of a group of piles

$$= 2(B + L)D\bar{c} + BLc_u N_c$$

where B and L are the breadth and length of the pile group

D is the depth of penetration of the piles into the cohesive stratum

$\bar{c}$ is the mean shear strength over the depth D

c_u is the shear strength of the soil below the base.

Even if the piles are sufficiently widely spaced for a 'block failure' not to occur, the bearing capacity of the group is reduced if the shear stresses induced in the soil by one pile affect the soil around and below adjacent piles. This reduction is clearly greatest for piles at the closest spacing. Whitaker has investigated this effect in a series of model tests [13.8]. Figure 13.3 shows typical results from his tests.

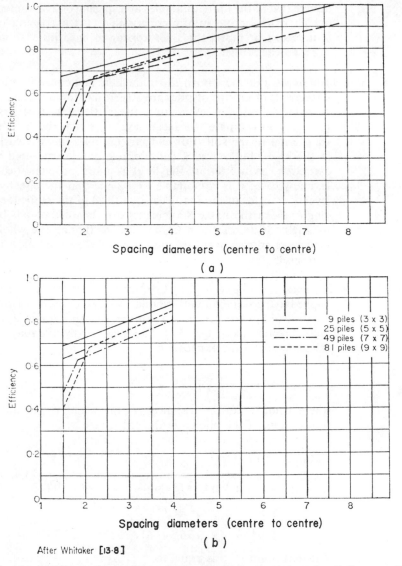

After Whitaker [13·8]

Fig. 13.3   Efficiency of model pile groups in clay. (a) Penetration equal to 48 diameters. (b) Penetration equal to 24 diameters.

13.15   *Groups of piles in cohesionless soils:*   Pile driving in clean sands and gravels causes considerable compaction of the soil between the piles—so much so that great difficulty may be experienced in driving the later piles in a group if the piles are closely spaced. This compaction of the soil increases the bearing capacity of the group, which may be nearly double that of the individual piles at a spacing of 2 diameters (which is about the minimum possible). At a spacing of 5 diameters or more, the piles appear to have little or no effect on each other [Kezdi, 13.9].

13.16 *Settlement of pile groups:* Both the immediate and consolidation settlements of groups of piles are larger than the settlements of a single pile. In estimating the settlement it is first necessary to estimate the mean level at which the load is transferred to the soil. For end bearing piles this is clearly at the bases, but where a considerable part of the load is carried by skin friction the level is more problematical. In many cases it is reasonable to assume that the effective level of transfer of the load is at one-third of the embedded length above the base. Settlements are then estimated, by the methods described in Chapter 12, as though the whole load were spread over the total area of the foundation at that level.

13.17 *Eccentric loads on pile groups:* Consider a vertical load (P) applied to a group of n vertical piles, with eccentricities from the centroid of the group of e_x in the x direction and e_y in the y direction (see Fig. 13.4). The load on any pile may be calculated from the expression

$$P\left(\frac{1}{n} + \frac{e_x x}{\sum x^2} + \frac{e_y y}{\sum y^2}\right)$$

where x and y are the distances, in the x and y directions respectively, of the pile from the centroid.

Example 13.1
Consider a pile group shown in plan in Fig. 13.4. A vertical load of 4 000 kN is applied at point G. Then

$$\Sigma x^2 = 8 \times 2^2 = 32$$

$$\Sigma y^2 = 6 \times 0 \cdot 75^2 + 6 \times 2 \cdot 25^2 = 33 \cdot 75$$

The maximum load is equal to

$$4\,000\left(\frac{1}{12} + \frac{0 \cdot 8 \times 2}{32} + \frac{0 \cdot 5 \times 2 \cdot 25}{33 \cdot 75}\right) = 667\,\text{kN}$$

13.18 *Pile groups carrying inclined loads:* Vertical piles can carry a limited horizontal component of load, as a result of the passive resistance of the soil against the shaft. However, the horizontal deflections under load may be very large, and cannot be satisfactorily predicted from the soil properties.

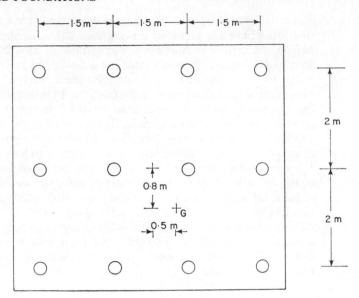

Fig. 13.4   Example 13.1.

Any substantial horizontal loads are therefore usually carried by raking piles (Fig. 13.5).

Where the piles are in three rows only, the load on each pile may be estimated by constructing the force diagram, as shown in Fig. 13.5. The piles are assumed to be axially loaded, and to be freely hinged at the top. Let the resultant of the loads Q_B and Q_C on the piles B and C be S. The line of action of S must pass through the intersection of Q_B and $Q_C(M)$ and also through the intersection (N) of P with Q_A. Thus the direction of S is known, and the force diagram may be completed.

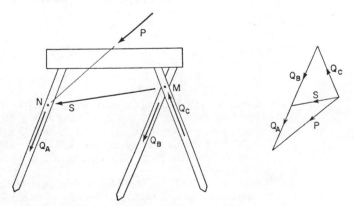

Fig. 13.5   Calculation of loads on raking piles.

13.19 General analysis of pile groups: A number of methods have been developed in recent years for the analysis of the distribution of forces and displacements in pile groups of any shape. Poulos [13.14] and his associates have developed methods based on the assumption that the piles are linearly elastic bodies embedded in another elastic medium. In principle, the finite element method could also be used to solve this problem, but the computational effort required seems to be excessive. Butterfield and Banerjee [13.15] have used the boundary element method to solve a similar problem, and Banerjee and Davies [13.16] have recently extended this to deal with soils whose elastic properties vary with depth.

Handled with care, these methods can yield satisfactory predictions of displacement under load, and of the distribution of load between piles in the group. A computer is required for the solution however, and there may be some difficulty in determining the soil parameters which should be used in the analysis.

REFERENCES

13.1 WHITAKER, T. and COOKE, R. W. 1961. A new approach to pile testing. *Proc. 5th Int. Conf. on Soil Mech. and Found. Eng.*

13.2 WHITAKER, T. 1967. Experiences with the Constant Rate of Penetration Test for piles. *Building Research Station Current Papers Engineering Series.*, **43**.

13.3 SKEMPTON, A. W. 1959. Cast-*in situ* bored piles in the London clay. *Géotechnique*, **9**.

13.4 TOMLINSON, M. J. 1977. *Pile design and construction.* Cement and Concrete Ass.

13.5 WHITAKER, T. and COOKE, R. W. 1966. An investigation of the shaft and base resistance of large bored piles in the London clay. *Proc. Symp. on Large Bored Piles.* Inst. Civil Eng.

13.6 TOMLINSON, M. J. 1970. The adhesion of piles in stiff clay. *CIRIA Res. Rep.*, **26**.

13.7 MEYERHOF, G. G. 1953. Recherche sur la force portante des pieux. *Ann. Inst. Tech. Batim*, **6**.

13.8 WHITAKER, T. 1957. Experiments with model piles in groups. *Géotechnique*, **7**.

13.9 KEZDI, A. 1957. Bearing capacity of piles and pile groups. *Proc. 4th Int. Conf. on Soil Mech. and Found. Eng.*

13.10 BEREZANTSEV, V. G., KHRISOFOROV, V. S. and GOLUBKOV, V. N. 1961. Load bearing capacity and deformation of piled foundations. *Proc. 5th Int. Conf. on Soil Mech. and Found. Eng.*

13.11 BURLAND, J. B. 1973. Shaft friction on piles in clay—a simple fundamental approach. *Building Research Station Current Paper CP 33/73.*

13.12 NORLUND, R. L. 1965. Bearing capacity of piles in cohesionless soils. *J. Soil Mech. and Found. Div. ASCE.*

13.13 BROMS, B. 1966. Methods of calculating the ultimate bearing capacity of piles. *Sols-Soil,* **5**.

13.14 POULOS, H. G. 1968. Analysis of the settlement of pile groups. *Géotechnique,* **18**.

13.15 BUTTERFIELD, R. and BANERJEE, P. K. 1971. The elastic analysis of compressible piles and pile groups. *Géotechnique,* **21**.

13.16 BANERJEE, P. K. and DAVIES, T. G. 1977. Analysis of pile groups embedded in Gibson soil. *Proc. 9th Int. Conf. on Soil Mech. and Found. Eng.*

CHAPTER 14

Geotechnical processes

IMPROVEMENT OF SOIL PROPERTIES

14.1 Purpose: In many cases, the engineering properties of soil in its natural state are so unfavourable for engineering construction that some method must be found for improving them if the engineer's freedom of choice in selecting the site and type of structure is not to be unreasonably restricted.

The purpose of such methods is generally to increase the shear strength, or to reduce the pore pressure, permeability or compressibility. In addition, it is often necessary to ensure that the improved soil properties do not deteriorate as a result of weathering, or of changes in the water content.

The methods used to improve the soil properties are collectively known as *geotechnical processes*. There is a wide choice of methods available, and the most suitable process in any particular case depends on the form of improvement required, and on the nature of the soil to be treated. Most of the techniques require considerable special knowledge and experience, and it is not possible here to do more than describe the processes in broad outline and to indicate the factors affecting the choice of method.

14.2 Methods available: The available methods for soil improvement may be conveniently grouped under the following heads:

(a) Compaction and preconsolidation.
(b) Drainage and ground water control.
(c) Injection grouting.
(d) Surface stabilisation.
(e) Soil freezing.

COMPACTION AND PRECONSOLIDATION

14.3 Soil compaction: Compaction rearranges the soil particles into a closer state of packing, and generally results in higher shear strength, lower compressibility, and a reduced susceptibility to water content changes. The principles underlying the compaction process were discussed in Chapter 1. In any particular case, it is necessary to decide on:

    (a)   The type and weight of plant to be used.
    (b)   The method of specification and control of the work.

14.4 Surface compaction: Fill materials for embankments and road bases are generally placed in thin layers, and are compacted by rolling. *Smooth-wheeled rollers* are made in a variety of weights from about 1 tonne to 20 tonnes. The costs of compaction with different types of machine are fully discussed in a series of technical reports issued by the Transport and Road Research Laboratory.

The usual type has a single drum at the front and two rolls of larger diameter at the rear. *Tandem rollers* have two (or exceptionally three) identical drums following in the same track. Smooth-wheeled rollers are generally most effective where a crushing action is required (as when compacting rubble or crushed rock), and for finishing the surfaces of road bases or pavements. Tandem rollers give a better surface finish, because the pressure is more evenly distributed between the rolls.

Pneumatic-tyred rollers have an open body, carried on a large number of pneumatic-tyred wheels. These wheels are independently sprung, so that they can accommodate irregularities in the surface to be rolled. The ground pressure may be altered by ballasting the machine with sand, or by altering the tyre pressures. The largest machines may have a ballasted weight of more than 40 tonnes and tyre pressures exceeding 800 kN/m^2. These machines are most effective in compacting fine-grained soils, including very fine sands. The maximum depth of soil compacted is about 0·25 to 0·3 m and the economic limit of compaction is usually achieved in from four to eight passes of the machine.

Sheeps-foot rollers are steel drums, about 1 m in diameter, fitted with a large number of 'feet' projecting about 0·2 m from the face of the drum. As the roller moves over the ground, the feet penetrate into the loose soil, compacting it from the bottom up, and the machine 'walks out of the ground'. These machines are most successful in compacting dry fine-grained soils, the best results being obtained where the water content is close to the optimum for the BS heavy hammer test. They are seldom suitable for British soils, which are almost always too wet. In suitable cases, the machine may be ballasted with water or sand in the drum, but excessive weight may cause shear failure of the soil beneath the feet so that the soil is churned up rather than compacted. The thickness of the soil layer to be compacted should not be more than 50 mm greater than the length of the foot. About 30 passes are usually required to achieve proper compaction.

Vibrating rollers are small smooth-wheeled machines fitted with a vibrating mechanism. This usually consists of one or more rotating eccentric weights, which impart an oscillating vertical force to the frame of the machine. The compaction of non-cohesive soils is greatly increased by this combination of rolling and vibration, and the machines can produce a compactive effect on these soils equal to that of a normal smooth-wheeled machine of many times their weight. They are generally ineffective on cohesive soils. Because of their small size, they are particularly useful for work in confined spaces, such as behind bridge abutments, where there would be insufficient space to manoeuvre a larger machine.

Power rammers compact the soil by imparting a blow to the surface. The energy for this blow is derived from the explosion of a petrol/air mixture in a cylinder containing a piston attached to the foot of the machine. Power rammers are most useful for work in confined spaces, such as the back-filling of trenches. The thickness of the soil layer which can be compacted depends mainly on the weight of the machine, but the largest sizes (up to 1 tonne in weight) can compact about 0·35 m in cohesive soils, and nearly double this depth in gravels.

14.5 Specification and control of surface compaction: The method of specification and control depends on the nature of the site. Where the works are concentrated in a relatively small area, such as the construction of an embankment dam, the soil conditions are likely to be fairly uniform, and the compaction operation reasonably easy to control. In such cases, it is usual to specify the method of operation—that is, the maximum thickness of soil to be compacted in each layer, the weight and type of roller, and the number of passes to be made over each layer. In important works, the method of operation is usually decided only after full-scale trials have been made with the actual soil and the actual plant to be used in the final construction.

Where the site is more dispersed and the soil more varied (as in a long length of road construction) it may be better to specify the results to be achieved, rather than to lay down the method to be used to achieve them. In the United States it is usual to specify the required density of the compacted soil as a percentage of the maximum dry density achieved in a compaction test, but this has disadvantages under British conditions. The water content of the soil is generally above the optimum as determined in the compaction test, is largely outside the contractor's control, and inevitably varies slightly, either as a result of different water contents in the borrow pit, or because of changes in the weather. If the specified dry density is made low enough to be possible in the wettest soil, it will permit a large air void ratio and poor compaction where the water content is low. It is generally better to specify the maximum air void ratio, with a further provision that the water content must lie within certain limits. Earlier Ministry of Transport specifications determined the required compaction in this way, but more recent practice [14.1] is to accept that the soil will be compacted at the natural water content, and to use a method specification which lays down the compactive effort to be applied to the soil and the manner of achieving it with different types of plant.

In a drier climate, the contractor may be reasonably expected to control the water content of the soil as placed. Specifications based on the dry density are then perfectly satisfactory. For sands and gravels with little or no fines, many authorities in the United States specify the compaction in terms of the relative density (RD) to be achieved.

14.6   *Compaction of deep beds of loose soil:*   Many water-borne deposits of fine sand are in a very loose state, with void ratios less than the critical value (e_c). Such loose soils in the foundations of structures are very dangerous, as they are unstable under the effects of quite small vibrations. Pile driving, blasting or the operation of heavy machinery nearby can cause compaction of the soil, resulting in a substantial decrease in volume, and settlement of the surface. Where such soils are saturated, vibration may cause a total loss of contact between the particles, and the liquefaction of the soil. This is frequently a major cause of damage and settlement of structures in severe earthquakes. Such loose deposits must therefore be compacted before any major structures are built on them.

14.7   *Compaction by pile driving:*   The effect of displacement piles in compacting cohesionless soils is well known. Where the sole purpose of the piles is to compact the soil, sand piles may be used. A steel casing is driven into the ground, the end being closed with a loose shoe or a plug of gravel. When the required depth has been reached, sand is placed and compacted in the hole so formed, while at the same time the casing is gradually withdrawn. The effect of this process is twofold. The vibration and displacement caused by driving the casing compacts the surrounding soil, while the 'pile' itself consists of a column of densely packed material.

14.8   Vibroflotation: This method is very effective in compacting deep beds of loose sand. The compactor consists of a vibrator, about 0·3 m in diameter and 1 m long, with water jets at the top and bottom. The vibrator is placed on the ground, water is supplied to the bottom set of jets, and the tool is 'jetted' into the ground. When the required depth has been reached, water is supplied to the top set of jets, and the tool is gradually withdrawn. The combination of vibration with the downward flow of water around the vibrator causes close compaction of the soil.

14.9   Preconsolidation: Normally consolidated alluvial soils frequently have high water contents, low shear strengths and high compressibilities. The shear strength may be increased, and the compressibility may be reduced, by temporarily pre-loading (and so consolidating) the soil with a heavy layer of fill before the permanent works are constructed. In suitable cases, sand drains (Section 14.14) may be used to increase the rate of consolidation.

DRAINAGE METHODS

14.10   Purpose of soil drainage: Drainage is generally undertaken to lower the water table, and so reduce the pore pressure in the soil. Permanent drainage systems are used to increase the stability of slopes and retaining walls, and to reduce the hydraulic uplift under structures. Drainage is also frequently necessary as a temporary measure to prevent the flow of water into excavations in water-bearing ground. Because of the prohibitive cost of pumping systems, long term soil drainage is practically restricted to flow by gravity.

14.11   De-watering excavations: If an excavation is to be made in water-bearing ground, it is generally necessary to lower the water level by pumping. However, if the water level in the excavation is lowered without first lowering the water table in the surrounding ground, instability may result, as the hydraulic gradient approaches the critical value.

   To prevent this, water is pumped from wells close to, but outside, the excavation. The water table is then lowered until it is below the bottom of the excavation, into which there is no flow. In very coarse sands and gravels, submersible pumps of high capacity are generally required to deal with the large volumes of water involved. In order to prevent internal erosion of the soil near the well, the pump inlets are protected with gravel filters.

In medium or fine sand, where the permeability is less, a 'well point' system may be used. The wells are simple tubes, perforated at the lower ends, and are usually sunk by jetting. The area to be excavated is surrounded with a curtain of these wells at fairly close spacing, and the heads of the wells are connected to a main from which water is pumped. Since the pumping is by suction, the water table can only be lowered about 5 to 6 m below the tops of the wells. Where excavations are to be made to greater depths, the water table must be lowered in stages, as shown in Fig. 14.1.

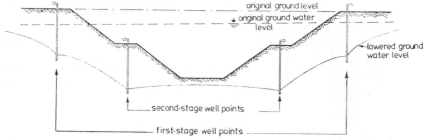

Fig. 14.1   Ground water lowering by well points.

14.12   Drainage behind retaining walls: The importance of draining the soil behind retaining walls was demonstrated in Section 9.12. A drain of crushed stone or clean gravel should be placed behind the wall, as shown in Fig. 14.2. Where the wall is to support natural ground, the drain must be placed immediately behind the wall (Fig. 14.2(a)), but where the wall is to support fill, it is usually more effective to install an inclined drain (Fig. 14.2(b)). Weep holes or some other system must be provided to carry water away from the base of the drain, which should be sealed to prevent percolation of water into the foundations. If the natural soil or fill material is very fine, an unprotected drain would soon be blocked by infiltration of fine soil into the larger pores. This can be prevented by protecting the drain with a graded filter, fine enough to prevent infiltration of the fine soil, but coarse enough to offer minimal resistance to the flow of water. Terzaghi suggested that a suitable filter material should have a grading curve generally similar to that of the soil, and that the D_{15} size of the filter (Section 1.19) should lie between 4 times the D_{15} size of the soil and 4 times its D_{85} size.

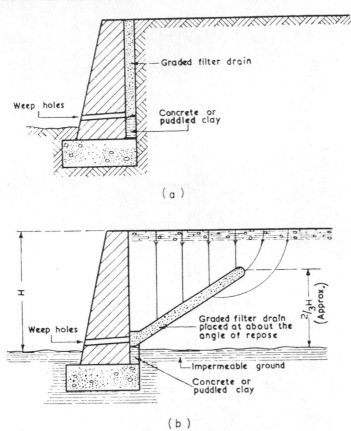

Fig. 14.2 *Drainage behind retaining walls. (After CECP No. 2 [14.2].)*

14.13 Drainage of embankment dams: The stability of an embankment dam is generally at its lowest:

(a) for the downstream face at the end of construction, and

(b) for the upstream face after rapid draw-down of the impounded water level.

Under both these conditions, it is very desirable that the dissipation of pore pressure should be as rapid as possible. If the permeability of the fill is low, it may be necessary to install drains within the embankment to assist this process. These drains usually consist of blankets of permeable material laid with a slight fall towards the face of the bank, as shown in Fig. 14.3. These blankets reduce the effective length of the drainage path and so increase the rate of

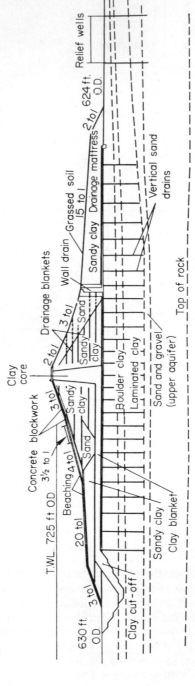

Fig. 14.3 Derwent dam cross-section [14.3].

dissipation. The vertical interval between the blankets is chosen to give a sufficient rate of dissipation to maintain stability under all conditions during the life of the structure.

14.14 *Sand drains:* Sand drains are used to shorten the drainage path in a consolidating stratum, with the object of

(a) increasing the rate of dissipation of pore pressure, and

(b) reducing the time required for consolidation settlement.

Piling equipment is used to sink a casing into the ground, and this is filled with a suitably graded sand or fine crushed chippings. The casing is then withdrawn. The upper ends of the drains are connected to a horizontal drainage blanket, as shown in Fig. 14.3.

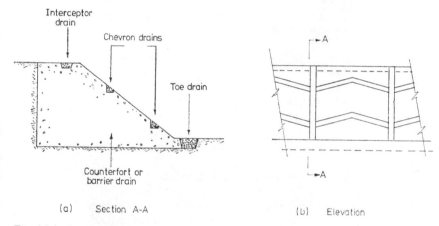

(a) Section A-A   (b) Elevation

Fig. 14.4 Drainage of cuttings.

14.15 *Drainage of natural slopes and cuttings:* The oldest and most effective method of draining slopes is the *counterfort drain,* shown in Fig. 14.4. The function of this drain is to reduce the pore pressure on the potential rupture surface within the bank. Toe drains may also be provided to prevent the development of seepage surfaces, while interceptor drains at the top of the slope and a system of chevron drains on the slope itself are used to prevent erosion by surface water. Grassing the slope also helps to prevent erosion.

14.16 *Electro-osmosis:* If an electrical potential is applied between two electrodes at either end of a capillary, the positively charged adsorbed layers are driven towards the cathode. In the

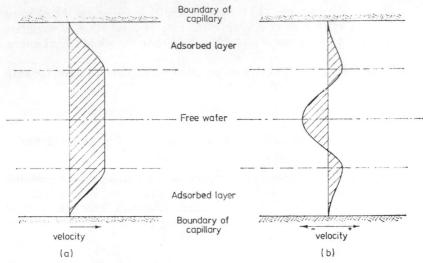

Fig. 14.5   Electro-osmosis. Velocity distribution in a circular capillary (a) with no hydraulic gradient, and (b) with no total flow.

absence of a contrary hydraulic gradient, they carry with them the plug of free water in the centre of the capillary. The resulting velocity distribution across the capillary is shown in Fig. 14.5(a). According to Helmholtz's law, the flow of water in a single circular capillary is

$$\Delta q_e = -\frac{Dr^2\zeta}{4\eta_w}\frac{dE}{dl}$$

where D is the dielectric constant
$\quad r$ is the radius of the capillary
$\quad \zeta$ is the zeta potential (that is, the potential existing between the positively charged adsorbed layers and the negative charges in the crystal face)
$\quad \eta_w$ is the kinematic viscosity of the water
$\quad E$ is the electrical potential
$\quad l$ is the length measured along the capillary.
Then, if all the capillaries in the soil were circular, the total flow through a section of area A would be

$$q_e = -A\frac{nD\zeta}{4\pi\eta_w}\frac{dE}{dl}$$

where n is the porosity of the soil.

Defining an electro-osmotic permeability k_e, and a potential gradient i_e, such that

$$k_e = \frac{nD\zeta}{4\pi\eta_w} \quad \text{and} \quad i_e = -\frac{dE}{dl}$$

the flow may be expressed in a form analogous to Darcy's law:

$$q_e = Ak_e i_e$$

Unlike the hydraulic permeability, however, k_e is independent of the capillary size. The zeta potential may be nearly zero for very high ion concentrations, and in some cases (particularly where calcium carbonate is a major constituent of the soil particles) may even be negative, so that the flow is towards the anode. For the majority of soils, however, k_e has a value of the order of 0.5×10^{-8} m^2/(s. volt) (Casagrande [14.4, 14.5]).

Where there is no flow in the capillaries, either because they are sealed or for other reasons, the electro-osmotic potential gradient in the adsorbed layers is superimposed on a hydraulic gradient in the opposite direction in the free water in the centre of the capillary. The resulting velocity distribution within the capillary is shown in Fig. 14.5(b).

Electro-osmosis may be used in practice for two purposes:

(a) The flow of water into an excavation may be controlled by applying an electro-osmotic potential away from the excavation to counteract the hydraulic gradient towards it. In this way, the seepage pressures are reduced, the effective stresses are increased, and the stability is improved.

(b) Permanent consolidation of fine-grained soils may be effected by sinking a number of perforated tube wells which act as cathodes, interspaced with rods forming anodes. A potential is applied to drive the water towards the wells, from which it is pumped. In certain soils, the application of an electro-osmotic potential causes base exchange, and results in a gain in shear strength much greater than would be expected from the effects of consolidation alone (Bjerrum *et al.* [14.6]).

INJECTION GROUTING

14.17 The purpose of injection grouting: The permeability of some soils may be greatly reduced by injecting suitable materials into the pores. In some cases the shear strength is also increased, although this is often of secondary importance. The material is injected in the form of a fluid grout which subsequently hardens, blocking the pores. The viscosity of the grout must be sufficiently low, and the size of any suspended particles must be sufficiently small, to allow the grout to penetrate into the pores and fissures in the soil. This is the most important factor in determining the most suitable type of grout in any particular case. The materials most commonly used are as follows:

    (a)   Cement and fly ash.
    (b)   Clay suspensions.
    (c)   Bituminous emulsions.
    (d)   Chemical grouts.

14.18   *Types of grout: Cement* is commonly used for grouting fissured rocks, gravels, and very coarse sands. It has been extensively used for forming the impervious curtains beneath dams, and for sealing faults in excavations for tunnels. Cement grout cannot be injected into any but the coarsest sand: in finer soils, the larger particles of cement become wedged across the pores, forming a filter. This removes water from the grout, and an almost impervious cake of cement builds up rapidly. Once this has happened, no further cement can enter the pore, however great the pressure of the grout.

A mixture of cement and *fly ash* has been successfully used by British Railways to stabilise slips in embankments and cuttings. The fly ash is added to the grout to allow easier mixing, to reduce the cost, and to improve the resistance to attack by sulphates in fills containing a high proportion of cinders.

Bentonite is a natural clay containing a high proportion of sodium montmorillonite, which has a great affinity for water, and forms stable suspensions which are highly thixotropic. Bentonite grout remains liquid during injection, but sets into a gel as soon as movement ceases. Since montmorillonite particles are very small, bentonite grout will penetrate fine sands which are impervious to cement. Small quantities of bentonite may be added to weak cement grouts to improve the stability (that is, to prevent the segregation of the cement particles before injection is complete).

Bitumen may be injected into medium and coarse sands in the form of finely dispersed suspension. Immediately before injection, a reagent—usually an organic ester—is added, which causes the emulsion to 'break' and to deposit the bitumen in the pores of the soil. The success of the grouting operation depends critically on the rate of breakdown of the emulsion. This is not always easy to control, since it is affected by the presence of dissolved salts in the soil water. It is also necessary to prevent the injection of large droplets of bitumen, since these would block the pores close to the point of injection, and prevent the entry of the emulsion.

Clay suspensions, in the concentrations normally used for grouting, offer some resistance to shear even when at rest. In order to inject very fine sands, it is necessary to use liquid grouts whose shearing resistance is almost purely viscous. A solution of *sodium silicate*, in the presence of weak acids or of polyvalent cations, deposits a stable gel of silica in the pores of the soil. In dilute solutions, the setting time of the silica gel is relatively long, and the materials may be mixed before injection. However, the resulting gel is soft, and is only useful for reducing the permeability of the soil. If greater shear strength is required, the concentration of the silicate must be increased, but this greatly shortens the setting time. In the Joosten process [14.7] a concentrated silicate solution is injected, followed by an injection of a reactive agent—usually calcium chloride. This results in an almost instantaneous formation of a strong silica gel. Owing to the high viscosity of the concentrated silicate solution, and the rapidity of the reaction, the effective radius of the grouting operation is small, and injection holes are required at about 0·7 m spacing. This, coupled with the high strength of the grouted   soil, can be a positive advantage where small volumes of soil are to be treated. On a large scale, however, the process tends to be expensive.

Silicate grout of high strength may be prepared for 'single shot' injection by mixing undiluted, or only slightly diluted, sodium silicate with an organic ester. The latter slowly decomposes, forming alcohol and acetic acid which attack the sodium silicate, depositing silica.

Ligno-chrome gels are formed by the action of a bichromate on ligno-sulphite, a by-product of the manufacture of paper. The setting time may be controlled over a wide range by varying the concentration of the bichromate. The strength of the grouted soil is greater than with soft silicate grouts, but is much less than that achieved with concentrated silicates. One disadvantage of the method is that the bichromate is toxic, and special precautions are required during mixing.

For injecting fine-grained soils, having permeabilities of the order of 10^{-7} m/s, it is necessary to use a grout having a viscosity close to that of water. Certain water-soluble monomers will polymerise at normal ground temperatures in the presence of a suitable catalyst, and these may be used for grouting such soils. Two of the best known are the American Cyanamide AM-9, and the synthetic resins of the *formaldehyde* group. AM-9 is a mixture of acrylamide with a methylene derivative with which it will cross-polymerise in the presence of a suitable acid catalyst. The setting time can be easily controlled by selecting a suitable concentration of the catalyst. *Resorcinol* polymerises with formaldehyde at normal ground temperatures in aqueous solutions in the presence of acid catalysts. Again, the setting time is closely dependent on the concentration of the catalyst.

14.19 *Methods of injection:* Injection is usually carried out in boreholes put down by rotary or percussion drilling. Single shot injections are generally started at the bottom of the boring, and proceed upwards towards the surface. In the Joosten process, the first injection is made working from the top downwards, while the second proceeds in the reverse direction. In practice, a single injection of grout is frequently insufficient to seal a stratum completely. For example, in a coarse sand it may be necessary to make a preliminary injection of cement to block the larger pores, followed by grout having a low viscosity to seal the finer pores which the cement cannot enter.

14.20 *Penetration of grout:* If the grout may be assumed to be a Newtonian liquid (that is, a material whose shearing resistance is purely viscous, and is directly proportional to the rate of shear strain), the rate of penetration into the soil may be estimated by Maag's formula [14.8].

Consider a Newtonian liquid injected into a homogeneous isotropic soil through an open-ended pipe having a radius r_0 which is small compared with the depth of penetration below the water table (Fig. 14.6). Assuming that the flow is spherical under these conditions, and that Darcy's law is valid, the flow of the liquid is

$$q = Ak_g i_g = 4\pi r^2 k_g \left(-\frac{dh}{dr} \right) \frac{\gamma_w}{\gamma_g}$$

where k_g is the permeability of the soil to grout
    γ_g is the unit weight of the grout.
The soil permeability to grout

$$k_g = k_w \frac{\eta_w \gamma_g}{\eta_g \gamma_w}$$

where k_w is the permeability of the soil to water
    η_w, η_g are the dynamic viscosities of the water and grout respectively.
Let the total head in the soil at the point of injection

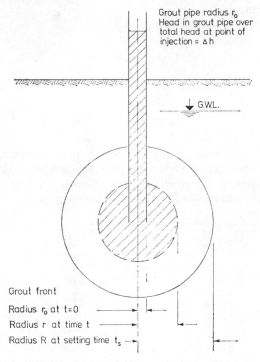

Grout pipe radius r_0
Head in grout pipe over
total head at point of
injection = Δh

G.W.L.

Grout front

Radius r_0 at $t=0$

Radius r at time t

Radius R at setting time t_s

Fig. 14.6   Penetration of a Newtonian grout into a uniform soil.

be initially h_0, and the head in the grout pipe be $h_0 + \Delta h$. Then

$$-\int_{h_0 + \Delta h}^{h_0} dh = \frac{q}{4\pi k_g} \int_{r_0}^{\infty} \frac{1}{r^2} \, dr \cdot \frac{\gamma_g}{\gamma_w}$$

and

$$\Delta h = \frac{q}{4\pi k_g r_0} \frac{\gamma_g}{\gamma_w}$$

But

$$n \frac{dr}{dt} = \frac{q}{4\pi r^2}$$

where n is the porosity of the soil. Then, if R is the radius of the grout front after the setting time t_s,

$$\int_{r_0}^{R} nr^2 \, dr = \frac{q}{4\pi} \int_{0}^{t_s} dt$$

and

$$\frac{n}{3}(R^3 - r_0{}^3) = \frac{q}{4\pi} t_s$$

Then

$$\Delta h = \frac{q}{4\pi k_g r_0} \frac{\gamma_g}{\gamma_w}$$

$$= \frac{n}{3k_g t_s r_0} \frac{\gamma_g}{\gamma_w} (R^3 - r_0^3)$$

so that

$$R = \left(\frac{3r_0}{n} k_w \frac{\eta_w}{\eta_g} \Delta h t_s + r_0^3 \right)^{\frac{1}{3}}$$

In practice, the conditions in the ground are never so simple as those assumed in deriving Maag's formula, but it serves to illustrate the factors which determine the penetration of the grout. These are as follows:

  (a)   The injection pressure.
  (b)   The viscosity of the grout.
  (c)   The porosity and permeability of the soil.
  (d)   The setting time of the grout.
  (e)   The diameter of the grout pipe.

The injection pressure should normally be kept below that which would cause the fissures or pores in the soil to open, and this requirement generally determines the pressure used. The viscosity of the grout is seldom constant in practice, but increases as the setting time approaches. Colloidal grouts are not Newtonian liquids, since they have some shear strength even when at rest (Fig. 14.7(b)). The size of the particles also limits the size of pore into which these grouts will penetrate.

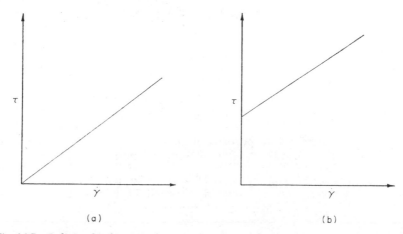

                (a)   (b)

Fig. 14.7   Relationship between shear stress and rate of strain (a) in a Newtonian liquid, and (b) in a Bingham material.

SURFACE SOIL STABILISATION

14.21 Mechanical stabilisation: For road and airfield subgrades and
bases, it is important that the soil should have adequate
resistance to lateral displacement under load, and a soil
which has such resistance is said to be stable. Well-graded
and well-compacted soils are naturally stable and generally
need no treatment. Where the grading is deficient, it may
often be improved by mixing another soil so that the
combined grading is nearer the ideal. This process is called
mechanical stabilisation.

The work of Fuller [14.9], Rothfuchs [14.10] and others
has shown that, for a material to have the greatest maximum
density, the grading should be approximately such that the
percentage passing any sieve is

$$\% \ passing = 100 \left(\frac{D}{D_{max}} \right)^{\frac{1}{2}}$$

where D is the aperture size of the sieve
D_{max} is the size of the largest particle in the soil.
Although Fuller's rule generally gives the greatest maximum
density, it is commonly found in practice that the most
stable soils have a greater proportion of fine particles than
the rule would suggest.

As with all methods of stabilisation, adequate mixing and
good compaction are essential if a stable condition is to be
achieved.

14.22 Additives for stabilisation: Many materials have been used as
additives to stabilise soils, including chlorides, lignins,
molasses and several of the synthetic resins. However, the
only materials now used on a large scale are cement, bitumen
and lime. Of these, ordinary Portland cement is much the
most frequently used, either to improve the subgrades of
roads or simply to provide a working platform to carry the
construction traffic and heavy slip-form pavers now
commonly used in highway construction.

14.23 Stabilisation with cement: Any soil which can be pulverised can
be stabilised with cement, but increasing proportions of silt
and clay-sized particles require larger cement contents and
more effort to pulverise the soil. Highly plastic fine-grained
soils can seldom be economically stabilised with cement.
The best soils for cement stabilisation are well-graded sand–
gravel mixtures with at least 10% of material passing the
75 μm sieve, and a uniformity coefficient of at least 5. The

presence of organic matter and sulphates in the soil may inhibit the setting of cement, and the presence of appreciable quantities of either of these materials may render the soil unsuitable for stabilisation.

14.24 *Specifications for cement-stabilised soils:* The maximum grading limits for the soil are determined by the need to avoid damage to the mixing plant. The British specification [14.1] recognises three different types of stabilised materials:

(a)   Soil–cement.
(b)   Cement-bound granular materials.
(c)   Lean concrete.

These are differentiated by increasingly stringent lower grading limits. For soil–cement, there is no lower grading limit. Cement-bound granular materials must not include more than 10% of material passing the $75\,\mu$m sieve. Materials for making lean concrete must not contain more than 6% of material passing the $150\,\mu$m sieve, and this practically restricts them to washed gravels or crushed rock.

Modern practice is to determine the cement content by the requirement that the stabilised material shall meet certain minimum strength requirements. The Department of Transport Specification for Road and Bridge Works [14.1] requires minimum strengths of $3\cdot5\,N/mm^2$ for cubes tested at 7 days for soil–cement and cement-bound granular materials, and $10\cdot0\,N/mm^2$ for cubes tested at 28 days for lean concrete.

14.25 *Construction methods:* There are two main types of plant used for soil stabilisation with cement:

(a)   Single-pass travel mixers.
(b)   Multi-pass mix-in-place machines.

Single-pass machines take up the soil and cement from the formation, mix it dry, add water, remix it wet, and deposit it on the formation ready for spreading and compaction. Multi-pass machines are generally similar to agricultural rotary cultivators. The soil is first pulverised by the mixer, and is shaped with a blade grader to the required final camber. Cement is then spread on the surface and the soil and cement are mixed dry. Water is next added in a number of stages, the soil and cement being remixed between each stage. When the required final water content has been reached, the soil is reshaped by grading, and compacted by rolling.

The single-pass machines are large and expensive, and operate most efficiently on long uninterrupted runs. The mix-in-place machines are generally smaller and lighter, and are convenient for use in confined spaces. It is necessary to complete compaction of the material as soon as possible after the water has been added, and certainly within two hours. Compaction can be completed much more rapidly with a single-pass machine.

Where the haul distance is not too large, or where the soil is to come from a borrow pit and not from the natural subgrade, mixing may be carried out in a central stationary mixer. This method is commonly used for lean concrete, and allows much closer control of the proportioning and mixing of the materials.

14.26   *Stabilisation with lime:*   *Lime* for soil stabilisation is prepared by calcining limestone or dolomite at about 900°C. The material cost of stabilising with quicklime is generally less than with hydrated lime, but the former is difficult and expensive to handle safely.

Pozzolanas are natural or artificial materials which react chemically with lime to produce a cement. The original pozzolana was a natural volcanic ash from Pozzuoli, on the Bay of Naples. Artificial materials with a pozzolanic action include burnt shale, blast furnace slag, and pulverised fuel ash (PFA). PFA has been used extensively to stabilise soils for highway construction in the United States, but so far has seldom been used for this purpose in Britain.

The method of application is generally similar to stabilisation with cement. However, hardening is much slower than that of cement-stabilised soil, and compaction need not be completed so rapidly.

14.27   *Stabilisation with bituminous materials:*   In suitable conditions, soil may be stabilised by the addition of bituminous materials, such as asphaltic bitumens, 'cut-back' bitumens, and bitumen emulsions. The bitumen seals the pores of the soil, reducing the permeability, and may also increase the shear strength considerably by binding the particles together. The extent of the latter effect depends mainly on the quantity of the stabilising agent used, and the nature of the soil.

The principal disadvantage of this method of stabilisation is that the *immediate* effect in saturated soils is to reduce the shear strength because of the increase in fluid content. For this reason, the method has not been extensively used in Britain.

14.28   Soil freezing: Water-bearing soils, too fine for injection grouting, may be temporarily stabilised by freezing. This is a slow and expensive method, but may be the only one possible in such soils. The method has been used for sinking mine shafts, particularly in Germany, but apart from these cases has not often been employed. In sinking shafts, a ring of vertical tubes is sunk into the ground surrounding the site of the excavation, and extending to the full depth of the strata to be stabilised. Low-temperature brine is circulated through these tubes, and the water in the soil surrounding the tubes is frozen. Eventually the site of the excavation is surrounded with a wall of frozen soil, which is strong enough to provide temporary support to the sides of the excavation and also prevents the inflow of water.

REFERENCES

14.1   Department of Transport. 1976. *Specification for Road and Bridge Works.* HMSO.

14.2   Civil Engineering Codes of Practice Joint Committee. 1951. *Earth retaining structures.* CECP No. 2. Inst. Struct. Eng.

14.3   RUFFLE, N. J. 1965. The Derwent Dam. *J. Inst. Water Eng.*, **19.**

14.4   CASAGRANDE, L. 1952. Electro-osmotic stabilisation of soils. *J. Boston Soc. Civil Eng.*

14.5   CASAGRANDE, L. 1959. Practical aspects of electro-osmosis in foundation engineering. *Pan American Conf. on Soil Mech. and Found. Eng.*

14.6   BJERRUM, L., MOUM, J., and EIDE, O. 1967. Application of electro-osmosis to a foundation problem in Norwegian quick clay. *Géotechnique,* **17.**

14.7   JOOSTEN, H. J. 1954. *The Joosten process for chemical soil solidification and its development from 1925 to date.* N.V. Amsterdamsche Ballast Maatschappij.

14.8   MAAG, E. 1938. Über die Verfestigung und Dichtung des Baugrundes (Injektionen). *Course notes in soil mechanics.* Zurich Tech. Hochsch.

14.9   FULLER, W. B. and THOMPSON, S. E. 1907. The laws of proportioning concrete. *Trans Amer. Soc. Civil Eng.*, **59.**

14.10 ROTHFUCHS, G. 1935. Particle size distribution to obtain the greatest density. *Zement,* **24.**

CHAPTER 15

Site investigation and *in situ* tests

PLANNING

15.1 The objects of a site investigation: A thorough investigation of the site is an essential preliminary to the construction of any civil engineering works. The British Standard Code of Practice defines the objects of such an investigation to be as follows:

    (a) To assess the general suitability of the site for the proposed works.

    (b) To enable an adequate and economical design to be prepared.

    (c) To foresee and provide against difficulties that may arise during construction due to ground and other local conditions.

    (d) To investigate the occurrence or causes of all natural or created changes of conditions and the results arising therefrom [15.1].

In addition, site investigations are necessary in reporting upon the safety of existing works and in investigating cases where failure has occurred.

The site investigation must cover all aspects of the site conditions, and the examination and testing of the soil often forms only a small part of a much more comprehensive study. This chapter is concerned only with investigations of the physical characteristics of the site in so far as they directly affect the design and construction of the works. The wider geographical, social and economic issues are beyond the scope of this book.

15.2 The information required from a site investigation: The information required from a site investigation falls broadly into three main classes:

    (a) *Information affecting the design of the structure,* such as the shear strength and compressibility of the soil.

    (b) *Information affecting the construction of the works,* such as the extent and properties of material to be

379

excavated, or to be used for fill or for road bases or concrete aggregates.

(c) *Information on ground water conditions*, including the level and seasonal variation of the water table, the pressures in the soil water, and the permeability of the soil.

15.3 Planning an investigation: In the earlier stages of an investigation, the information available is often inadequate to allow a firm and detailed plan to be made. The investigation must therefore proceed in three stages:

(a)   Collection of available information.
(b)   Preliminary reconnaissance.
(c)   Detailed exploration.

15.4 Collection of available information: It is important to collect all available information about the site before starting work. In Britain, much valuable data can be obtained from topographical and geological maps, Admiralty charts and tide tables, and meteorological and other records. A great deal of information can also be obtained locally. The offices of borough and district surveyors can often supply details of subsurface conditions in the area. Local public libraries frequently have collections of geological and historical publications. Local newspapers can sometimes provide old records and photographs showing previous uses of the site. Examination of existing structures in the same locality may indicate special foundation or other problems.

In more remote parts of the world, much of this information is not available, but it is most important that such records as do exist should be fully examined. If this is not done, a great deal of time and expense will be wasted.

Finally, to make sure that all the relevant features of the site are fully investigated, it is important to obtain at the earliest possible stage all available details of the nature and extent of the proposed works.

15.5 Preliminary reconnaissance: The principal purpose of the preliminary reconnaissance is to provde a general picture of the topography and geology of the site, so that the detailed investigation may be planned. A topographical survey should be made if this is not already available, and this may be supplemented by aerial photographs. The geological features should be thoroughly examined on the ground (preferably by walking over the site), and compared with the geological maps and records.

It is often necessary to make a number of exploratory pits or borings to determine the depth and succession of the strata, and their variability over the site. Some indication of the ground water conditions should also be obtained, both from surface observations of streams, ponds and springs, and from the borings.

The extent of the preliminary reconnaissance will vary greatly, depending on the nature of the site and the extent of the information previously available. Where the soil conditions are already fairly well known, a cursory inspection may be sufficient. On the other hand, for a large project in a remote area, it may be necessary to mount a major expedition.

15.6 Detailed exploration: The principal objects of the detailed soil survey are as follows:

(a) To determine in detail the geological structure of the site, including the thickness, sequence and extent of the strata.

(b) To determine the ground water conditions.

(c) To obtain disturbed and undisturbed samples for identification and laboratory testing.

(d) To carry out tests to determine the mechanical properties of the soil *in situ*.

15.7 The extent and depth of the investigation: The soil is examined in pits or headings, or by sinking boreholes from which samples are extracted. The number, location and depth of such pits or borings are decided as a result of the preliminary reconnaissance, and will depend both on the nature and variability of the soil strata, and on the form and extent of the works.

In uniform soils, borings may be 100 m apart or more, but in very erratic strata a spacing of 10 m or less may be necessary. Some borings should generally be made outside the limits of the proposed structures, to allow for a possible shift of alignment at a later stage in the design.

The preliminary borings should be continued to a sufficient depth to penetrate all the relevant strata. Subsequent borings may be of more limited depth. When the structure is to be founded in solid rock, borings may be stopped as soon as they have penetrated sufficiently to prove this stratum. It is important to ensure that the boring has reached the solid stratum and not just a detached boulder. All borings should therefore continue at least 1 m into the rock. In addition, at least one boring should penetrate at least 5–7 m into the rock to ensure that the stratum is of adequate thickness.

In compressible soils, borings should penetrate all strata whose compressibility would lead to significant settlement of the structure. Boring should continue to a depth below the foundation at least 50% greater than the width of the structure. At this level, for a square or circular foundation, the increase in vertical stress is about 20% of the value at foundation level: differential settlements due to compression of strata below this level will usually be small. However, if there are very compressible strata at greater depths, it may be necessary to continue the borings through them. As a rough guide, consolidation settlement in any stratum is usually negligible if the increase in vertical stress due to the structural loads does not exceed 10% of the effective overburden pressure.

EXCAVATION AND BORING METHODS

15.8 Trial pits and headings: Trial pits have a great advantage over borings, in that the succession of strata can be examined in the wall of the pit. Large samples can be taken for laboratory testing, or tests may be made *in situ*. In suitable soil, a shallow pit may be dug rapidly, using a mechanical excavator, but for depths exceeding 2–3 m the pit must be dug by hand. Excavation and timbering become increasingly slow and expensive as the depth increases, and trial pits are therefore seldom used for deep investigations.

Whatever the means used to dig the pit, the sides must be adequately supported before any man enters it. The collapse of the sides of pits and trenches is one of the commonest causes of fatal accidents on construction sites. These accidents are entirely avoidable.

15.9 Hand and powered augers: Hand driven post-hole augers may be used in clay soils to a depth of about 5 m. Disturbed samples may be obtained from the spoil brought up in the auger, and small undisturbed samples may be taken from the bottom of the hole.

For deeper investigations, powered augers may be used, and the larger machines are capable of drilling to depths exceeding 30 m. These machines are mainly used for the construction of bored piles, and the largest of them can make borings up to 1 m in diameter. In suitable soils, where no casing is required, the sides of the hole can be examined directly, as in a trial pit.

Although these machines are extremely fast, they are heavy and expensive, and there is often some difficulty in obtaining samples. Recently, continuous helix drills with hollow stems have been used in the United States. In these machines, sampling tools can be passed down through the hollow stem, and samples can be taken without removing the drilling tools from the borehole.

15.10 Wash boring: In this method, the soil is loosened by a high-pressure water jet from a pipe passing down the borehole. The washings are brought to the surface in the water which passes back up the outside of the jet pipe. The method is extremely cheap and rapid in suitable soils, and the disturbance of the soil below the bottom of the borehole is generally much less than with other methods of boring. On the other hand, the washings are usually so disturbed as to be valueless, even for identification. The method is widely used in the United States, but very rarely in Britain.

15.11 'Shell and auger' drilling: This is the method most commonly used in Britain, as it allows good progress to be made in widely differing soil types, by the use of suitable tools. The tools are operated at the bottom of solid drill rods, passing down the hole. The principal tools are as follows:

(a) *The 'shell' or bailer.* This is an open-ended cylinder with a cutting edge and flap valve at the bottom end. It is used for advancing the hole in sands and gravels, by alternately lifting and dropping it at the bottom of the hole. It is also used to remove slurry from the hole.

(b) *The auger.* This is commonly used for advancing the hole in cohesive soils, and for cleaning the bottom of the hole before sampling. It is generally turned by hand.

(c) *The chisel.* A variety of patterns are available for breaking up hard materials, the chippings being brought to the surface in the shell.

(d) *The clay cutter.* This is similar to the shell, but without the flap valve, and is used in the same manner. It is sometimes faster than the auger in clay soils, but causes much more disturbance at the bottom of the hole.

METHODS OF SAMPLING

15.12 Disturbed samples: Disturbed samples are generally required for two purposes:

(a) for identification, and
(b) for testing.

Samples for identification should be taken whenever a change of soil type is observed in the borehole. They should be about ½ kg in weight, and should be sealed in airtight glass or plastic jars, and securely labelled. Larger disturbed soil samples should be taken for testing from each of the main soil types encountered. These samples should be about 10 kg in weight and should be packed in polythene bags or in sealed airtight tins.

15.13 Undisturbed samples: Undisturbed samples are required for shear strength, consolidation and permeability tests. Undisturbed samples are usually taken from a borehole with a 'U4' sampler of the type shown in Fig. 15.1. The sampler consists of a steel tube about 0·1 m in diameter and 0·4 m long, fitted with a cutting shoe at the bottom and a cap with a non-return valve at the top. The sampler is fitted to the bottom of the drill rod, and is driven into the soil at the bottom of the borehole. In order to reduce the friction between the tool and the sample, the inner diameter of the tube is made slightly larger than that of the cutting shoe. This tool is far from ideal, and causes considerable disturbance of the sample, but it is easy to use and is generally adequate for routine tests, except in very sensitive soils.

15.14 Causes of sample disturbance: It is impossible to obtain a completely undisturbed sample, since the act of sampling must disturb the soil to some extent.

The principal causes of sample disturbance are:

(a) the boring process
(b) driving the sampling tool
(c) withdrawing the sampling tool, and
(d) the relief of stress in the soil.

The use of a shell or clay cutter causes considerable disturbance of the soil immediately below the bottom of the boring, and this soil is unsuitable for laboratory tests. Either this soil should be removed with an auger before sampling, or the upper part of the sample should be discarded.

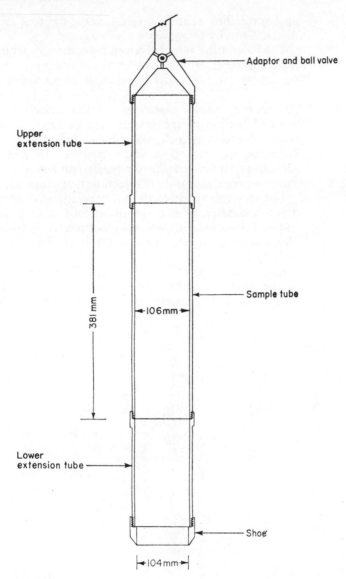

Fig. 15.1   A 'U4' sampler.

The disturbance caused in driving the sampling tool depends on the thickness of the tube and the manner in which it is driven. A thin-walled sampler causes much less disturbance, but is easily damaged. The least disturbance is caused when the sampler is driven in by steady pressure, but for stiff soils this requires the provision of a secure anchorage. For routine investigations, the tool is usually driven down by blows from a monkey.

The base of the sample is freed from the ground by rotating the sampler. This causes some disturbance of the lower part of the sample, which should also be discarded.

15.15 Methods of reducing sample disturbance: The disturbance of the top and bottom of the sample may be avoided by fitting extension pieces at each end of the sampling tube. The soil contained in these is carefully removed by hand. Only the centre section of the sample is contained within the sampling tube, and only this section is retained for testing.

For very soft sensitive alluvial deposits, it is better to use a piston sampler, of the type shown in Fig. 15.2. This is a thin-walled sampler in which the bottom end is temporarily closed with a piston. The sampler is fixed to the bottom end of a hollow shaft and the piston is fixed to a rod passing through the centre of this. When the sampler is in position at the bottom of the borehole, the piston is locked in position and the sampler is driven down past it into the

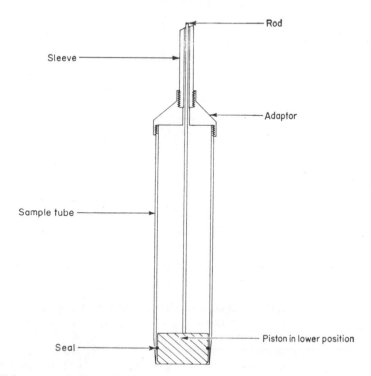

Fig. 15.2   A typical piston sampler.

soil below. When the sampler has been fully driven, piston and sampler are locked together and are withdrawn from the hole.

15.16 Handling and transport of samples: The loaded sampler is removed from the borehole, the cutting shoe and cap are removed, and the ends of the sample are sealed with wax to prevent changes in the water content. Undisturbed samples are still liable to damage after removal from the borehole. They should be protected from extremes of heat or cold, and from vibration or shock, and should be taken to the laboratory with the minimum of delay.

It is very important that all samples are accurately, clearly and securely labelled immediately they are taken from the ground. An unlabelled sample cannot be identified and is of no value.

IN SITU TESTS

15.17 Soil testing in situ: It is almost impossible to obtain satisfactory undisturbed samples of soft sensitive clays or of coarse' granular soils, since even the most sophisticated sampling techniques cause excessive disturbance of these soils. Methods have therefore been devised for estimating the soil properties from the results of tests carried out *in situ*. A considerable number of such tests have been developed, of which the most important are:

    (a) the vane shear test
    (b) the standard penetration test
    (c) the 'Dutch' static cone penetrometer
    (d) the plate bearing test, and
    (e) the pressuremeter.

15.18 The vane shear test [Cadling and Odenstad, 15.2]: This test was developed to measure the undrained shear strength of very soft and sensitive marine clays. A four-bladed vane (Fig. 15.3), about 50 mm in diameter and 100 mm long, is driven into the soil on the end of a rod. The vane is then rotated at a constant rate of 6·0 degrees per minute, and the cylinder of soil contained within the blades is sheared off. The torque required to effect this is measured. Then it may be easily shown that the torque required to shear the soil over the whole surface of the cylinder is

$$T = c_u \pi h d \frac{d}{2} + 2c_u \pi \frac{d^2}{4} \frac{d}{3}$$

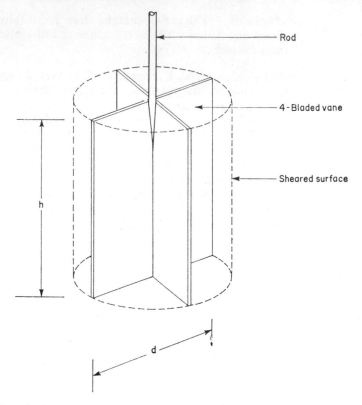

Fig. 15.3   Vane shear test apparatus.

so that

$$c_u = \frac{T}{\pi d^2 (h/2 + d/6)}$$

Example 15.1

A vane, 50 mm in diameter and 100 mm long, was driven into a soft marine clay, and the maximum torque required to rotate it was 5·9 N m.

Then

$$c_u = \frac{5\cdot9 \times 10^{-3}}{\pi \times 0\cdot05^2 [(0\cdot1/2) + (0\cdot05/6)]}$$

$$= 12\cdot8 \text{ kN/m}^2$$

The vane may be pushed down from the surface, measurements being made at regular intervals. The shaft on which the vane is mounted is enclosed within a sleeve, to prevent adhesion to the soil. In suitable soils, tests have been made in this way at depths exceeding 60 m. In Britain it is more usual to use the vane in a borehole. In this case, the vane must penetrate at least half a metre below the bottom of the hole to avoid the soil disturbed by boring.

15.19   The standard penetration test: This test has been used extensively in the United States and in Britain for estimating the relative density and angle of shearing resistance of coarse granular soils. A standard split spoon sampler, about 50 mm in diameter (Fig. 15.4), is driven into the ground by blows from

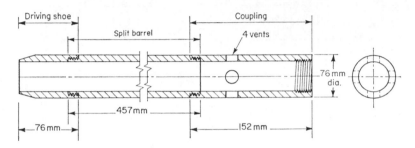

Fig. 15.4   Split spoon sampler (after BS 1377: 1975 [15.5]).

a drop-hammer weighing 64 kg (140 lb) and falling 0·76 m (30 in). The sampler is driven 0·15 m (6 in) into the soil at the bottom of the borehole, and the number of blows (N) required to drive it a further 0·3 m (12 in) is then recorded.

Although the test is entirely empirical, considerable experience with its use has enabled a reasonably reliable correlation to be established between the N value and certain soil properties. Figure 15.5 shows Peck, Hanson, and Thornburn's relationship between N and the relative density and angle of shearing resistance (φ') [15.3]. The use of the standard penetration test in predicting settlement of foundations was considered in Chapter 12.

Although the test is extremely simple, some difficulties arise in interpreting the results. In fine sands, the permeability is often relatively low. Where such soils are dilatant, a considerable reduction in pore pressure occurs under the sampler, at the instant when the blow is struck. This momentarily increases the effective stress and the shear strength of the soil, resulting in an abnormally high value for the penetration resistance.

The use of the split spoon sampler is satisfactory in sands, and allows disturbed samples to be obtained with each test.

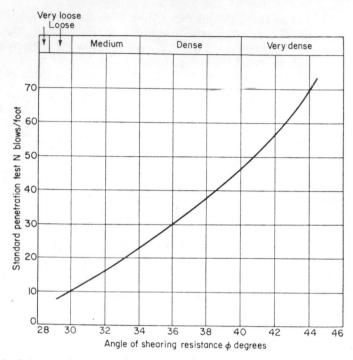

Fig. 15.5 Relation between RD, φ, and N obtained from the standard Penetration Test. (After Peck, Hanson and Thornburn [15.3].)

However, the tool is insufficiently robust for use in gravels. Palmer and Stuart [15.4] have developed a modified form of tool with a solid cone-shaped end for use in coarser soils. In sands, this tool gives results almost identical with those obtained using the spoon sampler, but there is evidence that the N value becomes increasingly conservative as the particle size increases.

The standard penetration test was originally developed for use in a wash boring only a little wider than the diameter of the spoon, and in the United States it is normally used in this way. In Britain, however, it is more usually used in a 'shell' boring, which may be 0·15 m or more in diameter. The greater disturbance of the soil at the bottom of the 'shell' boring, and the reduced lateral pressure due to the larger diameter of the boring, both tend to reduce the penetration resistance. It may be necessary to drive the tool more than the standard 0·15 m before beginning the count. Abnormally low values of penetration resistance have also been recorded near the ground surface, due to the lack of lateral support in the soil.

15.20 The 'Dutch' cone penetrometer: The estimation of the static shear strength of the soil by means of a dynamic test is inherently unsatisfactory, largely because the pore stresses differ in the two conditions, and the standard penetration test results cannot provide more than a rough guide to the soil properties. More reliable results are obtained from static penetration tests, of which the most commonly used is the 'Dutch' cone penetrometer. This device has been used extensively in Holland and Belgium, mainly in fine sands and silts.

The penetrometer is cone-shaped, with a maximum area of $1\,000\,mm^2$. The cone is attached to a rod, which is protected by an outer sleeve. The thrust needed to drive the cone and the sleeve into the ground may be measured independently, so that the end bearing resistance and the side friction may be separately determined.

This penetrometer was originally developed for the design of piles, but it has also been used successfully to estimate the bearing capacity and settlement of foundations on non-cohesive soils. This will be discussed in later chapters.

15.21 The plate bearing test: This is, in effect, a test on a model foundation. A trial pit is excavated to the required level, and a steel plate is set on the soil at the bottom of the pit. A static load is applied to the plate in successive increments, and the extent and rate of the settlement is recorded. The recovery of the settlement is then recorded as the load is reduced. Further increments of load are added, and the procedure is repeated until the soil below the plate fails.

Plate bearing tests are expensive, and are seldom used when the necessary information can be obtained in other ways. They are useful, however, in estimating the settlement and bearing capacity of foundations on such materials as hardcore, chalk, shale, and soft weathered rock, which cannot be easily investigated by any other form of test, and in the design of road and airfield pavements.

In some circumstances, the results of plate bearing tests may be very misleading. The depth to which the soil beneath a foundation is significantly stressed depends on the size of the loaded area. If the plate is much smaller than the eventual size of the foundation, a test carried out at foundation level will leave much of the soil untested (see Fig. 15.6). A number of tests at different depths are generally necessary. In any event, the width of the plate should be as near as possible to the width of the proposed foundation, and should be at least 0·3 m in diameter. The maximum size is usually limited by the

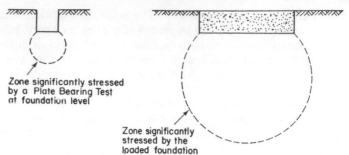

Zone significantly stressed
by a Plate Bearing Test
at foundation level

Zone significantly
stressed by the
loaded foundation

Fig. 15.6   Significance of foundation size in relation to the results of plate bearing tests.

available load, and plates exceeding 1·0 m in diameter are
seldom used.

Great care is needed when installing the plates in the
ground. All loose soil must be removed from the bottom of
the pit, and the plate must be set in mortar or plaster of Paris.
If this is not done with great care, the recorded settlements of
the plate will be much greater than the real deformation of
the soil.

15.22   The pressuremeter: The pressuremeter [15.5] was first developed
by Menard in France during the 1950s. In its original form, it
consists of a cylindrical cell covered with a rubber membrane
which is inserted into an unlined borehole. The membrane is
inflated against the side of the borehole by pumping in fluid,
and the pressure and volume of the fluid are continuously
monitored. Two guard cells, above and below the main cell,
are similarly inflated to ensure that the expansion of the main
cell is in a radial direction only. From the curves of volume
change against pressure, it is possible to estimate the shear
strength and stress/strain behaviour of the soil.

Modern pressuremeters are considerably more sophisticated than Menard's original instrument. Direct measurement of pressure and radial expansion, by means of electrical
transducers within the cell, have greatly improved their
accuracy. Automatic control of the pressure applied to the
cell allows a wide variety of test procedures to be used in
different ground conditions and for different purposes. In
soft clays, self-boring pressuremeters may be used. These
greatly reduce the disturbance of the soil during insertion
[15.6].

GEOPHYSICAL METHODS

15.23 Geophysical investigations: Direct methods of exploration by
boring and sampling may be supplemented in suitable cases

by indirect geophysical methods. These methods depend on our ability to locate the boundaries between strata of different composition, by detecting changes in the electrical resistivity or wave velocity in the soil, or in the magnetic or gravitational field.

Since these indirect methods give little or no information about the engineering properties of the soils or rocks encountered, they must always be correlated with borings from which the succession of strata can be established.

Gravitational and magnetic methods are generally most useful on a very large scale, and are seldom very suitable for the detailed studies required for civil engineering works, although proton magnetometers have been used successfully to detect swallow holes and cavities in limestone.

15.24 Seismic methods: The waves propagated in the ground by an explosion or impact are of three types;

    (a)   surface waves (Rayleigh waves)
    (b)   compression waves, and
    (c)   shear waves.

Compression and shear waves travel outwards in every direction from the point of generation. The compression waves have the greatest velocity of travel, and are therefore the first to arrive at any given point. The velocity of travel depends on the nature of the material through which the wave is passing, and is greatest in the most dense and rigid materials. If the velocity of the wave changes at an interface between two strata composed of different materials, the depth at which this change occurs may be determined by the methods of *seismic refraction* or *seismic reflection*.

15.25 Seismic refraction methods: Shock waves are set up in the ground, by exploding a small charge in a shallow boring. The time taken for the first shock wave to reach any point on the surface is determined by a sensitive detector, called a 'geophone'.

Consider a site, at which there are three successive strata with horizontal interfaces. Let the compression wave velocities be V_1, V_2, and V_3 in the upper, middle, and lower strata respectively. Let $V_1 < V_2 < V_3$.

When the shot is fired, compression waves move outwards in every direction through the upper stratum, at velocity V_1. At a point close to the shot-hole, the first wave to arrive is a compression wave which has passed directly through the upper stratum. As this direct compression wave moves outwards, it will eventually be overtaken by a second wave which passes down through the upper stratum, and is

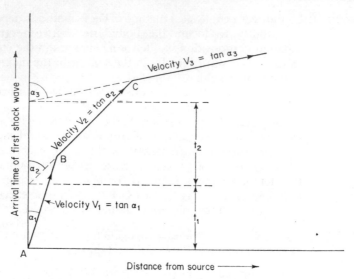

Fig. 15.7 Seismic refraction tests. Curve showing time of arrival of the first shock wave plotted against distance from the source.

refracted parallel to the interface. This wave then passes through the middle stratum with a velocity V_2 ($> V_1$), and is eventually refracted back through the upper stratum to the surface. Since the wave must be refracted parallel to the interface, the angle of incidence (β_1) is $\sin^{-1}(V_1/V_2)$.

If, for various points on the surface, we plot the time of arrival of the first compression wave against the distance of the point from the shot-hole, we obtain a curve of the type shown in Fig. 15.7. The line AB shows the first wave arriving directly through the upper stratum. The velocity of this wave (V_1) is $\tan \alpha_1$. Between B and C this direct wave has been overtaken by a wave which has passed through the middle stratum with a velocity $V_2 = \tan \alpha_2$. Then the time for the two journeys through the upper stratum is t_1, and the thickness of this stratum is

$$H_1 = \frac{t_1 V_1}{2 \cos \beta_1}$$

The curve beyond C shows the arrival of waves refracted through the lower stratum. The time for the double journey through the middle stratum is t_2, and the thickness of this stratum is

$$H_2 = \frac{t_2 V_2}{2 \cos \beta_2} \quad \text{where} \quad \beta_2 = \sin^{-1} \frac{V_2}{V_3}$$

Seismic refraction methods cannot detect an interface at

which the velocity in the lower stratum is less than that in the upper material. In this case, the compression waves, on reaching the interface, are refracted away from the surface. Interpretation is also complicated if the wave velocity increases gradually with depth, or if the interface is not parallel to the ground surface.

15.26 Seismic reflection methods: For deeper strata, the depth may be determined by recording the time of return of shock waves reflected from the interface. For investigations under the sea bed, a 'sparker' technique is commonly used. A continuous series of high voltage (10 000 to 12 000 volts) discharges are made at the end of a cable towed behind a launch. The resulting shock waves are reflected from the sea bed, and from the interfaces of the strata below. The time of return of the reflected wave from each interface is registered on a hydrophone towed behind the launch, and is automatically recorded in such a way as to give a continuous trace of the relative depths of the interfaces between the strata.

15.27 Electrical resistivity methods: These methods consist of detecting differences in the electrical resistivity of the strata. Most soil and rock minerals are poor conductors of electricity, and the conductivity of the ground is mainly due to the dissolved salts in the pore water. There may, therefore, be considerable differences between the resistivities of different strata. Dense unfissured rocks generally have much higher resistivities than loose soils. There may also be considerable differences in the resistivity above and below the water table.

In the usual method of measurement, a potential is applied between two electrodes buried near the ground surface (Fig. 15.8). Measurements are made of the current flowing between these electrodes, and of the potential drop between a pair of intermediate electrodes. The spacing of the intermediate electrodes may be varied to suit the site conditions. For most purposes, a satisfactory arrangement is the Wenner configuration (Fig. 15.8) in which the four electrodes are equally spaced. For this arrangement, and for a soil of uniform resistivity ρ, it may be shown that

$$\rho = 2\pi a \, . \, V/I$$

where a is the electrode spacing
V is the potential drop between the inner pair of electrodes
I is the current flowing between the outer pair of electrodes.

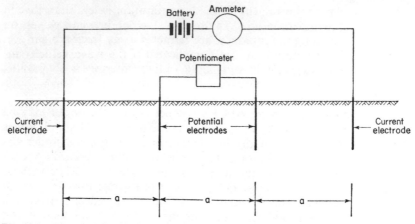

Fig. 15.8   *Electrical resistivity methods. The Wenner configuration.*

The interfaces between the strata may be located by comparing the apparent values of the soil resistivity obtained from a series of measurements.

The depth of horizontal or nearly horizontal interfaces may be determined by using an expanding electrode system. The centre point of the electrode system is fixed, but the spacing between the electrodes is progressively increased. Figure 15.9 shows a typical curve obtained by plotting apparent resistivity against electrode spacing. The depth to

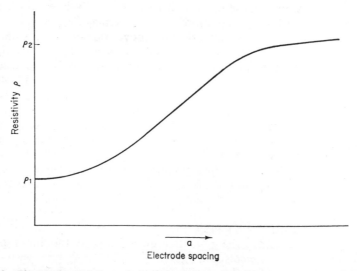

Fig. 15.9   *Electrical resistivity methods. Apparent resistivity plotted against electrode spacing for two strata of resistivity ρ_1 and ρ_2.*

the interfaces between the strata is approximately ⅔ of the electrode spacing at which the corresponding point of inflection appears on the curve. A more precise estimate of the depth may be obtained by comparing the measured values of apparent resistivity with a series of standard curves.

Vertical or nearly vertical interfaces may be detected by traversing. The electrode spacing is kept constant, but the whole electrode system is moved along between each set of readings. A sharp change in the apparent resistivity may be observed as each electrode passes over the interface.

RECORDS AND REPORTS

15.28 Borehole records: During all boring operations, the foreman maintains a log, from which a comprehensive series of borehole records is later prepared. A typical record is shown in Fig. 15.10. These borehole records are incorporated in the final report on the investigation, and each should contain the following information:

    (a)   Details of the boring, including the location and original ground level, the diameter and method of boring, and the date of commencement and completion.

    (b)   A description of each soil type encountered, and the level at which each change in soil type was observed.

    (c)   The serial number and type of each sample from the borehole, and the level from which each was taken.

    (d)   The levels at which standard penetration tests, or other *in situ* tests, were carried out, and the results of these tests.

    (e)   The level of any water table, and the level at which any seepage into the boring was observed.

    (f)   The level at which boring was stopped.

15.29 Soil survey reports: The borehole records are incorporated in a comprehensive report on the investigation, which should comprise the following:

    (a)   A brief description of the proposed works, and a statement of the purpose of the investigation.

    (b)   A general description of the site, including all surface features which are relevant to the design or construction of the works.

BOREHOLE RECORD

Location: Highfield green
Job No.: 709/8
Borehole No.: 4
Date commenced: 17-7-68
Date completed: 18-7-68

Method of boring: Shell and Auger
Diameter of boring: 0·2 m to 8·9 m
Casing: 0·2 m to 6·0 m
Ground level: 28·9 m OD

Changes in strata			Samples		Water level		In-situ tests	
Depth (m)	Legend	Description	Depth (m)	Type and No.	Level (m)	Depth of boring (m)	Depth (m)	Type
0·8		Made ground mainly brick rubble and cinders						
		Firm yellow-brown sandy clay	1·0 1·3 1·8	D(21) U4 (72)				
2·1								
2·4		Dense brown sand with some clay						
		Dense brown gravel with some sand and occasional layers of sandy clay	2·6	D(22)	2·9m	3·0	3·0	SPT (N=24)
							5·0	SPT (N=14)
5·6								
		Medium to soft brown silty clay	5·8	D(23)	Sealed	6·0		
6·2			6·4 6·9	U4 (73)				
		Stiff fissured grey clay	7·2	D(24)				
			7·9 8·4	U4 (74)				
8·9								
	Boring stopped at 8·9 m							

Fig. 15.10 A typical borehole record.

    (c)   A clear and concise statement of all work carried out, including the laboratory tests.

    (d)   A site plan showing:
        (i)   all relevant surface features,
        (ii)   an outline of the proposed works, and
        (iii)   the location of all trial pits and borings.

    (e)   The borehole records described above.

    (f)   The results of any *in situ* tests not included in the borehole records.

    (g)   The records of geophysical surveys.

    (h)   The information obtained from laboratory tests.

    (j)   Comments and recommendations.

The extent and content of the final section of the report depends on the purpose of the investigation. In many cases, a client employs a firm of site investigation specialists to obtain certain specific information about the site. The subsequent interpretation of this information is undertaken by his own engineers. In such a case, the final section of the report serves only to summarise the information obtained, and to draw attention to features of particular importance. However, many of the firms engaged in this work are able, from their special knowledge and experience, to advise on the design and construction of the works in the light of the information obtained. If the client asks for such advice, it forms the final section of the report. It is important that, if such advice is required, this fact should be made clear at the outset. Those carrying out the investigation should then be left reasonably free to decide what information is needed, and how it is to be obtained.

It is important to remember that soil survey reports must often be read by persons who have not visited the site. Clear maps and diagrams of the site are essential, with all work carried out clearly marked and indexed in the report, and giving the exact position of all borings. Geological names for soil materials should be used if this can be done reliably.

REFERENCES

15.1   British Standards Institution. Site investigations. CP 2001: 1957.

15.2   CADLING, L. and ODENSTAD, S. 1950. The vane borer. *Proc. Roy. Swedish Geotech. Inst.* **2.**

15.3   PECK, R. B., HANSON, W. E. and THORNBURN, T. H. 1953. *Foundation engineering.* Wiley (New York).

15.4   PALMER, D. J. and STUART, J. G. 1957. Some observations on the standard penetration test, and a correlation of the test with a new penetrometer. *Proc. 4th Int. Conf. Soil Mech. and Found. Eng.*

15.5   BAGUELIN, F., JEZEQUEL, J-F. and SHIELDS, D. H. 1978. *The pressuremeter and foundation engineering.* Trans. Tech. Publ. (Basle).

15.6   WINDLE, D. and WROTH, C. P. 1977. The use of a self-boring pressuremeter to determine the undrained properties of clays. *Ground Engineering,* **10.**

Index